LYMPHOCYTE ACTIVATION AND IMMUNE REGULATION IX

ADVANCES IN EXPERIMENTAL MEDICINE AND BIOLOGY

Recent Volumes in this Series

Volume 503
INTEGRATING POPULATION OUTCOMES, BIOLOGICAL MECHANISMS AND RESEARCH METHODS IN THE STUDY OF HUMAN MILK AND LACTATION
Edited by Margarett K. Davis, Charles E. Isaacs, Lars Å. Hanson, and Anne L. Wright

Volume 504
MYCOTOXINS AND FOOD SAFETY
Edited by Jonathan W. DeVries, Mary W. Trucksess, and Lauren S. Jackson

Volume 505
FLAVONOIDS IN CELL FUNCTION
Edited by Béla A. Buslig and John A. Manthey

Volume 506
LACRIMAL GLAND, TEAR FILM, AND DRY EYE SYNDROMES 3: Basic Science and Clinical Relevance
Edited by David A. Sullivan, Michael E. Stern, Kazuo Tsubota, Darlene A. Dartt, Rose M. Sullivan, and B. Britt Bromberg

Volume 507
EICOSANOIDS AND OTHER BIOACTIVE LIPIDS IN CANCER, INFLAMMATION, AND RADIATION INJURY 5
Edited by Kenneth V. Honn, Lawrence J. Marnett, Santosh Nigam, and Charles Serhan

Volume 508
SENSORIMOTOR CONTROL OF MOVEMENT AND POSTURE
Edited by Simon C. Gandevia, Uwe Proske, and Douglas G. Stuart

Volume 509
IRON CHELATION THERAPY
Edited by Chaim Hershko

Volume 510
OXYGEN TRANSPORT TO TISSUE, VOLUME 23
Edited by David Wilson, Sydney M. Evans, John Biaglow, and Anna Pastuszko

Volume 511
PEDIATRIC GENDER ASSIGNMENT: A Critical Reappraisal
Edited by Stephen A. Zderic, Douglas A. Canning, Michael C. Carr, and Howard McC. Snyder

Volume 512
LYMPHOCYTE ACTIVATION AND IMMUNE REGULATION IX: Homeostasis and Lymphocyte Traffic
Edited by Sudhir Gupta, Eugene Butcher, and William Paul

LYMPHOCYTE ACTIVATION AND IMMUNE REGULATION IX

Homeostasis and Lymphocyte Traffic

Edited by

Sudhir Gupta
University of California, Irvine
Irvine, California

Eugene Butcher
Stanford University School of Medicine
Stanford, California

and

William Paul
National Institute of Allergy and Infectious Diseases
National Institutes of Health
Bethesda, Maryland

Kluwer Academic / Plenum Publishers
New York, Boston, Dordrecht, London, Moscow

Library of Congress Cataloging-in-Publication Data

International Conference on Lymphocyte Activation and Immune Regulation (9th: 2002 Newport Beach, Calif.)

Lymphocyte activation and immune regulation IX: homeostasis and lymphocyte traffic /edited by Sudhir Gupta, Eugene Butcher, and William Paul.

p. cm. — (Advances in experimental medicine and biology; v. 512)

"Proceedings of the Ninth International Conference on Lymphocyte Activation & Immune Regulation held in Newport Beach, California on February 8–12, 2002."

Includes bibliographical references and index.

ISBN 0-306-47395-X

1. Lymphocyte transformation—Congresses. 2. Immune response—Regulation—Congresses. 3. Cells—Motility—Congresses. I. Gupta, Sudhir. II. Butcher, Eugene. III. Paul, William E. IV. Title. V. Series.

QR185.8.L9 I553 2002
616.07′9—dc21

2002027574

Proceedings of the 9th International Conference on Lymphocyte Activation and Immune Regulation held in Newport Beach, California on February 8–10, 2002.

ISBN 0-306-47395-X

233 Spring Street, New York, N.Y. 10013

http://www.wkap.nl/

10 9 8 7 6 5 4 3 2 1

A C.I.P. record for this book is available from the Library of Congress

Printed in the United States of America

11/26/02
cw

PREFACE

During the past 5 years major progress has been made in the understanding the basic mechanisms of lymphocyte homeostasis and in the developmental relationship between different memory T subsets and their traffic patterns and functional significance. This volume highlights the current concepts of lymphocyte development, factors regulating lymphocyte trafficking and development, and specialized characteristics and functional properties of naïve and memory subsets.

This volume is divided into 3 sections. Section I deals with factors that regulate the development and maturation of T cells and B cells and lymphocyte traffic. The significance of C-kit , Bcl-6, IL-7 and Vav in the development of T and B lymphocytes is discussed. A role of lymphotoxins and VAP-1 in trafficking of leucocytes is reviewed. Finally, the trafficking and homing characteristics of T cell and B cell subsets, and the regulation of these processes during the immune response, is presented.

Section II discusses various aspects of naïve and memory T cell biology, including clonal expansion, reprogramming of genes including those encoding cytokines and cytotoxic granules, changes in the expression of cell surface proteins involved in cell-cell adhesion, homing of naïve and memory T cells, the role of MHC and cytokines in the maintenance of naïve and memory T cells, and the characterization and differentiation of virus-specific memory T cell heterogeneity in mice and humans.

Novel methods of visualization of immune cells and immune synapses are reviewed in Section III. This includes tracking of dendritic cells *in vivo*, monitoring arterial smooth muscle-specific T cells in the inflamed vasculature, imaging of molecular migrations in immune synapses, and visualization of various immune cells in intact lymphoid tissues by two photon confocal imaging.

This volume should be of interest to immunologists, molecular biologists, microbiologists, pathologists, academic physicians, cell biologists, and scientists and clinicians interested in vaccine development.

Sudhir Gupta
Eugene Butcher
William Paul

CONTENTS

LYMPHOCYTE DEVELOPMENT AND LYMPHOCYTE TRAFFIC

NAÏVE AND MEMORY T CELLS

TRACKING IMMUNE CELLS

LYMPHOCYTE DEVELOPMENT IN NEONATAL AND ADULT C-KIT-DEFICIENT (C-KIT$^{W/W}$) MICE

Claudia Waskow and Hans-Reimer Rodewald*

ABSTRACT

Hematopoietic stem cells and lymphocyte progenitors express the receptor tyrosine kinase c-Kit. In fetal and neonatal life, c-Kit plays a redundant role in T, and no apparant role in B cell development. In neonatal mice deficient for both c-Kit and the common γ chain (γ_c), a component of the interleukin-7 (IL-7) receptor, the thymus is alymphoid, and therefore lacks T cell receptor (TCR) β, γ, and δ rearrangements. Thus, a critical role for c-Kit in T cell development around birth is well established. More recently, it has become possible to examine the impact of c-Kit deficiency under conditions of steady state lymphopoiesis in adult life. Such analysis has been made possible by the identification of a viable adult c-Kit-deficient (c-Kit$^{W/W}$) variant, termed the Vickid mouse. The Vickid mouse arose by outcrossing c-Kit^{W}-bearing mice of the WB strain, in which lack of c-Kit is lethal, to a mixed genetic background. In adult Vickid mice, mainstream $\alpha\beta$ TCR$^+$ thymocyte development, and B cell development in the bone marrow are severely c-Kit-dependent with progressive age. Analysis of other pathways of developing T cells, i.e. CD4$^-$CD8$^-$ (double neagative [DN]) $\alpha\beta$ TCR$^+$ and DN $\gamma\delta$ TCR$^+$ thymocytes revealed that the development of both lineages is also severely affected by lack of c-Kit. However, numbers of $\gamma\delta$ TCR$^+$ T cells decline before numbers of $\alpha\beta$ TCR$^+$ T cells in the thymus. In contrast to T and B cell development, generation of NK cells is not affected in adult c-Kit$^{W/W}$ mice.

INTRODUCTION

The receptor tyrosine kinase c-Kit is a growth factor receptor important for the development and function of stem and progenitor cells. C-Kit is expressed in

*Dept. for Immunology, University of Ulm, D-89070 Ulm, Germany, claudia.waskow@medizin.uni-ulm.de, hans-reimer.rodewald@medizin.uni-ulm.de, tel: 49-731-5002 3360, fax: 49-731-5002 3367

hematopoietic cells, germ cells, melanocytes, intestinal pacemaker cells, and neuronal cells (Bernstein *et al.*, 1990; Besmer, 1991). According to this expression pattern, c-Kit deficiency causes anemia, sterility, depigmentation (dominant white spotting), and intestinal pacemaker dysfunction. Numerous naturally occurring c-Kit (W alleles) or Kit-ligand (Kitl; formerly termed stem cell factor [SCF], or mast cell growth factor [MGF]) (steel [Sl] alleles) mutants have been identified (Russell, 1979; Galli *et al.*, 1994). Mutants entirely lacking cell surface expression of c-Kit (c-Kit$^{W/W}$) are lethal within a few days after birth (Russell, 1979). This lethality can be overcome by transplantation of wild type hematopoietic stem cells (Mintz and Fleischman, 1979). The most prominent pathological symptom in c-Kit$^{W/W}$ mice is a severe macrocytic anemia (reduced red blood cell [RBC] numbers with increased RBC volume), and this anemia is the putative cause of death (Russell, 1979). Hence, until now, analyses of adult c-Kit mutants have been limited to mild W or steel alleles. The c-Kit$^{W/W}$ genotype was known to be lethal in the WB mouse strain background (as well as in all other tested backgrounds). We have identified a white, black-eyed c-Kit$^{W/W}$ mouse, surviving to adulthood, in out-crosses of WB-c-Kit$^{W/+}$ to mixed genetic backgrounds. This novel mutant has been termed the "Vickid" mouse (Waskow et al. unpublished). The molecular nature of the putative c-Kit suppressor mutation is unknown. Here, we summarize our detailed analyses of lymphocyte development in adult Vickid mice, and compare these findings to the role previously recognized for c-Kit in lymphocyte development during fetal and neonatal life.

Thymocyte Development in Neonatal c-Kit$^{W/W}$ Mice

Analyses of thymocyte development in both neonatal c-Kit$^{W/W}$ mice, as well as in adult wild type mice grafted with either wild type or c-Kit-ligand-deficient fetal thymic transplants (Sl/Sl) revealed that intrathymically expressed c-Kit ligand is a major factor driving expansion of very immature thymocytes *in vivo*. Transplantation of Sl/Sl grafts was important to separate effects due to lack of Kitl at the hematopoietic stem cell (HSC) level (as in the fetal liver of c-Kit$^{W/W}$ mice) from intrathymic development (as in Sl/Sl thymi grafted into wild type mice). In the neonatal c-Kit$^{W/W}$ thymus, and in Sl/Sl thymus grafts, the size of the CD3$^-$CD4$^-$CD8$^-$ (triple-negative [TN]) thymocyte compartment was reduced ~40 fold and ~12 fold, respectivly. This reduction in TN cell numbers was due to diminished proliferation in the thymus as shown by *in vivo* bromodeoxyuridine labeling of thymocytes in wildtype or Sl/Sl mutant grafts. Collectively, lack of Kitl in the thymic environment reduces the expansion rate of TN, but not overall thymocytes, by ~50%. Apart from these clear alterations at early stages of thymocyte development, thymocyte development was permissive in both neonatal c-Kit$^{W/W}$ mice and in Sl/Sl thymus grafts. Thus, this study (Rodewald *et al.*, 1995), together with earlier *in vivo* work (Asamoto and Mandel, 1981), has established a role, albeit not an essential one, for c-Kit/Kitl interactions in the expansion of very early thymic immigrants.

Thymocyte Development in Neonatal Mice Double-Deficient for c-Kit and the Common γ Chain (γ_c) (c-Kit$^{W/W}\gamma_c^-$ Mice)

Given that thymocyte development was affected but permissive in neonatal c-Kit$^{W/W}$ mice, we pursued the idea that a second growth factor might be redundant with Kitl at this stage of development, and thus compensate for the lack of c-Kit at this stage. Since interleukin-7 (IL-7) was known to be an important, albeit again not essential growth factor driving thymocyte proliferation (Conlon *et al.*, 1989; Murray *et al.*, 1989; von-Freeden-Jeffry *et al.*, 1995), we tested the roles of these two growth factor receptor pathways *in vivo* by generating and analyzing mice lacking either c-Kit (c-Kit$^{W/W}\gamma_c^+$ mice), or the common γ chain (γ_c) (DiSanto *et al.*, 1995), a component of the IL-7 receptor (reviewed in Akashi *et al.*, 1998; DiSanto and Rodewald, 1998) (c-Kit$^+\gamma_c^-$ mice), or both c-Kit and γ_c (c-Kit$^{W/W}\gamma_c^-$ mice). Indeed, these double mutant mice were the first mutants in which T cell development was completely abrogated as a result of growth factor deficiencies (Rodewald *et al.*, 1997). Lack of γ_c alone reduced thymic cellularity, but, as shown for c-Kit$^{W/W}\gamma_c^+$ mice, was permissive for thymocyte development (DiSanto *et al.*, 2000; DiSanto and Rodewald, 1998). In c-Kit$^{W/W}\gamma_c^-$ mice, thymocytes, including all progenitors, were absent by flow cytometric and histological analysis. Absence of thymocytes was also evident by lack of diverse T cell receptor (TCR) rearrangements at the TCR β (Rodewald *et al.*, 1997), TCR γ and TCR δ loci (Rodewald and Fehling, 1998; Rodewald and Haller, 1998). In contrast, immunoglobulin junctions were diverse in B lymphocytes from c-Kit$^{W/W}\gamma_c^-$ mice (Rodewald *et al.*, 1997). These experiments revealed that c-Kit and IL-7Rγ_c provide essential, overlapping and synergistic signals for two distinct receptor-mediated signaling pathways in thymocyte development in neonatal life.

In subsequent studies, we have utilized c-Kit$^{W/W}\gamma_c^-$ mice to analyze whether or not a normal pro-T cell compartment is required to generate thymic dendritic cells, a pathway of DC development proposed some time ago (Ardavin *et al.*, 1993). Our results provided clear evidence that thymocyte development can be dissociated from thymic dendritic cell development, both for so called myeloid as well as lymphoid dendritic cells (Rodewald *et al.*, 1999). These findings were in principle corroborated by the report that common myeloid progenitors (CMP) can give rise to so called lymphoid dendritic cells (Traver *et al.*, 2000).

The molecular basis for the double growth factor receptor dependency at birth, each receptor is potentially regulating proliferation, survival, or differentiation, is only poorly understood. In attempts to rescue the defects in thymocyte development in growth factor receptor mutants by overexpression of the anti-apoptotic protein bcl-2, other investigators (Akashi *et al.*, 1997; Kondo *et al.*, 1997a; Maraskovsky *et al.*, 1997), and we (Rodewald *et al.*, 2001), have obtained different results, at least with regard to the potential rescue of γ_c deficiency by enforced expression of bcl-2. Transgenic bcl-2, biologically active at the appropriate stages of development (Eμ-bcl-2-25; expressed in the T cell lineage) (O'Reilly *et al.*, 1997), was introduced into the following genotypes: c-Kit and γ_c wild type (c-Kit$^+\gamma_c^+$bcl$^+$), or c-Kit-deficient (c-Kit$^{W/W}\gamma_c^+$bcl$^+$), or γ_c-deficient (c-Kit$^+\gamma_c^-$bcl$^+$), or c-Kit and γ_c double-deficient mice (c-Kit$^{W/W}\gamma_c^-$bcl$^+$). In our hands, introduction of transgenic bcl-2 promoted survival of *ex vivo* isolated thymocytes, but failed to release T cell precursors from their phenotypic block, and no increase in progenitor or total thymocyte cellularity was observed in any of the mutants studied. Thus, we propose that bcl-2 does not act as a key mediator in signaling pathways driving early thymocyte development (Rodewald *et al.*, 2001). Our experiments (Rodewald *et al.*, 2001) agree with some (Blom *et al.*, 1997) but not other reports (Kondo and Weissman, 2000), and such differences might be accounted for by modifier genes present in the genetic background of the various mutants studied.

B Cell Development in Lethal c-Kit$^{W/W}$ Mice, and Following Transfer of c-Kit$^{W/W}$ Fetal Liver Stem Cells into Rag-2-Deficient Hosts

Several lines of evidence originally suggested a role for c-Kit in B cell development: (1) Kitl and IL-7 can synergize in the proliferation of precursor B cells *in vitro* (Martin *et al.*, 1990; McNiece *et al.*, 1991; Williams *et al.*, 1992; Era *et al.*, 1994; Yasunaga *et al.*, 1995); (2) blocking anti-c-Kit antibody inhibits the growth of B cell precursors *in vitro* (Rolink *et al.*, 1991); (3) In c-Kit$^{W/Wv}$ mice, numbers of B cell precursors in bone marrow are weakly reduced (Mackarehtschian *et al.*, 1995); (4) B lymphocyte development is blocked following injection of anti-c-Kit antibody into wildtype mice (Era *et al.*, 1994), However, the latter experiment cannot distinguish between inhibition by anti-c-Kit antibody at the level of the HSC, or at the level of the committed B lymphocyte progenitor.

The role of c-Kit in B lymphocyte development was analyzed in day 16 and day 18 fetal liver, and in bone marrow and spleen in 5 days old c-Kit$^{W/W}$ mice. B cell generation was also examined following transfer of c-Kit$^{W/W}$ or wild-type fetal liver cells from day 13 of gestation into lymphocyte-deficient Rag-2$^{-/-}$ mice. In contrast to the evidence summarized above, these experiments indicated that fetal liver-derived HSC can develop normally into precursor B cells in the bone marrow and into mature B1 and B2 type B cell in peripheral lymphoid sites (peritoneal cavity and spleen) in the absence of c-Kit.

Identification of Viable Adult c-Kit$^{W/W}$ Mice

Given the critical role for c-Kit in T cell development around birth, it was of interest to examine the impact of c-Kit deficiency under conditions of steady state lymphopoiesis in adult life. However, homozygous null mice (c-Kit$^{W/W}$) suffer from severe macrocytic anemia, and die by 10 days of age (Russell and Lawson, 1959; Mintz and Fleischman, 1979). Thus, no information was available on c-Kit function in adult hematopoiesis.

In the process of generating c-Kit$^{W/W}\gamma_c^-$ mice (see above), WB-c-Kit$^{W/+}$ were mated to γ_c^- mice which were on a mixed genetic background (BALB/c, 129P2/OlaHsd and C57BL/6) (DiSanto *et al.*, 1995). In 1997, in offspring from these c-Kit$^{W/+}$ x c-Kit$^{W/+}$ crosses, we noticed, in addition to lethal c-Kit$^{W/W}$ mice, a white, black-eyed mouse surviving to adulthood. The "survivor" trait has been continued by inter-crossing fertile c-Kit$^{W/+}$ littermates from Vickid mice, a strategy which has yielded more than 50 viable c-Kit$^{W/W}$ mice in the last four years. Genetic and biochemical analyses have shown that these mice, termed Vickid for *vi*able *c-Kit-d*eficient, are "true" c-Kit null mice (Waskow et al. unpublished). The W-mutant c-Kit protein lacks the transmembrane domain (Hayashi *et al.*, 1991; Nocka *et al.*, 1990). Therefore, in Vickid mice, genetically c-Kit$^{W/W}$, the c-Kit protein cannot be expressed on the cell surface. This fact is underscored by the complete absence of c-Kit on lineage-marker (lin) (B220, Mac-1, Gr-1, Ter119, CD3, CD4, CD8)$^-$ bone marrow cells. Roughly 75% of this population which contains mostly uncommitted progenitors, including hematopoietic stem cells defined as lin$^-$c-Kit^{high}Sca-1$^+$, express c-Kit. In clear contrast, c-Kit expression is undetectable in lin$^-$ cells from Vickid bone marrow. In addition, Vickid-derived hematopoietic cell are c-Kit-ligand (Kitl) unresponsive. Overall bone marrow cellularity is reduced by ~50% in Vickid, when compared to c-Kit$^{W/+}$ mice. Vickid mice have an ameliorated bone marrow phenotype since bone marrow cellularity is reduced only two-fold compared to wild type mice. In contrast, in lethal neonatal c-Kit$^{W/W}$ mice, or in fetal Kitl-deficient mice, numbers of bone marrow cells are reduced 10fold compared to wild type mice (Russell, 1979; Ikuta and Weissman, 1992; Rodewald *et al.*, 1997). The putative c-Kit suppressor mutation underlying the viability of Vickid has not been identified yet. Analysis of erythropoiesis suggests that Vickid may have overcome the lethal anemia of WB c-Kit$^{W/W}$ by a partial rescue of red blood cell (RBC) numbers. In Vickid, we find 3.7 x 10^6 (~30% of the normal number) of RBC, whereas, on postnatal day two, lethal WB c-Kit$^{W/W}$ have only 0.8 x 10^6 RBC, ~15% of the normal number) (Waskow et al. unpublished).

Thymocyte Development is Blocked at the Pro-T Cell Stage in Adult c-Kit$^{W/W}$ Mice

Early intrathymic stages of T cell development can be resolved using the expression of c-Kit, CD44, and CD25 on CD3$^-$CD4$^-$CD8$^-$ (triple negative [TN]) thymocytes (developmental order: TN 1 to TN 4 stages (c-Kit$^+$CD25$^-$CD44$^+$ [TN 1]; c-Kit$^{+/-}$ CD44$^+$CD25$^+$ [TN 2]; c-Kit$^-$CD44$^-$CD25$^+$ [TN 3]; and c-Kit$^-$CD25$^-$CD44$^-$ [TN 4] (Pearse *et al.*, 1989; Godfrey *et al.*, 1993). TN give rise to CD4$^+$CD8$^+$ (double positive [DP]), and mature CD3$^+$CD4$^+$CD8$^-$ or CD3$^+$CD4$^-$CD8$^+$ (single positive [SP]) cells. Strikingly, normal percentages of DP are only present in wild type and c-Kit$^{W/W^V}$ thymi but DP are missing in the old Vickid thymus. Heterozygous c-Kit$^{W/W^V}$ mice carry the c-Kit^W allele together with the kinase-weak c-Kit$^{W^V}$ allele. In these mice, very low levels of c-Kit are expressed on the cell surface. A search to identify the stage of this block, shows that (1) CD25$^+$ (TN 2 + TN 3) are absent, and that (2) TN 1 from Vickid, but not wild type mice, fail to express crucial stage-specific (Rag-1, Rag-2, pTα, c-Kit) but not the T lineage-specific genes (CD3ε, CD3δ, Gata-3, Tcf-1), or (IL-7Rα, γ_c) (Waskow et al., unpublished). We propose that the failure to detect mRNAs encoding Rag-1, Rag-2, pTα and c-Kit in TN 1 correlates with absence of pro-T cells in aged Vickid mice.

Finally, cell transfer experiments of lineage-marker negative bone marrow cells from Vickid mice show that (1) c-Kit null HSC can engraft in adoptive hosts, and that (2) the observed defects in B cell development in the bone marrow (see below), and $\alpha\beta$ T cell development in the thymus are cell-intrinsic.

Severe Block in Early B Cell Development in the Bone Marrow in Adult c-Kit-Deficient Mice

In Vickid mice, numbers of peripheral leukocytes (in the blood and in the spleen) are normal and these leukocytes contain normal percentages of mature B ($B220^+IgM^+$) and T ($CD4^+$ and $CD8^+$) lymphocytes. Whether ratios of naïve and memory T and B cell populations – in the absence of *de novo* lymphopoiesis – are normal in aged Vickid mice, remains to be determined. B cell production in Vickid bone marrow is unaffected by lack of c-Kit in young mice but a striking block in B cell development is apparent in mice older than 3 month. In old Vickid mice, B cell development is phenotypically arrested at the transition from fraction B (nomenclature according to Hardy *et al.*, 1991), defined as $B220^+CD43^+CD24^+BP\text{-}1^-$ cells, to fraction C, defined as $B220^+CD43^+CD24^+BP\text{-}1^+$. Consequently, subsequent stages of immature B cells, fraction D ($B220^+CD43^-IgM^-$), and E ($B220^+CD43^-IgM^+$) are also absent. In contrast, recirculating mature B cells (fraction F; $B220^{high}CD43^-IgM^+$) are present in old Vickid bone marrow. This phenotypic alteration, caused by the c-Kit null mutation, was not seen in bone marrow from old $c\text{-}Kit^{W/W^V}$ mice.

Further analyses of progenitor B cells from fraction B, the stage which appears to be phenotypically unaffected according to the "Hardy markers", showed that several developmental defects are already apparent at this stage. (1) Percentages of $CD19^+$ fraction B progenitors are reduced by roughly 50% in Vickid compared to wild type. (2) $CD19^+$ and $CD19^-$ cells from Vickid mice are incapable to develop into B cell colonies *in vitro* compared to the corresponding populations from wild type mice. (3) RT-PCR for expression of stage- and lineage-specific genes, as found in c-Kit null pro T cells, shows a marked difference between wild type and c-Kit null B progenitor cells. mRNAs encoding Igα, Igβ, CD19, E47, VpreB and IL-7Rα are expressed in both wild type and Vickid pro-B cells. Surprisingly, mRNAs encoding Rag-1, Rag-2, and λ5, expressed in wild type pro-B cells, are lacking in Vickid pro-B cells.

In summary, B cell development is strongly c-Kit-dependent in adult bone marrow. Fraction B $CD19^+$ cells are developmentally compromised *in vitro* and *in vivo*. Failure to express a pre B cell receptor due to lack of Rag-1, Rag-2 and λ5 mRNA could explain the affected stage and the magnitude of the observed block. This c-Kit-dependent block in adult mice is in striking contrast to the data obtained from fetal and neonatal B cell development *in situ*. Likewise, initial B cell development in Rag-deficient mice transplanted with c-Kit-deficient (W/W) fetal liver stem cells is c-Kit independent (Takeda *et al.*, 1997). We conclude that B cell development occurs independent from c-Kit signals early in life, but becomes progressively c-Kit-dependent in adult life.

Development of $CD4^-CD8^-$ αβ TCR^+ and $CD4^-CD8^-$ DN γδ TCR^+ Thymocytes in Adult $c\text{-}Kit^{W/W}$ Mice

In Vickid mice, $CD25^+$ TN and and DP thymocytes are lost over time, and thus development of "mainstream" αβ TCR^+ T cells in the thymus is interrupted. Are other intrathymic pathways of T cell development also c-Kit-dependent? In addition to the mature $CD3^+CD4^+CD8^-$ and $CD3^+CD4^-CD8^+$ populations, mature $CD3^+$ thymocytes are found among $CD4^-CD8^-$ thymocytes (negative for B220, CD19, DX5, Gr-1, Mac-1, Ter119). $CD3^+CD4^-CD8^-$ can be divided into αβ TCR^+ and γδ TCR^+ DN cell populations. Analyses of these compartments revealed that, in absolute numbers, γδ TCR^+ DN thymocytes are reduced over 20 fold in Vickid compared to wild type mice already in young adult mice (Waskow et al, unpublished). Thus, we find an early onset of the dependency on c-Kit for proper γδ TCR^+ thymocyte generation. This decline in αβ follows the decline in γδ thymocyte development, and, in old Vickid mice, numbers of both αβ TCR^+ and γδ TCR^+ thymocytes are reduced more than 60fold compared to age-matched wild type mice.

A dependency of γδ TCR^+ T cells on Kitl/c-Kit interactions has previously been shown for γδ TCR^+ intraepithelial lymphocytes (IELs). In mice expressing weak alleles of the c-Kit receptor (c-Kit^{W/W^v}) or Kit ligand ($Kitl^{Sl/Sl^d}$; only soluble Kitl is produced (Brannan *et al.*, 1991)), γδ TCR^+ IELs declines with age (Puddington *et al.*, 1994). These findings demonstrated a role for c-Kit in the development of γδ TCR^+ IELs. However, lack of c-Kit may cause defects in extrathymic and/or intrathymic γδ TCR^+ T cell development, both of which contribute to of γδ TCR^+ IELs. The early decline of γδ TCR^+ thymocytes in Vickid mice now shows that the intrathymic pathway of γδ T cell development is strongly c-Kit-dependent.

Development of Natural Killer Cells in Adult c-$Kit^{W/W}$ Mice

C-Kit expression is a "common theme" in hematopoietic progenitors, and may therefore also be involved in the development of natural killer cells. This lineage is HSC-derived (Haller *et al.*, 1977), and its development may follow common pathways with other lymphoid lineages, i.e. the common lymphocyte progenitors (CLP) can generate NK cells in addition to T and B cells (Kondo *et al.*, 1997b), and fetal thymocytes include a common T/NK progenitor (Rodewald *et al.*, 1992). Growth factors and receptors known to be involved in NK development/function are IL-15 (Kennedy *et al.*, 2000), the IL-15Rα chain (Lodolce *et al.*, 1998), CD122 (IL-2R β chain) (Suzuki *et al.*, 1997) and γ_c (DiSanto *et al.*, 1995). A committed NK cell precursor (NKP) population (lin^- $CD122^+DX5^-NK1.1^-$) has been reported in the bone marrow (Rosmaraki *et al.*, 2001). Using this phenotypic resolution, we analyzed bone marrow in Vickid mice for NK precursors. In contrast to pro T and pro B cells, bone marrow from Vickid mice contains normal percentages of $lin^-NK1.1^-$ cells (0.6% of total bone marrow), and percentages of $CD122^+$ NKP within the $lin^-NK1.1^-$ population were similar in wild type (3.8%±1.7; n = 3), and Vickid mice (2.8%±1.4; n = 5). We conclude that, in contrast to T and B cell development which are strongly c-Kit-dependent in adult life, NK cell development occurs normally in the absence of c-Kit. Colucci and DiSanto who analyzed NK cell development and function in Rag-$2^-\gamma_c^-$ mice transplanted with c-Kit-deficient (W/W) fetal liver stem cells have suggested that c-Kit provides a critical signal for survival, expansion, and maturation of mouse natural killer cells. Our analysis of Vickid mice does not support a role of c-Kit in NK cell development. Whether or not NK cell function, primarily cytotoxicity, is also normal in adult c-Kit null mice remains to be determined.

ACKNOWLEDGMENTS

We thank Dr. H.J. Fehling for critical reading of the manuscript.

REFERENCES

Akashi, K., Kondo, M., von Freeden-Jeffry, U., Murray, R., and Weissman, I. L., 1997, Bcl-2 rescues T lymphopoiesis in interleukin-7 receptor-deficient mice. *Cell* **89**, 1033-1041.

Akashi, K., Kondo, M., and Weissman, I. L., 1998, Role of interleukin-7 in T-cell development from hematopoietic stem cells. *Immunol Rev* **165**, 13-28.

Ardavin, C., Wu, L., Li, C.-L., and Shortman, K., 1993, Thymic dendritic cells and T cells develop simultaneously in the thymus from a common precursor population. *Nature* **362**, 761-763.

Asamoto, H., and Mandel, T. E., 1981, Thymus in mice bearing the steel mutation: Morphologic studies on fetal, neonatal, organ-cultured, and grafted fetal thymus. *Lab Invest* **45**, 418-426.

Bernstein, A., Chabot, B., Dubreuil, P., Reith, A., Nocka, K., Majumder, S., Ray, P., and Besmer, P., 1990, The mouse W/c-kit locus. *Ciba Found Symp* **148**, 158-166.

Besmer, P., 1991, The kit ligand encoded at the murine Steel locus: a pleiotropic growth and differentiation factor. *Curr Opin Cell Biol* **3**, 939-946.

Blom, B., Spits, H., and Krimpenfort, P., 1997, *The role of the common gamma chain of the IL-2, IL-4, IL-7 and IL-15 receptors in development of lymphocytes. Constitutive expression of bcl-2 does not rescue the developmental defects in gamma common deficient mice.* Kluwer Academic Publishers, Dordrecht.

Brannan, C. I., Lyman, S. D., Williams, D. E., Eisenman, J., Anderson, D. M., Cosman, D., Bedell, M. A., Jenkins, N. A., and Copeland, N. G., 1991, Steel-Dickie mutation encodes a c-kit ligand lacking transmembrane and cytoplasmic domains. *Proc Natl Acad Sci* **88**, 4671-4674.

Conlon, P. J., Morrissey, P. J., Nordan, R. P., Grabstein, K. H., Prickett, K. S., Reed, S. G., Goodwin, R., Cosman, D., and Namen, A. E., 1989, Murine thymocytes proliferate in direct response to interleukin-7. *Blood* **74**, 1368-1373.

DiSanto, J. P., Muller, W., Guy-Grand, D., Fischer, A., and Rajewsky, K., 1995, Lymphoid development in mice with a targeted deletion of the interleukin 2 receptor gamma chain. *Proc Natl Acad Sci U S A* **92**, 377-381.

DiSanto, J. P., Radtke, F., and Rodewald, H. R., 2000, To be or not to be a pro-T? *Curr Opin Immunol* **12**, 159-165.

DiSanto, J. P., and Rodewald, H. R., 1998, *In vivo* roles of receptor tyrosine kinases and cytokine receptors in thymocyte development. *Curr Opin Immunol* **10**, 196-207.

Era, T., Nishikawa, S., Sudo, T., Wang, F. H., Ogawa, M., Kunisada, T., Hayashi, S.-I., and Nishikawa, S.-I., 1994, How B-precursor cells are driven to cycle. *Immunol Reviews* **137**, 35-51.

Galli, S. J., Zsebo, K. M., and Geissler, E. N., 1994, The kit ligand, stem cell factor. *Adv Immunol* **55**, 1-96.

Godfrey, D. I., Kennedy, J., Suda, T., and Zlotnik, A., 1993, A developmental pathway involving four phenotypically and functionally distinct subsets of CD3-CD4-CD8- triple-negative adult mouse thymocytes defined by CD44 and CD25 expression. *J Immunol* **150**, 4244-4252.

Haller, O., Kiessling, R., Orn, A., and Wigzell, H., 1977, Generation of natural killer cells: an autonomous function of the bone marrow. *J Exp Med* **145**, 1411-1416.

Hardy, R. R., Carmack, C. E., Shinton, S. A., Kemp, J. D., and Hayakawa, K., 1991, Resolution and characterization of pro-B and pre-pro-B cell stages in normal mouse bone marrow. *J Exp Med* **173**, 1213-1225.

Hayashi, S., Kunisada, T., Ogawa, M., Yamaguchi, K., and Nishikawa, S., 1991, Exon skipping by mutation of an authentic splice site of c-kit gene in W/W mouse. *Nucleic Acids Res* **19**, 1267-1271.

Ikuta, K., and Weissman, I. L., 1992, Evidence that hematopoietc stem cells express mouse c-kit but do not depend on steel factor for their generation. *Proc Natl Acad Sci USA* **89**, 1502-1506.

Kennedy, M. K., Glaccum, M., Brown, S. N., Butz, E. A., Viney, J. L., Embers, M., Matsuki, N., Charrier, K., Sedger, L., Willis, C. R., Brasel, K., Morrissey, P. J., Stocking, K., Schuh, J. C., Joyce, S., and Peschon, J. J., 2000, Reversible defects in natural killer and memory CD8 T cell lineages in interleukin 15-deficient mice. *J Exp Med* **191**, 771-780.

Kondo, M., Akashi, K., Domen, J., Sugamura, K., and Weissman, I. L., 1997a, Bcl-2 rescues T lymphopoiesis, but not B or NK cell development, in common gamma chain-deficient mice. *Immunity* **7**, 155-162.

Kondo, M., and Weissman, I. L., 2000, Function of cytokines in lymphocyte development. *Curr Top Microbiol Immunol* **251**, 59-65.

Kondo, M., Weissman, I. L., and Akashi, K., 1997b, Identification of clonogenic common lymphoid progenitors in mouse bone marrow. *Cell* **91**, 661-672.

Lodolce, J. P., Boone, D. L., Chai, S., Swain, R. E., Dassopoulos, T., Trettin, S., and Ma, A., 1998, IL-15 receptor maintains lymphoid homeostasis by supporting lymphocyte homing and proliferation. *Immunity* **9**, 669-676.

Mackarehtschian, K., Hardin, J. D., Moore, K. A., Boast, S., Goff, S. P., and Lemischka, I. R., 1995, Targeted disruption of the flk2/flt3 gene leads to deficiencies in primitive hematopoietic progenitors. *Immunity* **3**, 147-161.

Maraskovsky, E., O'Reilly, L. A., Teepe, M., Corcoran, L. M., Peschon, J. J., and Strasser, A., 1997, Bcl-2 can rescue T lymphocyte development in interleukin-7 receptor-deficient mice but not in mutant rag-1-/- mice. *Cell* **89**, 1011-1019.

Martin, F. H., Suggs, S. V., Langley, K. E., Hsieng, S. L., Ting, J., Okino, K. H., Morris, C. F., McNiece, I. K., Jacobsen, F. W., Mendiaz, E. A., Birkett, N. C., Smith, K. A., Johnson, M. J., Parker, V. P., Flores, J. C., Patel, A. C., Fisher, E. F., Erjavec, H. O., Herrera, C. J., Wypch, J., Sachdev, R. K., Pope, J. A., Leslie, I., Wen, D.-Z., Lin, C.-H., Cuppies, R. L., and Zsebo, K. M., 1990, Primary structure and functional expression of rat and human stem cell factor DNAs. *Cell* **63**, 203-211.

McNiece, I. K., Langley, K. E., and Zsebo, K. M., 1991, The role of recombinant stem cell factor in early B cell development. *J Immunol* **146**, 3785-3790.

Mintz, B., and Fleischman, R. A., 1979, Prevention of genetic anemias in mice by microinjection of normal hematopoietic stem cells into the fetal placenta. *Proc Natl Acad Sci U S A* **76**, 5736-5740.

Murray, R., Suda, T., Wrighton, N., Lee, F., and Zlotnik, A., 1989, IL-7 is a growth and maintenance factor for mature and immature thymocyte subsets. *Int Immunol* **1**, 526-531.

Nocka, K., Tan, J. C., Chiu, E., Chu, T. Y., Ray, P., Traktman, P., and Besmer, P., 1990, Molecular bases of dominant negative and loss of function mutations at the murine c-kit/white spotting locus: W37, Wv, W41 and W. *Embo J* **9**, 1805-1813.

O'Reilly, L., Harris, A., and Strasser, A., 1997, bcl-2 transgene expression promotes survival and reduces proliferation of CD3-CD4-CD8- T cell progenitors. *Int Immunol* **9**, 1291-1301.

Pearse, M., Wu, L., Egerton, M., Wilson, A., Shortman, K., and Scollay, R., 1989, A murine early thymocyte developmental sequence is marked by transient expression of the interleukin 2 receptor. *Proc Nat Acad Sci USA* **86**, 1614-1618.

Puddington, L., Olson, S., and Lefrancois, L., 1994, Interactions between stem cell factor and c-Kit are required for intestinal immune system homeostasis. *Immunity* **1**, 733-739.

Rodewald, H. R., and Fehling, H. J., 1998, Molecular and cellular events in early thymocyte development. *Adv Immunol* **69**, 1-112.

Rodewald, H. R., and Haller, C., 1998, Antigen receptor junctional diversity in growth factor receptor deficient mice. *Dev Comp Immunol* **22**, 351-365.

Rodewald, H. R., Kretzschmar, K., Swat, W., and Takeda, S., 1995, Intrathymically expressed c-kit ligand (stem cell factor) is a major factor driving expansion of very immature thymocytes *in vivo*. *Immunity* **3**, 313-319.

Rodewald, H. R., Moingeon, P., Lucich, J. L., Dosiou, C., Lopez, P., and Reinherz, E. L., 1992, A population of early fetal thymocytes expressing FcγRII/III contains precursors of T lymphocytes and natural killer cells. *Cell* **69**, 139-150.

Rodewald, H. R., Ogawa, M., Haller, C., Waskow, C., and DiSanto, J. P., 1997, Pro-thymocyte expansion by c-kit and the common cytokine receptor γ chain is essential for repertoire formation. *Immunity* **6**, 265-272.

Rodewald, H. R., Brocker, T., and Haller, C., 1999, Developmental dissociation of thymic dendritic cell and thymocyte lineages revealed in growth factor receptor mutant mice. *Proc Natl Acad Sci USA* **96**, 15068-15073.

Rodewald, H. R., Waskow, C., and Haller, C., 2001, Essential requirement for c-kit and common γ chain, γc) in thymocyte development cannot be overruled by enforced expression of Bcl-2. *J Exp Med* **193**, 1431-1437.

Rolink, A., Streb, M., Nishikawa, S.-I., and Melchers, F., 1991, The c-kit-encoded typrosine kinase regulates the proliferation of early pre-B cells. *Eur J Immunol* **21**, 2609-2612.

Rosmaraki, E. E., Douagi, I., Roth, C., Colucci, F., Cumano, A., and DiSanto, J. P., 2001, Identification of committed NK cell progenitors in adult murine bone marrow. *Eur J Immunol* **31**, 1900-1909.

Russell, E., 1979, Hereditary anemias of the mouse : A review for geneticists. *Adv Genetics* **20**, 357-459.

Russell, E. S., and Lawson, F. A., 1959, Selection and inbreeding for longevity of a lethal type. *J Hered* **50**, 19-25.

Suzuki, H., Duncan, G. S., Takimoto, H., and Mak, T. W., 1997, Abnormal development of intestinal intraepithelial lymphocytes and peripheral natural killer cells in mice lacking the IL-2 receptor beta chain. *J Exp Med* **185**, 499-505.

Takeda, S., Shimizu, T., and Rodewald, H. R., 1997, Interactions between c-kit and stem cell factor are not required for B cell development *in vivo*. *Blood* **89**, 518-525.

Traver, D., Akashi, K., Manz, M., Merad, M., Miyamoto, T., Engleman, E. G., Weissman, I. L., 2000, Development of CD8α-positive dendritic cells from a common myeloid progenitor. *Science*, **290**, 2152-2154.

von-Freeden-Jeffry, U., Vieira, P., Lucian, L. A., McNeil, T., Burdach, S. E., and Murray, R., 1995, Lymphopenia in interleukin (IL)-7 gene-deleted mice identifies IL-7 as a nonredundant cytokine. *Journal of Experimental Medicine*, **181**, 1519-1526.

Williams, D. E., de Vries, P., Namen, A. E., Widmer, M. B., and Lyman, S. D., 1992, The Steel Factor. *Dev Biology* **151**, 368-376.

Yasunaga, M., Wang, F., Kunisada, T., Nishikawa, S., and Nishikawa, S., 1995, Cell cycle control of c-kit+IL-7R+ B precursor cells by two distinct signals derived from IL-7 receptor and c-kit in a fully defined medium. *J Exp Med* **182**, 315-323.

DEVELOPMENT OF ANTIGEN-SPECIFIC HELPER T CELL RESPONSES IN VIVO

Antigen-specific Th cell subsets

Louise McHeyzer-Williams, Laurent Malherbe, Michael Eisenbraun, David Driver and Michael McHeyzer-Williams*

1. INTRODUCTION

Helper T cell-regulated B cell immunity constitutes a major component of the immune response to many pathogens. Spatially and temporally regulated changes in lymphocyte function and physiology underpin the primary adaptive response to antigen. We have developed flow-cytometric and single-cell RT-PCR based strategies to quantify the antigen-specific Th cell and B cell responses in non-transgenic animals directly ex vivo. Here, we will summarize our studies in the B10.BR murine response to pigeon cytochrome c (PCC) with some emphasis on the cellular complexity of the Th cell-dependent B cell response to the hapten (4-hydroxy-3-nitrophenyl)acetyl (NP).

Formation of immune synapsis in vivo is presented as the central regulating event in the development of high-affinity humoral immunity. A series of three quantitatively and qualitatively distinct immune synapses are proposed to act as checkpoints in the development of the primary immune response to antigen. Synapse I occurs between activated antigen presenting cells (APC) and naïve antigen-specific Th cells. This synapse communicates the inflammatory context of initial antigen encounter to the adaptive immune system and serves to recruit T cell help with TCR of the 'best fit'. Synapse II occurs between clonally-expanded specific Th cells and antigen-activated B cells. This immune synapse constitutes the framework for the delivery of cognate T cell help to B cells and most likely represents different sets of inter-cellular molecular activities between functionally discrete cellular subsets. Synapse III occurs within the confines of the germinal center (GC) reaction that focuses on the development of affinity-matured B cell memory. GC Th cells and GC B cells are phenotypically-distinct from

The Scripps Research Institute, La Jolla, California, 92037 and Duke University Medical Center, Durham, North Carolina, 27710 their pre-GC counterparts, hence we also consider synapse III molecular interactions as unique and distinct.

The development of Th cell and B cell memory is one outcome of the primary immune response and will be discussed in a final section on memory responses. Cellular heterogeneity in the rapid memory B cell response to antigen recall is emphasized.

2. IMMUNE SYNAPSIS

Specific recognition of peptide-MHC (pMHC) complexes by the TCR of helper T cells is central to initiating adaptive immunity. Successful cognate interactions occur at the cellular interface of antigen-specific Th cells and antigen-activated APC [1, 2]. In Th cell-regulated B cell responses, the most likely APC are dendritic cells and antigen-specific B cells at temporally separate phases of immune response development. In general, adhesion molecule interactions establish initial cell contact and then assort to the periphery of the contact area when TCR-pMHC interactions of sufficient avidity occur. TCR concentrate centrally at the interface together with co-stimulators such as CD28 and co-receptors such as CD4. Signaling intermediates also focus at the intracellular face of these contact sites now commonly referred to as the immunological synapse. These molecular interactions organize too slowly to account for the earliest signaling events associated with antigen recognition [3]. Nevertheless, they appear indispensable to cell fate determination and are required for cell cycle entry and development of effector Th cell function.

Immune synapse formation can be considered a critical regulating checkpoint during the complex developmental program that underlies humoral immunity [4]. While the organized molecular interactions of synapsis have not been demonstrated in vivo, the quality of the information exchange can substantially impact the cellular outcome. The strength and duration of TCR-pMHC interactions together with the extent and ever-growing range of co-stimulation provides a flexible mechanism for influencing cell fate decisions. Using this notion, we have divided the developing primary immune response into three separate but overlapping phases of cellular development, each focused around the formation of an immune synapse between cells at distinct phases of their antigen-driven development.

3. CLONAL SELECTION

We study the I-E^k restricted B10.BR response to the model protein antigen PCC [5-7]. The majority of the PCC-specific Th cell response to this antigen is focused to a single peptide epitope recognized by Th cells that express a restricted pair of V region genes, Vα11 and Vβ3 [8]. Not all Vα11Vβ3 expressing Th cells express peptide-binding junctional regions (CDR3) hence to identified the PCC-responsive component we use the modulation of cell surface phenotype to indicate antigen experience in vivo (Figure 1)[5, 9]. We developed pMHC tetramers to allow direct ex vivo analysis of antigen-specificity that focused on peptide specificity and MHC restriction as the defining characteristics for cellular quantification [10]. An example of class II restricted pMHC tetramer labeling is present in Figure 1B [11, 12]. Hence, we had two complementary flow-cytometric based strategies for quantifying and then isolating PCC-specific Th cells directly ex vivo. This approach allowed us to monitor the emergence of the antigen-specific Th cell arm of the adaptive immune response to a T dependent protein antigen that induces mainly humoral immunity.

Aggravation of APC at the site of antigen entry is a necessary and formative aspect of the adaptive immune response. We use whole protein antigen in the RAS adjuvant system that acts to activate APC using the Ribi lipopolysaccharide (LPS) derivative. Dendritic cells are the most likely APC in these responses and will become activated non-specifically due to the influence of the adjuvant and then uptake, process and present peptides of PCC. These activated DC migrate to the T cell zones of secondary lymphoid tissue over the first ~3 days after immunization. We primarily focus on the periaortic and inguinal LNs after base of tail subcutaneous injection with antigen.

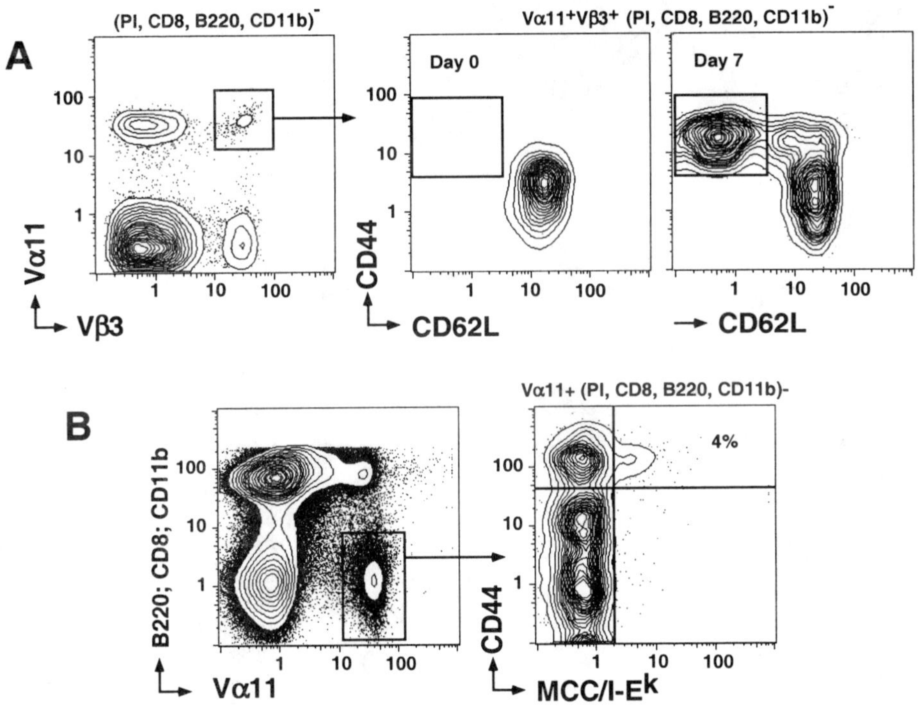

Figure 1. The PCC specific Th cell response in B10.BR mice. (A) The standard V region-specific strategy for identifying PCC specific Th cells. LN cells are labeled with anti-B220, CD8, CD11b antibodies and propidium iodide (PI) to exclude B cells, CD8 T cells, macrophage and dying cells from analysis. Vα11 and Vβ3 levels are displayed in the first panel to identify Th cells that express both Vα11 and Vβ3 (the insert box). PCC responsive cells will up-regulated CD44 and down-regulate CD62L in response to antigen and form a sub-population of Vα11+Vβ3+ cells clearly discernible at day 7 after immunization (insert box third panel). (B) pMHC tetramer binding stratègy to isolate PCC-specific Th cells. Similar labeling is used to focus analysis on CD4 Th cells that express Vα11 V region genes (insert box first panel). CD44 versus pMHC tetramer labeling identifes a small subpopulation of MCC/I-Ek binding Th cells in the draining LN at day 7 after primary immunization. These cells are also mostly CD62L low (data not shown) as predicted from the V region-specific strategy in section A.

We detect the presence of antigen-specific Th cells at numbers significantly higher than adjuvant only controls by day 3 after priming [9]. To confirm the specificity of these cells, we amplify the TCR junctional regions for evidence of canonical features previously associated with PCC-binding in hybridoma analyses. These analyses are conducted with single cell resolution to provide a secondary assessment of responder cell frequency in vivo.

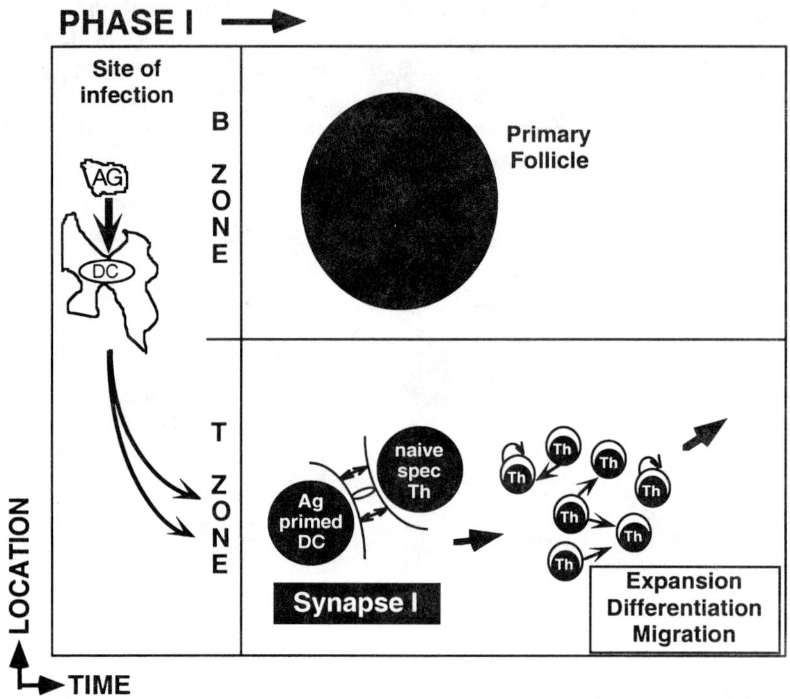

Figure 2. Phase I: Recruitment and expansion of antigen-specific T cell help. The B cell response begins at the time of initial antigen entry with activation of local APC, antigen uptake and migration of the APC to the T cell zones of the draining lymph node. Synapse I occurs between activated APC presenting foreign pMHC complexes to naïve Th cells with receptors of sufficient avidity to trigger a signaling response. TCR signals together with the approproiate costimulation induce cell cycle entry, clonal expansion and effector Th cell differentiation. Th cells with preferred TCR are selectively expanded and migrate to the T/B borders of these secondary lymphoid tissues. While the precise details are still unclear, naïve antigen-speciificB cells also encountering antigen during this early phase in development but will require cognate T cell help to clonally-expand and differentiate.

At day 3, ~40% of the PCC-specific Th cells express >6/8 of the preferred junctional features associated with this response. By day 5 the percentage reaches ~80% and does not change substantially thereafter (to day 14 of the primary response). Upon antigen recall there is a further focusing of the TCR usage reaching 96% of PCC-specific Th cells expressing >6 of the preferred junctional features. These data suggest that clonal maturation in the PCC response is very rapid and largely over by day 5 after priming. Thus, PCC-specific Th cells are recruited early in the response but those with TCR of the'best fit' are preferentially-expanded over the course of the first week to dominate the responder cell compartment in vivo.

Preferential clonal expansion is accompanied by effector Th cell differentiation. Analyzing cytokine mRNA expression directly ex vivo with single cell resolution, we demonstrate that the appearance of specialized functions occurs in progressive waves over time and across subsets of the PCC-specific Th cell compartment [12]. Antigen-driven activation and clonal selection is also associated with the down-regulation of responsiveness to T zone localized chemokines and the up-regulation of CXCR5 expression that confers responsiveness to B zone chemokines [13]. This change in chemokine balance drives the responsive Th cells to the T/B borders of these LNs to promote contact with antigen-activated B cells. This clonal expansion of PCC-specific Th cells and their movement towards B cell zones constitutes the end of phase I of the developing humoral response.

4. HELPER T CELL SUBSETS

To receive cognate T cell help, B cells must have encountered and recognized the same antigen, processed and presented the dominant PCC epitope in the context of MHC class II. The precise details regarding when, where and how this happens still remains unclear. While B cells can clearly bind soluble antigen, cell-associated antigen is far more efficient at activating specific B cells [14]. Nevertheless, using a protein antigen, or hapten-protein carrier model antigen, T cell help is required for clonal expansion, isotype switch recombination and affinity maturation of the antigen-specific B cell compartment. The initial recruitment and activation of the specific B cells should be considered a Phase I activity in the development of the humoral response. Cyster and colleagues have recently demonstrated a similar chemokine balancing mechanism that actively relocates antigen-activated B cells to the T/B borders of LNs [15]. This relocation increases the likelihood for cognate interchange with the antigen-responsive Th cells.

Phase II begins with the formation of immune synapse II. This synapse occurs between the antigen-responsive Th cells (most reasonably considered the effector Th cells) and the antigen-activated B cells. These interactions can be visualized directly ex vivo using adoptive transfer of antigen receptor transgenic Th cells and B cells [16]. The frequency of the responding B cells before expansion is generally too low in non-transgenic animals to allow reliable detection by immunhistochemistry. While cytokine production is considered a key and critical Th cell function, it is now clear that the expressed cell surface molecules exert an equally critical role in B cell differentiation. CD40L was the earliest of these molecules characterized on activated Th cells. In the absence of CD40L, B cell differentiation was grossly impaired with no isotype-switch or GC formation in the resultant B cell response. TNF/TNFR family members [17] and B7/CD28 family of molecules [18] also have substantial impact on the quality of the subsequent B cell response.

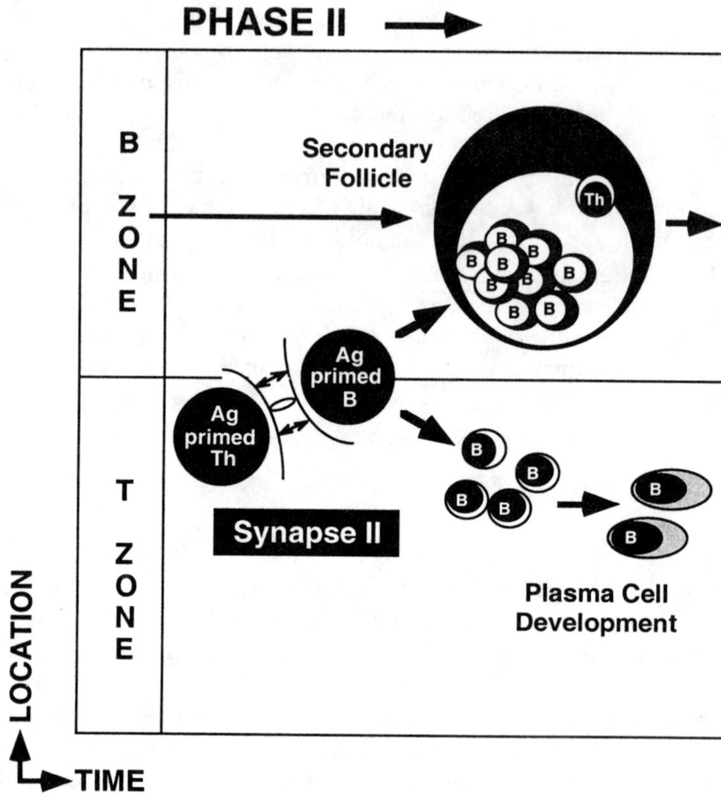

Figure 3. Phase II: The delivery of cognate T cell help to B cells. The formation of immune synapse II constitutes the beginning of Phase II of the humoral immune response. Synapse II occurs between antigen-responsive and expanded Th cells and antigen-activated B cells at the T/B borders of secondary lymphoid tissue. B cell fate bifurcates at this point. One pathway proceeds in the T cell zones as plasma cell differentiation with isotype-switch recombination and no somatic diversification of germline encoded BCR. The second pathway proceeds in the B cell zones with the formation of the secondary follicles as the first step in the germinal center reaction.

We have also recently demonstrated that the nature of the signal through the TCR is substantially altered at this stage of the primary immune response (day 7-9 after priming). While naïve Th cells transduce a robust intracellular calcium response upon ligation of the CD3 complex in vitro, PCC-specific Th cells at the peak of clonal expansion in vivo (day 7) display blocked TCR-mediated calcium responsiveness coupled to substantial reductions in TCR-independent capacitative calcium entry (CCE) [19]. While the significance of this altered physiology is still speculative, it clearly demonstrates that the reception of pMHC recognition changes substantially as a consequence of antigen experience. Thus, the rules that govern the outcome of synapse II interactions promise to be quite unique and distinct to those more closely studied at synapse I.

Cell surface phenotype is also not evenly distributed at this stage of the Th cell response. We have detected subset distribution for variety of TNFR family receptors on the PCC-specific Th cells such as CD27, CD30, CD40L. CD69 is also expressed on a substantial subset of day 7 cells and appears to have a role to play in the regulation of TCR-independent CCE in these cells [19]. Thus, PCC-specific effector Th cells that deliver help to B cells are not all the same at this stage in the developing humoral response.

The GC reaction is the next phase in the development of humoral immunity. The secondary follicle polarizes into a T zone proximal region (dark zone) of highly proliferating sIg^- B cells that somatically diversify the BCR V region genes and a zone of sIg^+ B cells at the opposite pole (light zone). The sIg^+ B cells in the light zone display evidence of somatic hypermutation and undergo selection based on increased affinity for antigen binding. The mechanism that controls this selection process remains unclear. The light zone is filled with follicular dendritic cell (FDC) processes with evidence for antigen-specific Th cells in this region. Blocking antigen trapping by the FDC appears to have little effect on selection. However, interfering with T cell help at this stage of the response alters the efficiency antigen-driven selection and can terminate the GC microenvironment. Further, it is presumed that the delivery of T cell help in this microenvironment is also antigen-specific and requires the formation of a qualitatively and quantitatively distinct immune synapse III for this specialized function. Antigen-specific GC Th cells are thought to display unique cell surface phenotype from their T zone counterparts [20] however, their precise role in the GC cycle of activity needs further analysis. GC B cells also display a phenotype that is distinct from their pre-GC counterparts with a unique constellation of co-stimulatory molecules that could impact the quality of synapse III interactions.

5. MEMORY B CELL SUBSETS

The outcome of the GC reaction is a long-lived memory B cell compartment. This post-GC compartment comprises at least two broadly different cellular products. Long-lived terminally-differentiated plasma cells are one of these products. These plasma cells display evidence of somatic hypermutation and antigen-driven selection in the sequence of their expressed V region genes [21]. They preferentially home to the bone marrow where they persist secreting antibody for extended periods of time [22, 23]. These cells can be considered memory B cells as they contribute the persisting serum presence of high affinity specific antibody that is a first line of defense against re-exposure to the same antigen.

The second type of memory B cell is a non-secreting memory response precursor. These post-GC B cells also display the same level of somatic hypermutation and affinity maturation as the long-lived plasma cells, however, they do not spontaneously secrete antibody in the absence of rechallenge. These cells may be programmed for accelerated antigen recall due to their increased affinity and also changes in basic response physiology as a consequence of initial antigen experience. These memory response precursors rapidly expand upon antigen rechallenge and differentiate into plasma cells in a memory Th cell dependent manner. Typically, these memory response precursors are thought to have isotype switched and are IgM/IgD negative and express the pan B cell marker B220. We have recently demonstrated that there are multiple varieties of affinity matured memory response precursors (Figure 4) [24, 25]. In the memory response to the hapten NP on a protein carrier, we demonstrate the presence of the typical B220$^+$ memory response precursor and a second substantial compartment of atypical NP-specific B cells that were B220$^-$. These atypical B220$^-$ memory responders expanded rapidly upon antigen recall, expressed predominantly the lambda light chain genes with evidence for somatic mutation and expanded and efficiently produced NP-specific plasma cells upon transfer with antigen into Rag deficient hosts.

The B220$^-$ memory cells further sub-divided into two major fractions. The first subset expressed high levels of CD11b and IgG while the second subset labeled with anti-IgE (and not IgG) with lower levels of CD11b. Both subsets dominated the bone marrow compartment of NP-specific cells as late as 42 days post antigen recall. While the CD11b high subset expressed high levels of CD79b, the BCR co-receptor, the IgE+ subset expressed very low levels of this co-receptor molecule (similar to those of plasma cells). All three subsets of memory response precursor develop during the primary response to antigen and appear to be the product of the primary response GC reaction [26].

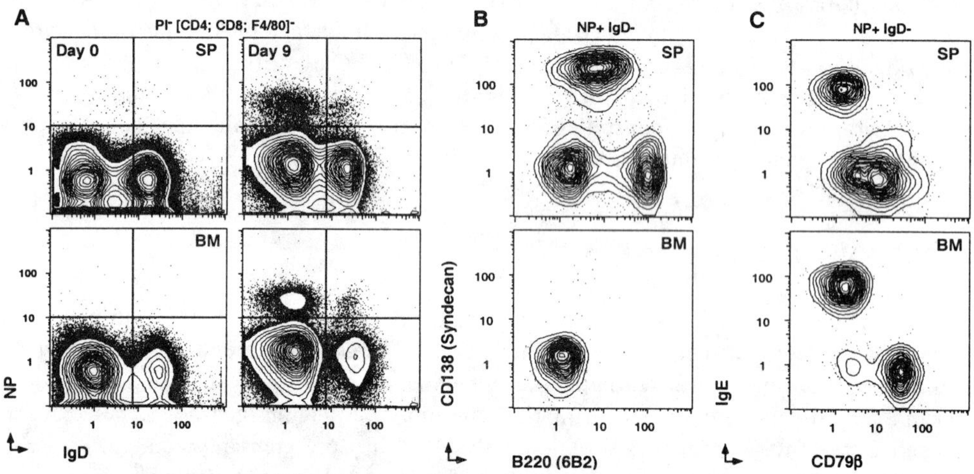

Figure 4. NP-specific memory B cell response, day 5 antigen recall. (A) Spleen cells have been labeled with anti-CD4, CD8 and F4/80 and PI to exclude T cells, macrophage and dying cells from analysis. Displayed are the profiles for IgD expression and NP-binding on the remaining cells in the spleen (upper) or bone marrow (lower) either 56 days after primary injection (labeled day 0 of the memory response) or 9 days after a secondary injection (day 9 memory). The B220 versus CD138 profiles highlight the major division in the NP-binding cells and the IgE versus CD79b profiles indicate the further sub-division of the B220- B cells (easiest to visualize in the BM).

The precise division of labor among these different memory response precursors is still unclear. Proliferative capacity and differentiative potential of the different subsets upon adoptive transfer suggests that the B220- group display less plasticity than their

B220+ counterparts but more rapidly produce plasma cells upon antigen recall. The differential isotype expression among the memory response precursors is intriguing and may suggest a different set of memory Th cell requirements for re-activation on antigen recall. This level of cellular heterogeneity in the memory B cell compartment suggests a more complex organization of the recall response and long humoral protection than has been previously appreciated.

REFERENCES

1. C. R. Monks, B. A. Freiberg, H. Kupfer, N. Sciaky and A. Kupfer. Three-dimensional segregation of supramolecular activation clusters in T cells. *Nature*. (395): 82-6. (1998)
2. A. Grakoui, S. K. Bromley, C. Sumen, M. M. Davis, A. S. Shaw, P. M. Allen and M. L. Dustin. The immunological synapse: a molecular machine controlling T cell activation. *Science*. (285): 221-7. (1999)
3. K. H. Lee, A. D. Holdorf, M. L. Dustin, A. C. Chan, P. M. Allen and A. S. Shaw. T cell receptor signaling precedes immunological synapse formation. *Science*. (295): 1539-42. (2002)
4. M. G. McHeyzer-Williams, L. J. McHeyzer-Williams, J. Fanelli Panus, G. Bikah, R. R. Pogue-Caley, D. J. Driver and M. D. Eisenbraun. Antigen-specific immunity. Th cell-dependent B cell responses. *Immunol Res*. (22): 223-36 (2000)
5. M. G. McHeyzer-Williams and M. M. Davis. Antigen-specific development of primary and memory T cells in vivo. *Science*. (268): 106-11. (1995)
6. M. G. McHeyzer-Williams. Visualizing immune responses in vivo. *Curr Opin Immunol*. (8): 321-6. (1996)
7. M. G. McHeyzer-Williams, J. D. Altman and M. M. Davis. Tracking antigen-specific helper T cell responses. *Curr Opin Immunol*. (8): 278-84. (1996)
8. F. W. Lu, K. Yasutomo, G. B. Goodman, L. J. McHeyzer-Williams, M. G. McHeyzer-Williams, R. N. Germain and J. D. Ashwell. Thymocyte resistance to glucocorticoids leads to antigen-specific unresponsiveness due to "holes" in the T cell repertoire. *Immunity*. (12): 183-92. (2000)
9. L. J. McHeyzer-Williams, J. F. Panus, J. A. Mikszta and M. G. McHeyzer-Williams. Evolution of antigen-specific T cell receptors in vivo: preimmune and antigen-driven selection of preferred complementarity-determining region 3 (CDR3) motifs. *J Exp Med*. (189): 1823-38. (1999)
10. J. D. Altman, P. A. Moss, P. J. Goulder, D. H. Barouch, M. G. McHeyzer-Williams, J. I. Bell, A. J. McMichael and M. M. Davis. Phenotypic analysis of antigen-specific T lymphocytes. *Science*. (274): 94-6. (1996)
11. P. A. Savage, J. J. Boniface and M. M. Davis. A kinetic basis for T cell receptor repertoire selection during an immune response. *Immunity*. (10): 485-92. (1999)
12. J. F. Panus, L. J. McHeyzer-Williams and M. G. McHeyzer-Williams. Antigen-specific T helper cell function: differential cytokine expression in primary and memory responses. *J Exp Med*. (192): 1301-16. (2000)
13. K. M. Ansel, L. J. McHeyzer-Williams, V. N. Ngo, M. G. McHeyzer-Williams and J. G. Cyster. In vivo-activated CD4 T cells upregulate CXC chemokine receptor 5 and reprogram their response to lymphoid chemokines. *J Exp Med*. (190): 1123-34. (1999)
14. F. D. Batista, D. Iber and M. S. Neuberger. B cells acquire antigen from target cells after synapse formation. *Nature*. (411): 489-94. (2001)
15. D. C. Hargreaves, P. L. Hyman, T. T. Lu, V. N. Ngo, A. Bidgol, G. Suzuki, Y. R. Zou, D. R. Littman and J. G. Cyster. A coordinated change in chemokine responsiveness guides plasma cell movements. *J Exp Med*. (194): 45-56. (2001)
16. P. Garside, E. Ingulli, R. R. Merica, J. G. Johnson, R. J. Noelle and M. K. Jenkins. Visualization of specific B and T lymphocyte interactions in the lymph node. *Science*. (281): 96-9. (1998)
17. G. A. Bishop and B. S. Hostager. B lymphocyte activation by contact-mediated interactions with T lymphocytes. *Curr Opin Immunol*. (13): 278-85. (2001)
18. A. H. Sharpe and G. J. Freeman. The B7-CD28 superfamily. *Nature Rev Immunol*. (2): 116-26. (2002)
19. G. Bikah, R. R. Pogue-Caley, L. J. McHeyzer-Williams and M. G. McHeyzer-Williams. Regulating T helper cell immunity through antigen responsiveness and calcium entry. *Nat Immunol*. (1): 402-12.
20. B. Zheng, S. Han and G. Kelsoe. T helper cells in murine germinal centers are antigen-specific emigrants that downregulate Thy-1. *J Exp Med*. (184): 1083-91. (1996)

21. M. G. McHeyzer-Williams, M. J. McLean, P. A. Lalor and G. J. Nossal. Antigen-driven B cell differentiation in vivo. *J Exp Med.* (178): 295-307. (1993)
22. M. K. Slifka, R. Antia, J. K. Whitmire and R. Ahmed. Humoral immunity due to long-lived plasma cells. *Immunity.* (8): 363-72. (1998)
23. M. G. McHeyzer-Williams and R. Ahmed. B cell memory and the long-lived plasma cell. *Curr Opin Immunol.* (11): 172-9. (1999)
24. L. J. McHeyzer-Williams, M. Cool and M. G. McHeyzer-Williams. Antigen-specific B cell memory: expression and replenishment of a novel b220(-) memory b cell compartment. *J Exp Med.* (191): 1149-66. (2000)
25. L. J. McHeyzer-Williams, D. J. Driver and M. G. McHeyzer-Williams. Germinal center reaction. *Curr Opin Hematol.* (8): 52-9. (2001)
26. D. J. Driver, L. J. McHeyzer-Williams, M. Cool, D. B. Stetson and M. G. McHeyzer-Williams. Development and maintenance of a B220- memory B cell compartment. *J Immunol.* (167): 1393-405. (2001)

BCL-6 UNCOUPLES B LYMPHOCYTE PROLIFERATION FROM DIFFERENTIATION

Douglas T. Fearon, Peter M. Manders, and Simon D. Wagner[1]

1. INTRODUCTION

The effective stimulation of lymphocytes by antigen has three potential outcomes for antigen-specific clones, their replication leading to net expansion of cell numbers, their differentiation into effector cells and their differentiation into memory cells. The coordination of these cellular responses is important because, as in other cellular systems, differentiation to the effector stage of development may be associated with an inability to replicate further. For B lymphocytes, terminal differentiation is coupled to cessation of mitosis since non-transformed plasma cells cannot re-enter the cell cycle; it is not clear if terminally differentiated effector T cells have a similar limitation. Therefore, in the immune sytem where the number of antigen-specific clones in a naïve individual is low, premature terminal differentiation occurring before sufficient cellular replication has been achieved would limit the effectiveness of an immune response. Yet, the need for clonal expansion must be appropriately balanced with the obvious benefits of early control and elimination of the infectious process by differentiated lymphocytes. Finally, control of effector cell differentiation and function in the immune system may protect the individual from autoimmune or allergic disease.

2. THE CONTRASTING OUTCOMES OF T-INDEPENDENT AND -DEPENDENT RESPONSES OF B CELLS TO ANTIGEN

Understanding the means by which lymphocytes coordinate their replicative and developmental responses is readily studied with B cells than T cells because, as indicated above, the former resemble most other developmental systems in which replication occurs only with relatively undifferentiated cells and ceases upon terminal differentiation, whereas this may not be the situation with T cells which may revert from an effector stage of development to a pre-effector stage[1]. When B cells respond to TI-2 antigens, or are stimulated *in vitro* by crosslinking of their membrane Ig in the presence of a cytokine, such as IL-4, they rapidly enter cell cycle and divide approximately 8-10 times over a period of

---

[1] All authors are at the Wellcome Trust Immunology Unit, MRC Centre, Hills Road, Cambridge CB2 2SP, UK

3-4 days. They then differentiate to plasma cells by a process involving the expression of Blimp-1[2], which not only regulates the expression of genes associated with the plasma cell stage of development[3], but also represses the expression of c-myc[4], linking terminal differentiation to exit from the cell cycle. No B cells remain for a secondary response, indicating that all antigen-stimulated cells have terminally differentiated or died. Once stimulated, the response of an individual B cell appears to be programmed in that varying the strength of the antigen receptor signal may only changes the number of B cells entering cycle, but does not alter the proliferation and eventual differentiation of those B cells that have reponded[5], an observation that is similar to recent studies of CD8 T cells[6,7]. In short, with T-independent B cell reactions, proliferation is inexorably linked to either differentiation or death.

The B cell response to a T-dependent antigen differs in that the number of cell cycles that a B cell may traverse is relatively unlimited, and it does not necessarily lead to plasma cell differentiation, with the memory B cell being an additional developmental option. These two differences reflect the process of affinity maturation of antibody that is unique to a T-dependent B cell response. Affinity maturation requires that B cells undergo an unpredictable number of iterations of proliferation, mutation of expressed Ig genes and selection. Proliferation is not only for the purpose of clonal expansion, as in a T-independent response and that can be achieved by a pre-determined number of cell cycles, but also for an unpredictable, step-wise acquisition of mutations leading to higher affinity antibodies. That memory B cells persist after the primary response to T-dependent antigen for future interactions with antigen probably reflects the advantage of retaining these highly selected B cells that carry unique, non-germline V region sequences in their H and L chain genes.

The processes of somatic mutation of Ig genes and the development of memory B cells, both of which represent a form of pre-terminal developmental arrest, share a common origin in that they occur in B cells participating in the germinal center reaction. Therefore, in considering models for understanding how lymphocytes might control terminal differentiation, the germinal center B cell is a logical place to look for potential mechanisms. Among the genes that are expressed and required for this phase of B cell development, we chose to examine BCL-6 because of its homology to Tramtrack69 (TTK69), a zinc finger transcriptional repressor that inhibits cellular differentiation another developmental system, that of suppression by Notch of neuronal differentiation during asymmetric cell division of the sensory organ precursor[7a].

3. THE MOLECULAR BIOLOGY OF BCL-6

The *BCL6* gene was discovered through its involvement in recurrent breakpoint sites of chromosomal translocations in diffuse large B-cell lymphomas[8-12]. Although the partners of the chromosomal translocations are variable, each of these genes formed a fusion transcript with BCL6 by removing the first *BCL6* noncoding exon. Thus, transcription of the *BCL6* gene in these B cell lymphomas is under heterologous control, while the oncoprotein is intact. The structure of BCL-6 indicates that it is one of approximately 100 proteins proteins that contain an N-terminal POZ (poxvirus and zinc finger domain) or BTB (Broad-Complex, Tramtrack, and Bric-a-brac) domain and a C-terminal region comprised usually of several zinc fingers. The POZ/BTB domain of BCL-6 and other members of this group of transcriptional repressors mediates homo- and heterodimerization with other POZ/BTB domains, and interacts with a co-repressor proteins, Sin3A, SMRT (silencing mediator for retinoid and thyroid hormone receptors) and N-Cor (nuclear receptor corepressor), which recruit histone deacetylase-1 (HDAC-1) to effect transcriptional repression[13-16]. The structural basis for these protein-protein interactions is suggested by the crystallographic analysis of the POZ/BTB domain of another member of this family of proteins, promyelocytic leukemia zinc finger (PLZF) protein[17,18]. This revealed a tightly intertwined dimer with an extensive hydrophobic interface wherein approximately one-quarter of the monomer surface area was involved in the dimer intermolecular contact, a feature characteristic of obligate homodimers. Mutations that block dimerization abolish the function of these proteins, indicating that this structural feature is essential. A surface-exposed groove lined with conserved amino acids was formed at the dimer interface has been suggested to be the site of interaction of the POZ/BTB domain with nuclear corepressors, although mutational studies have not supported this possibility.

The C-terminal zinc fingers confer sequence specific binding of these proteins to DNA. Cys2His2 zinc fingers are one of the most common DNA-binding motifs found in eukaryotic transcription factors. These proteins typically contain several fingers that make tandem contacts along the DNA. Each finger has a conserved β–β–α structure, and amino acids on the surface of the α-helix contact bases in the major groove, with each finger contacting three nucelotides.[19,20] This simple, modular structure of zinc finger proteins enable them to recognize a wide variety of DNA sequences, and six individual zinc fingers, as are present in BCL-6, linked together would recognise a DNA sequence of up to 18 bp in length, sufficiently long to constitute a rare address in the human genome.

Several groups have determined consensus binding sequences for BCL-6 and, although these vary in length from nine to twenty nucleotides, a common core sequence, TCC(NNN)GAA, is apparent[21-26]. This was recognized as being identical to the canonical palindromic sequence that is characteristic of STAT binding elements. Accordingly, BCL-6 has been shown to bind to the STAT elements of several genes, including the STAT6 element of the *CD23b* gene[25, 27], the germline immunoglobulin ε gene[26], and the *IL-4*gene[27], and the STAT3 element of the α2-macroglobulin gene[28], but not to the STAT5 element of the IL-2Rα chain. The presence of six zinc fingers in BCL-6 makes likely that nucleotides outside the core STAT binding element of 9-10 nucleotides contribute to DNA recognition, and this may be the basis for its ability to bind more than one type of STAT element. Most importantly with respect to its biological function, BCL-6 suppressed transcription from reporter constructs containing these STAT3 and STAT6 binding elements[24-28], reflecting its ability to recruit HDAC1. Therefore, BCL-6 is a sequence-specific transcriptional repressor that can modify signal transduction events involving STATs.

Consensus BCL-6 Binding Sites	Consensus STAT Binding Site
(T/A)NCT**TTC**NAG**G(A/G)A** T[22]	**TTC**($N_{3\text{-}4}$)**GAA**
AAANNNCNNTAN**GAAA**[23]	
GAAAA**TTC**CTA**GAA**AGCATA[24]	
TC(T/C)T(C/A)**GAA**(A/T)G[25]	
TTC(C/T)T(A/C)**GAA**[26]	
GAAAA**TTC**CTA**GAA**AGCATA[27]	

4. THE IMMUNOLOGICAL FUNCTION OF BCL-6

BCL-6 mRNA is found in many cell types, but protein has been reported in only three sites: the germinal center B[29-31] and T cell, and an occasional marginal zone B cell[29]. This finding not only indicates that BCL-6 expression is subject to some form of post-transcriptional regulation, which has been confirmed by the finding that MAPkinase-dependent phosphorylation of serines in a PEST domain of BCL-6 causes accelerated degradation of protein[32], but also that BCL-6 function may be critical during T-dependent immune responses of B cells. Three separate groups have interrupted the *BCL-6* gene in mice and all reported the absence of germinal centers following immunization of BCL-6-/- mice with protein antigens[25,33,34]. No abnormalities in the development of mature B cells were found, and T-independent immune responses were intact. An additional phenotype was reported by two groups that involved Th2-type inflammatory reaction with eosinophilic infiltration of multiple tissues[25,33]. The latter may relate to the ability of BCL-6 to regulate STAT6-dependent transcription, although the phenotype persists in *IL-4-/-* and *STAT6-/-* mice[35], and it is not mediated by *BCL-6-/-* lymphocytes[34,36]. It has been suggested that the inflammatory phenotype of BCL-6-/- mice is mediated by increased chemokine expression by macrophages[36], although macrophages have not yet been shown to express BCL-6 protein. The third group, which found that the absence of germinal centers was B cell autonomous, also found growth retardation and severe myocardial injury, but have not characterized these additional abnormalities further[34]. In summary, the analysis of *BCL-6-/-* mice have demonstrated a non-redundant function for the transcriptional repressor in the germinal center B cell; its less well described functions in other cell types are not the subject of this review.

Two distinctly different approaches to understanding BCL-6 function in the germinal center B cell have been made. The first took advantage of expression profiling using customized microarrays that had been developed to characterize various B cell lymphomas[37], while the second emanated from the hypothesis that the germinal center B cell required some means by which plasma cell differentiation could be suppressed[28]. In the former, RAJI B lymphoblastoid B cells, which display a germinal center phenotype with expression of BCL-6 protein, were transfected with a construct encoding a fusion protein between the hormone binding domain of the estrogen receptor and the BCL-6 zinc fingers. This enabled the inducible expression of a dominant negative form of BCL-6 in these cells, thereby relieving genes from transcriptional suppression by their endogenous BCL-6. One of the newly transcribed genes in "de-repressed" Raji cells was Blimp-1, which drives terminal differentiation of B cells and is highly expressed in plasma cells, but not in earlier stages of B cell development. Although the Blimp-1 expressing Raji cells did not differentiate into antibody-secreting plasma cells, this finding suggested that BCL-6 may repress transcription of Blimp-1 to prevent the terminal differentiation of germinal center B cells. Curiously, this induced Blimp-1 was also associated with a lower expression of c-myc in the Raji cells, which would not be anticipated, despite the ability of Blimp-1 to suppress transcription of a wild type c-myc allele[4], because all Burkitt lymphomas, including Raji cells, have a translocated c-myc allele under control of Ig transcriptional elements that should have increased activity in plasma cells.

The second approach examined the effect of BCL-6 on the differentiation of the BCL-1 murine B cell lymphoma line, which has been a model for analyzing cytokine-induced plasma cell development for many years[38]. Treatment of BCL-1 cells with IL-2 and IL-5 causes them to express Blimp-1 and J chain, to secrete IgM, and to cease growing and undergo apoptosis. When these cells were made the express BCL-6 by tranduction with a BCL-6 encoding retrovirus, all three effects of the cytokines were blocked. Furthermore, introducing ectopic Blimp-1 into the BCL-6 expressing BCL-1 cells caused terminal differentiation of the cells, formally proving that BCL-6 is downstream of the IL-2 and IL-5 receptors, and upstream of Blimp-1. Since the IL-2 and IL-5 receptors signal through STAT3 and STAT5, but not STAT6, these findings also suggested that BCL-6 can regulate transcription from additional STAT elements. This was confirmed by the findings that BCL-6 bound to the STAT3 element of the α2-macroglobulin promoter and suppressed transcription from a reporter gene having this site, and that a dominant negative form of STAT3, when expressed by BCL-1 cells, replicated the effects of BCL-6 on cytokine-induced terminal differentiation. Finally, BCL-6 was shown to suppress plasma cell differentiation of primary murine B cells stimulated through CD40 and cytokine receptors sharing the common γ_c, indicating that the effects were not limited to transformed cells.

In summary, then, two quite different studies have both found that BCL-6 can affect the terminal differentiation of B cells. Therefore, BCL-6 expression may be required for the germinal center B cell to be able to undergo multiple cycles of mutation, proliferation, and selection to create high affinity variants of Ig during affinity maturation of the humoral immune response. BCL-6 accomplishes this is by suppressing STAT3-dependent transcription. The ultimate target in the B cell appears to be Blimp-1, but it is not known whether there are STAT elements in its promoter, or whether the effect of STAT3 in the BCL-1 system is indirect. BCL-6 clearly can bind to different STAT elements, and we would propose that it "fine tunes" cytokine signaling, allowing cytokine driven cellular proliferation to proceed, while suppressing cytokine induced terminal differentiation. As indicated in the introductory paragraphs, a need for this function in the immune system may be more widespread than the germinal center B cell, and we anticipate that other examples of this general phenomenon of uncoupling proliferation from differentiation to effector functions will be found, and that these will be mediated by transcriptional repressors similar to BCL-6.

REFERENCES

1. J.T. Opferman, B.T. Ober, and P.G. Ashton-Rickardt, 1999, Linear differentiation of cytotoxic effectors into memory T lymphocytes, *Science* **283,** 1745-1748.
2. C.A. Turner, Jr., D.H. Mack, M.M. Davis, 1994, Blimp-1, a novel zinc finger-containing protein that can drive the maturation of B lymphocytes into immunoglobulin-secreting cells, Cell **77,** 297-306.
3. J.F. Piskurich, K.I. Lin, Y. Lin, Y. Wang, J.P. Ting, and K. Calame, 2000, BLIMP-1 mediates extinction of major histocompatibility class II transactivator expression in plasma cells, *Nat. Immunol.* **1,** 526-532.
4. Y. Lin, K. Wong, and K. Calame, 1997, Repression of c-myc transcription by Blimp-1, an inducer of terminal B cell differentiation, *Science* **276,** 596-599.
5. P. D. Hodgkin, J. H. Lee, and A.B. Lyons, 1996, B cell differentiation and isotype switching is related to division cycle number, *J. Exp. Med.* **184,** 277-281.
6. S.M. Kaech, and R. Ahmed, 2001, Memory CD8+ T cell differentiation: initial antigen encounter triggers a developmental program in naive cells, *Nat. Immunol.* **2,** 415-422.

7. M.J. van Stipdonk, E.E. Lemmens, and S.P. Schoenberger, 2001, Naive CTLs require a single brief period of antigenic stimulation for clonal expansion and differentiation, *Nat. Immunol.* **2,** 423-429.

7a. B.J. Dickson, 1998, Photoreceptor development: breaking down the barriers, *Curr. Biol.* **8,** R90-R92.

8. B.W. Baron, G. Nucifora, N. McCabe, R. Espinosa III, M.M. Le Beau, and T.W. McKeithan, 1993, Identification of the gene associated with the recurring chromosomal translocations t(3;14)(q27;q32) and t(3;22)(q27;q11) in B-cell lymphomas, *Proc. Natl. Acad. Sci. USA* **90,** 5262-5266.

9. J.P. Kerckaert, C. Deweindt, H. Tilly, S. Quief, G. Lecocq, and C. Bastard, 1993, LAZ3, a novel zinc-finger encoding gene, is disrupted by recurring chromosome 3q27 translocations in human lymphomas. *Nature Genet.* **5,** 66-70.

10. B.H. Ye, F. Lista, F. Lo Coco, D.M. Knowles, K. Offit, R.S. Chaganti , and R. Dalla-Favera, 1993, Alterations of a zinc finger-encoding gene, BCL-6, in diffuse large-cell lymphoma, *Science* **262,** 747-750.

11. C. Deweindt, J.P. Kerckaert, H. Tilly, S. Quief, V.C. Nguyen, and C. Bastard , 1993, Cloning of a breakpoint cluster region at band 3q27 involved in human non-Hodgkin's lymphoma. *Genes Chromosomes Cancer* **8,** 149-154.

12. T. Miki, N. Kawamata, A. Arai, K. Ohashi, Y. Nakamura, A. Kato, S. Hirosawa, and N. Aoki, 1994, Molecular cloning of the breakpoint for 3q27 translocation in B-cell lymphomas and leukemias, *Blood* **83,** 217-222.

13. P. Dhordain, O. Albagli, R.J. Lin, S. Ansieau S, S. Quief S, A. Leutz A, J.P.Kerckaert, R.M. Evans, and D. Leprince, 1997, Corepressor SMRT binds the BTB/POZ repressing domain of the LAZ3/BCL6 oncoprotein, *Proc. Natl. Acad. Sci. USA* **94,** 10762-10767.

14. P. Dhordain, R.J. Lin, S. Quief, D. Lantoine, J.P. Kerckaert, R.M. Evans, O. Albagli, 1998, The LAZ3(BCL-6) oncoprotein recruits a SMRT/mSIN3A/histone deacetylase containing complex to mediate transcriptional repression, *Nucleic Acids Res.* **26,** 4645-4651.

15. K.D. Huynh, and V.J. Bardwell, 1998, The BCL-6 POZ domain and other POZ domains interact with the co-repressors N-CoR and SMRT, *Oncogene* **17,** 2473-2484.

16. C.W. Wong, and M.L. Privalsky, 1998, Components of the SMRT corepressor complex exhibit distinctive interactions with the POZ domain oncoproteins PLZF, PLZF-RARalpha, and BCL-6, *J. Biol. Chem.* **273,** 27695-27702

17. K.F. Ahmad, C.K. Engel, and G.G. Prive GG, 1998, Crystal structure of the BTB domain from PLZF, *Proc. Natl. Acad. Sci. USA* **95,** 121231212-8

18. X. Li, H. Peng, D.C. Schultz, J.M. Lopez-Guisa, F.J. Rauscher, 3rd, and R. Marmorstein, 1999, Structure-function studies of the BTB/POZ transcriptional repression domain from the promyelocytic leukemia zinc finger oncoprotein, *Cancer Res.* **59,** 5275-5282.

19. A. Klug, 1999, Zinc finger peptides for the regulation of gene expression, *J. Mol. Biol.* **293,** 215-218.

20. S.A. Wolfe, L. Nekludova, and C.O. Pabo, 2000, DNA recognition by Cys2His2 zinc finger proteins, *Annu. Rev. Biophys. Biomol. Struct.* **29,** 183-212.

21. N. Kawamata, T. Miki, K. Ohashi, K. Suzuki, T. Fukuda, S. Hirosawa, and N. Aoki, 1994, Recognition DNA sequence of a novel putative transcription factor, BCL6, *Biochem. Biophys. Res. Commun.* **204,** 366-374.

22. B.W. Baron, R.R. Stanger, E. Hume, A. Sadhu, R. Mick, J.P. Kerckaert, C. Deweindt, C. Bastard, G. Nucifora, and N. Zeleznik-Le, 1995, BCL6 encodes a sequence-specific DNA-binding protein. *Genes Chromosomes Cancer* **13,** 221-224.
23. C.C. Chang, B.H. Ye, R.S. Chaganti, and R. Dalla-Favera, 1996, BCL-6, a POZ/zinc-finger protein, is a sequence-specific transcriptional repressor, *Proc. Natl. Acad. Sci. USA* **93,** 6947-6952.
24. V.L. Seyfert, D. Allman, Y. He, and L.M. Staudt, 1996, Transcriptional repression by the proto-oncogene BCL-6, *Oncogene* **12,** 2331-2342.
25. A.L. Dent, A.L. Shaffer, X. Yu, D. Allman, and L.M. Staudt, 1997, Control of inflammation, cytokine expression, and germinal center formation by BCL-6, *Science* **276,** 589-592.
26. M.B. Harris, C.C. Chang, M.T. Berton, N.N. Danial, J. Zhang, D. Kuehner, B.H. Ye, M. Kvatyuk, P.P. Pandolfi, G. Cattoretti, R. Dalla-Favera, and P.B. Rothman, 1999, Transcriptional repression of Stat6-dependent interleukin-4-induced genes by BCL-6: specific regulation of iepsilon transcription and immunoglobulin E switching, *Mol. Cell Biol.* **19,** 7264-7275.
27. T. Hartatik, S. Okada, S. Okabe, M. Arima, M. Hatano, and T. Tokuhisa, 2001, Binding of BAZF and Bc16 to STAT6-binding DNA sequences, *Biochem. Biophys. Res. Commun.* **284,** 26-32.
28. R. Reljic, S.D. Wagner, L.J. Peakman, and D.T. Fearon DT, 2000, Suppression of signal transducer and activator of transcription 3-dependent B lymphocyte terminal differentiation by BCL-6, *J. Exp. Med.* **192,** 1841-1848.
29. T. Onizuka, M. Moriyama, T. Yamochi, T. Kuroda, A. Kazama, N. Kanazawa, K. Sato, T. Kato, H. Ota, and S. Mori, 1995, BCL-6 gene product, a 92- to 98-kD nuclear phosphoprotein, is highly expressed in germinal center B cells and their neoplastic counterparts, *Blood* **86**, 28-37.
30. G. Cattoretti, C.C. Chang, K. Cechova, J. Zhang, B.H. Ye, B. Falini, D.C. Louie,K. Offit, R.S. Chaganti, and R. Dalla-Favera, 1995, BCL-6 protein is expressed in germinal-center B cells, *Blood* **86,** 45-53.
31. L. Flenghi, B.H. Ye, M. Fizzotti, B. Bigerna, G. Cattoretti, S. Venturi, R. Pacini, S. Pileri, F. Lo Coco, and E. Pescarmona, 1995, A specific monoclonal antibody (PG-B6) detects expression of the BCL-6 protein in germinal center B cells, *Am. J. Pathol.* **147,** 405-411.
32. H. Niu, B.H. Ye, and R. Dalla-Favera, 1998, Antigen receptor signaling induces MAP kinase-mediated phosphorylation and degradation of the BCL-6 transcription factor, *Genes Dev.* **12,** 1953-1961.
33. B.H. Ye, G. Cattoretti, Q. Shen, J. Zhang, N. Hawe, R. de Waard, C. Leung, M. Nouri-Shirazi, A. Orazi, R.S. Chaganti, P. Rothman, A.M. Stall, P.P. Pandolfi, and R. Dalla-Favera, 1997, The BCL-6 proto-oncogene controls germinal-centre formation and Th2-type inflammation, *Nat. Genet.* **16,** 161-170.
34. T. Fukuda, T. Yoshida, S. Okada, M. Hatano, T. Miki, K. Ishibashi, S. Okabe, H. Koseki, S. Hirosawa, M. Taniguchi, N. Miyasaka, and T. Tokuhisa, 1997, Disruption of the Bcl6 gene results in an impaired germinal center formation, *J. Exp. Med.* **186,** 439-448.
35. A.L. Dent, J. Hu-Li, W.E. Paul, and L.M. Staudt, 1998, T helper type 2 inflammatory disease in the absence of interleukin 4 and transcription factor STAT6, *Proc. Natl. Acad. Sci. USA* **95,** 13823-13828.
36. L.M. Toney, G. Cattoretti, J.A. Graf, T. Merghoub, P.P. Pandolfi, R. Dalla-Favera,

B.H. Ye, and A.L. Dent, 2000, BCL-6 regulates chemokine gene transcription in macrophages. *Nat. Immunol.* **1,** 214-220.

37. A.L. Shaffer, X. Yu, Y. He, J. Boldrick, E.P. Chan, and L.M. Staudt, 2000, BCL-6 represses genes that function in lymphocyte differentiation, inflammation, and cell cycle control, *Immunity* **13,** 199-212.
38. M.A. Blackman, M.A. TiggeS, M.E. Minie, and M.E. Koshland, 1986, A model system for peptide hormone action in differentiation: interleukin 2 induces a B lymphoma to transcribe the J chain gene, *Cell* **47,** 609-617.

THE ROLE OF VAV PROTEINS IN B CELL RESPONSES

Martin Turner

Laboratory of Lymphocyte Signaling and Development
Molecular Immunology Programme, The Babraham Institute
Babraham, Cambridge CB2 4AT, UK

1. INTRODUCTION

B and T lymphocyte activation mediated through the antigen receptor orchestrates integration of the cell cycle, transcription and differentiation. Furthermore antigen receptor elicited changes in the actin cytoskeleton contribute to the formation of lipid rafts, the movement and directional release of secretory granule contents and the movement of lymphocytes themselves [1, 2]. Small GTPases of the Ras HOmolog (Rho) family are widely recognised as regulators of the cytoskeleton, transcription and cell cycle. Rho proteins, including RhoA, Rac-1, Rac-2 and Cdc42 have been shown to regulate the survival, proliferation and differentiation of lymphocytes[3]. Biochemically, Rho GTPases function as molecular switches, shuttling between an "inactive" GDP-bound state and an "active" GTP-bound state and GTP-bound Rho proteins bind to and control the activity of a great number of effector proteins[4, 5].

Vav proteins, are guanine nucleotide exchange factors (GEFs) that activate Rho proteins by promoting exchange of GDP for GTP. Vav proteins are rapidly phosphorylated and activated upon antigen receptor engagement, and provide the major, and perhaps only, link from the antigen receptor to Rho family of GTPases[6]. Vav-1 is a critical regulator of T cell development, particularly positive and negative selection in the thymus[7, 8]. Vav-1 function is necessary for the T-cell antigen receptor (TCR) to mediate inositol tris-phosphate production, calcium flux and the formation of actin filaments as well as to activate the ERK and NF-κB pathways[7, 9-11]. B cell development in Vav-1-deficient mice is normal, with the exception of the B1 subset that is Vav-1-dependent[12-15].

Vav-1-deficient B cells respond normally to thymus-independent antigens and to thymus-dependent antigens if T cell help is provided[14, 15]. B and T lymphocytes thus rely on Vav-1 to different extents for their development and function.

Vav-2 and Vav-3 are also phosphorylated on tyrosine in lymphocytes following antigen receptor crosslinking suggesting that they may participate in antigen receptor-mediated signaling[16-18]. Overexpression of Vav-2 augments B cell receptor (BCR) stimulated calcium flux and promotes transcription from an NFAT reporter in B cells[18], but not in T cells[16-18]. These findings suggest that Vav-2 may play an important role in signaling from the BCR. To clarify the role of Vav-2 in lymphocyte development and the immune response we created a Vav-2-deficient mouse. Vav-$2^{-/-}$ mice contain normal numbers of the major subsets of B and T lymphocytes. However, immunization with thymus-independent and thymus-dependent antigens revealed a severe immunodeficiency in terms of immunoglobulin class switching and the germinal center reaction. B-lymphocytes lacking both Vav-1 and Vav-2 showed a defect in development and signaling similar to Btk-deficient *xid* mice.

2. RESULTS AND DISCUSSION

Vav-2 was deficient mice were generated by gene targeting in embryonic stem cells which were used to establish a null mutation in the mouse germline[19]. A gene cassette containing the bacterial neomycin resistance gene under the control of the phosphoglycerate kinase promoter was inserted into an exon encoding an essential part of the Dbl domain (amino acid 253 lies within Conserved Region 2 of the Dbl homology domain[20]. Targeted ES cell clones were used to generate chimaeric mice which transmitted the mutation to their offspring. Homozygous mutant animals were established that were healthy and fertile and western blotting confirmed the absence of Vav-2. The numbers of T and B cells in the peripheral lymphoid organs of Vav-2-deficient mice were not different from littermate controls suggesting that Vav-2 was not required for lymphocyte development. To determine whether the lymphocytes generated in Vav-2 deficient mice were functional, we immunized mice intraperitoneally *(i.p.)* with the T-independent-type II (TI-2) antigen, DNP-ficoll. The hapten-specific IgM response of Vav-$2^{-/-}$ mice following immunization was impaired. Similarly, the levels of DNP-specific IgG3 produced were significantly lower in Vav-$2^{-/-}$ mice. TI-2 antigens, which are typically derived from polysaccharides, usually consist of complex repeating units that drive B cell responses by extensively crosslinking specific membrane Ig. As the responses of Vav-$2^{-/-}$ mice to TI-2 antigens are impaired, we suggest that Vav-2 is a component of the surface Ig signaling cascade necessary for a TI-2 response. This result contrasts with the normal responsiveness of Vav-1-deficient mice to TI-2 antigens[14, 15]. Recently, marginal zone B cells have been identified as the population principally responsible for the TI-2 response[21]. In Vav-2-deficient mice the MZ was identified in histological sections and no reduction was seen in the population of $CD21^{high}CD23^{dull}B220^{+}$ MZ B cells. Our results favor the suggestion that there is a defect in the ability of Vav-2-deficient B cells to respond to TI-2 antigens.

The primary response to TD antigens was measured seven days after the *i.p.* injection of DNP-KLH. Wild-type littermate mice produced IgM, IgG1 IgG2a, IgG2b and IgG3 antibodies to DNP (anti-DNP). By contrast, although vav2$^{-/-}$ mice produced levels of anti-DNP antibody of the IgM and IgG2a subclasses similar to control mice, they produced far less anti-DNP antibody of the IgG1, IgG2b and IgG3 subclasses. Reimmunization of control animals with DNP-KLH led to a robust secondary response, however, while mice lacking Vav-2 produced antigen-specific IgM and IgG2a, there was a significant defect in the production of antigen-specific IgG1, IgG2b and IgG3. Enumeration of germinal centers by immunohistological analysis showed an 80% reduction in the number of PNA^{+} germinal centers compared with the control mice. Thus, following immunization with T-dependent antigens Vav-2$^{-/-}$ mice fail to mount normal primary or secondary antibody responses. This phenotype is similar to that of CD19-deficient mice that are also able to produce antigen specific IgG2a but show a defect in the switch to other Ig subclasses, particularly IgG1, the major isotype produced in response to DNP-KLH immunization[22, 23]. These defects do not appear to be as severe as those we had previously reported in Vav-1$^{-/-}$ mice; in Vav-1$^{-/-}$ mice the B cell defect was attributed to a lack of T cell help[14]. This may reflect the ability of helper T cells to function in the absence of Vav-2. In support of this, we have found that in addition to the normal development of CD4 and CD8 T cells in Vav-2-deficient mice, *in vitro* stimulation of *vav-2*-deficient T cells with antibodies to CD3 and CD28 yields normal levels of proliferation and antibodies to CD3 stimulates a normal calcium flux. Furthermore, *in vitro* studies that show Vav-2 can regulate Ca^{2+} flux in B cells, but not T cells [18] and a recent study using mice deficient in both Cbl-b and Vav-1, suggested that Vav-2 could not compensate for loss of Vav-1 in T cells[24]. Taken together these data add further weight to the possibility that the defect in humoral immune responses in Vav-2$^{-/-}$ mice may be primarily a B cell defect. Transfer experiments to produce chimaeric animals with mixtures of wild-type T cells and Vav-2-deficient B cells will be necessary to address this issue but will require the mice to be on an inbred genetic background.

As the initial phenotypic analysis of lymphocyte subsets in Vav-2-deficient mice failed to reveal any developmental abnormalities in the B and T cell lineages, we generated mice lacking both Vav-1 and Vav-2 to eliminate any compensatory functions of these molecules. FACS analysis of the bone marrow of Vav-$^{1-/-}$Vav-2$^{-/-}$ mice did not reveal any major blocks in B cell development as assessed by the numbers and proportions of B220^{+} cells which could be classified into the pro-, pre- or immature B cell stages[25]. We did observe, however, a 50% reduction in the absolute number of mature B cells in the bone marrow of Vav-1$^{-/-}$Vav-2$^{-/-}$ mice. This reduction in the number of B cells was also seen in the spleen of Vav-1Vav-2 double-deficient mice, which contain one third of the number of B cells found in control mice. The number of mature (IgD^{hi}) B cells is reduced between 80-90 % in the spleens of Vav-1$^{-/-}$Vav-2$^{-/-}$ mice when compared to either wild-type, Vav-1$^{-/-}$ or Vav-2$^{-/-}$ mice. Combination of the Vav-1 and Vav-2 mutations had little additional effect upon T cell development, which was already severely impaired in mice lacking Vav-1. By contrast, the absence of both Vav-1 and Vav-2 led to a clear defect in B cell maturation. Thus Vav-1$^{-/-}$Vav-2$^{-/-}$ mice had only 30-50% the number of total B cells. The majority of IgM^{hi}IgD^{lo} B cells [26-28] are recent immigrants from the bone marrow which have been termed transitional B cells. The numbers of transitional B cells is reduced in Vav-1$^{-/-}$Vav-2$^{-/-}$ mice, but this reduction is not as severe as that seen for mature B cells. Thus the majority of B cells in Vav-1$^{-/-}$Vav-2$^{-/-}$ mice are phenotypically immature. Furthermore a much higher proportion of B cells from Vav-1$^{-/-}$Vav-2$^{-/-}$ mice were CD24^{high}, which is consistent with this suggestion. Transitional B cells require

BCR-mediated signals to progress to the more mature IgM^{lo}IgD^{hi} RF stage[28,29]. A partial block at this stage of B cell development could explain the dramatic drop in the numbers of mature peripheral B cells in mice lacking Vav-1 and Vav-2. However, our findings could also reflect defective selection of immature B cells in the bone marrow or defective emigration of immature B cells from the bone marrow. Alternatively, the survival of mature B cells, which is dependent upon low levels of "tonic" BCR signaling[30], may be defective in Vav-1$^{-/-}$Vav-2$^{-/-}$ mice.

One of the hallmarks of BCR-induced signaling is the mobilization of intracellular calcium (Ca^{2+}) and this can be shown in mature B-lymphocytes following addition of antibodies against IgM. Calcium fluxes have been implicated in the selection events that regulate B cell maturation and in antigen-receptor triggered proliferation[31,32]. To examine the role of Vav-1 and Vav-2 in BCR-elicited calcium flux, B cells from wild-type, Vav-1$^{-/-}$, Vav-2$^{-/-}$ or Vav-1$^{-/-}$Vav-2$^{-/-}$ mice were preloaded with the Ca^{2+}-sensitive dye Indo-1 and changes in intracellular Ca^{2+} concentrations following crosslinking of the BCR complex were measured. These studies showed that Ca^{2+} fluxes were reduced in the Vav-1 or Vav-2-deficient B cells. This was particularly evident in the latency period when a low dose of anti-IgM was used. At higher doses, B cells lacking Vav-1 or Vav-2 responded at near normal levels, however, B cells lacking both Vav-1 and Vav-2 were still unresponsive. Thus, both Vav-1 and Vav-2 appear to be required for the full coupling of BCR activation to Ca^{2+} mobilization.

The phenotype of Vav-1$^{-/-}$Vav-2$^{-/-}$ mice resembles that of mice lacking a functional Bruton's tyrosine kinase gene[33]. Mutation of *btk* affects maturation of peripheral B cells without seriously compromising B cell development in the bone marrow. Those mature B cells that do develop are defective in their *in vitro* proliferative and Ca^{2+} responses. Additionally *btk*$^{-/-}$ mice lack peritoneal B cells and have defective immune responses to both TI-2 antigens and suboptimal, but normally immunogenic, doses of TD antigen. Recently, a number of mutations of signal transduction molecules that affect peripheral B cell maturation have been identified including the p85 regulatory subunit of phosphoinositide 3-kinase, phospholipase C γ-2 and B cell linker protein, BLNK (also called SLP65 or BASH)[34]. All of these mutations yield similar phenotypes and it has been suggested that these proteins are components of a signaling complex named the "signalosome". The "signalosome" concept envisions a molecular complex that is regulated both temporally and spatially and whose outputs include quantitative and qualitative effects on the levels of intracellular Ca^{2+}. Both peak threshold and sustained increases in intracellular Ca^{2+} are required for the activation of the MAP, JNK and p38 kinase cascades and the Ca^{2+} dependent transcription factors NFAT and NF-□B[32]. Our work provides genetic evidence to suggest that both Vav-1 and Vav-2 are likely to be components of this complex.

3. ACKNOWLEDGEMENTS

Work in the authors laboratory is supported by Biotechnology and Biological Sciences Research Council, the Leukaemia Research Fund and the Cancer Research Campaign [CRC].

REFERENCES

1. Acuto, O. & Cantrell, D. A. T cell activation and the cytoskeleton. *Ann. Rev. Immunol.* **18**, 165-184 (2000).
2. Dustin, M. L. & Cooper, J. A. The immunological synapse and the actin cytoskeleton: molecular hardware for T cell signaling. *Nat. Immunol.* **1**, 23-29 (2000).
3. Henning, S. W. & Cantrell, D. A. GTPases in antigen receptor signalling. *Curr. Opin. Immunol.* **10**, 322-329 (1998).
4. Van Aelst, L. & D'Souza-Schorey, C. Rho GTPases and signalling networks. *Gen. Dev.* **11**, 2295-2322 (1997).
5. Bishop, A. B. & Hall, A. Rho GTPases and their effector proteins. *Biochem. J.* **348**, 241-255 (2000).
6. Bustelo, X. R. Regulatory and Signalling Properties of the Vav Family. *Mol. Cell. Biol.* **20**, 1461-1477 (2000).
7. Turner, M. *et al.* A requirement for the Rho-Family GTP Exchange Factor Vav in Positive and Negative Selection of Thymocytes. *Immunity* **7**, 451-460 (1997).
8. Kong, Y. Y. *et al.* Vav regulates peptide-specific apoptosis in thymocytes. *J. Exp. Med.* **188**, 2099-2111 (1998).
9. Fischer, K. D. *et al.* Vav is a regulator of cytoskeletal reorganisation mediated by the T-cell receptor. *Curr. Biol.* **8**, 554-562 (1998).
10. Holsinger, L. J. *et al.* Defects in actin-cap formation in Vav-deficient mice implicate an actin requirement for lymphocyte signal transduction. *Curr. Biol.* **8**, 563-572 (1998).
11. Costello, P. S. *et al.* The Rho-family GTP exchange factor Vav is a critical transducer of T cell receptor signals to the calcium, ERK, and NF-kappaB pathways. *Proc. Natl. Acad. Sci. U.S.A.* **96**, 3035-3040 (1999).
12. Tarakhovsky, A. *et al.* Defective antigen receptor-mediated proliferation of B and T cells in the absence of Vav. *Nature* **374**, 467-470 (1995).
13. Zhang, R., Alt, F. W., Davidson, L., Orkin, S. H. & Swat, W. Defective signalling through the T- and B-cell antigen receptors in lymphoid cells lacking the vav proto-oncogene. *Nature* **374**, 470-473 (1995).
14. Gulbranson-Judge, A. *et al.* Defective immunoglobulin class switching in Vav-deficient mice is attributable to compromised T cell help. *Eur. J. Immunol.* **29**, 477-487 (1999).
15. Bachmann, M. F. *et al.* The guanine-nucleotide exchange factor Vav is a crucial regulator of B cell receptor activation and B cell responses to nonrepetitive antigens. *J. Immunol.* **163**, 137-142 (1999).
16. Billadeau, D. D., Mackie, S. M., Schoon, R. A. & Leibson, P. J. The Rho family Guanine Nucleotide Excahnge factor Vav-2 regulates the Development of Cell-mediated Cytotoxicity. *J. Exp. Med.* **192**, 381-391 (2000).
17. Moores, S. L. *et al.* Vav Family Proteins couple to Diverse Cell Surface Receptors. *Mol. Cell. Biol.* **20**, 6364-6373 (2000).
18. Doody, G. M. *et al.* Vav-2 controls NFAT-dependent transcription in B-but not T-lymphocytes. *EMBO. J.* **19**, 6173-6184 (2000).
19. Doody, G. M. *et al.* Signal transduction through Vav-2 participates in humoral immune responses and B cell maturation. *Nat. Immunol.*, 542-547 (2001).
20. Soisson, S. M., Nimnual, A. S., Uy, M., Bar-Sagi-D. & Kuriyan, J. Crystal structure of the Dbl and Pleckstrin Homology Domains from the Human Son of Sevenless Protein. *Cell* **95**, 259-268 (1998).
21. Guinamard, R., Okigaki, M., Schlessinger, J. & Ravetch, J. V. Absence of marginal zone B cells in Pyk-2-deficient mice defines their role in the humoral response. *Nat. Immunol.* **1**, 31-36 (2000).
22. Sato, S., Steeber, D. A., Jansen, P. J. & Tedder, T. F. CD19 expression levels regulate B lymphocyte development: human CD19 restores normal function in mice lacking endogenous CD19. *J. Immunol.* **158**, 4662-4669 (1997).
23. Gardby, E. & Lycke, N. Y. CD19-deficient mice exhibit poor responsiveness to oral immunisation despite evidence of unaltered total IgA levels, germinal centre and IgA-isotype switching in peyer's patches. *Eur. J. Immunol.* **30**, 1861-1871 (2000).
24. Krawczyk, C. *et al.* Cbl-b is a negative regulator of receptor clustering and raft aggregation in T cells. *Immunity* **13**, 463-473 (2000).
25. Hardy, R. R., Carmack, C. E., Shinton, S. A., Kemp, J. D. & Hayakama, K. Resolution and characterisation of Pro-B and Pre-Pre-B cell stages in normal mouse bone marrow. *J. Exp. Med.* **173**, 1213-1225 (1991).

26. Allman, D. M., Ferguson, S. E. & Cancro, M. P. Peripheral B cell maturation. I. Immature peripheral B cells in adults are heat-stable antigen^{hi} and exhibit unique signaling characteristics. *J. Immunol.* **149**, 2533-2540 (1992).
27. Allman, D. M., Ferguson, S. E., Lentz, V. M. & Cancro, M. P. Peripheral B cell maturation. II. Heat-stable antigen$^{(hi)}$ splenic B cells are an immature developmental intermediate in the production of long-lived marrow-derived B cells. *J. Immunol.* **151**, 4431-4444 (1993).
28. Loder, F. *et al.* B cell development in the spleen takes place in discrete steps and is determined by the quality of B cell receptor-derived signals. *J. Exp. Med.* **190**, 75-89 (1999).
29. Meffre, E., Casellas, R. & Nussenzweig, M. C. Antibody regulation of B cell development. *Nat. Immunol.* **1**, 379-385 (2000).
30. Lam, K. P., Kuhn, R. & Rajewsky, K. *In vivo* ablation of surface Ig on mature B cells by inducible gene targeting results in rapid cell death. *Cell* **90**, 1073-1083 (1997).
31. Bijsterbosch, M. K., Meade, C. J., Turner, G. A. & Klaus, G. G. B. B lymphocyte receptors and polyphosphoinositide degradation. *Cell* **41**, 999-1006 (1985).
32. Healy, J. I. & Goodnow, C. C. Positive versus negative signaling by lymphocyte antigen receptors. *Ann. Rev. Immunol.* **16**, 645-670 (1998).
33. Satterthwaite, A. B., Li, Z. & Witte, O. N. Btk function in B cell development and response. *Semin. Immunol.* **10**, 309-316 (1998).
34. Fruman, D. A., Satterthwaite, A. B. & Witte, O. N. Xid-like Phenotypes: A B cell signalosome takes shape. *Immunity* **13**, 1-3 (2000).

TRAFFIC PATTERNS OF B CELLS AND PLASMA CELLS

Jason G. Cyster, K. Mark Ansel, Vu N. Ngo, Diana C. Hargreaves and Theresa T. Lu

1. INTRODUCTION

B cells at various stages of development and activation undergo specific migration events that are important for their function. A majority of the immature B cells emerging from the bone marrow home to spleen and after a short period of further maturation become competent for migration into B cell follicles. Most B cells join the conventional or B2 subset and recirculate between lymphoid follicles of secondary lymphoid organs. Smaller numbers of cells take a different path, differentiating to B1 cells which preferentially lodge within the body cavities. Following encounter with antigen, conventional B cells redistribute from follicles to T cell zones where they interact with helper T cells. Antigen-stimulated B cells that differentiate to IgM- or IgG-secreting plasma cells leave the T and B cell areas and migrate to splenic red pulp, lymph node medullary cords or bone marrow. Each of these migration events contributes to regulate B cell fate and understanding the factors controlling B cell migration has been a major goal of our work. Here we describe our recent findings showing the roles played by chemokines (Cyster, 1999; Zlotnik and Yoshie, 2000) in controlling the positioning of B cells and plasma cells.

2. B CELL HOMING TO LYMPHOID FOLLICLES

Lymphoid follicles constitute a major compartment in all secondary lymphoid organs. B cell homing to lymphoid follicles occurs in the absence of antigenic stimulation and a principal role for the follicle is to promote surveillance of antigen-presenting follicular dendritic cells (FDCs) by naïve B cells. Follicles also support the germinal center reaction, a site of intense T-cell dependent B cell proliferation and V-gene somatic mutation. Follicles lacking germinal centers are termed primary follicles and when a germinal center forms the follicle is renamed as a secondary follicle. FDCs are present within both primary and secondary follicles, although these specialized stromal cells undergo major changes during the germinal center reaction, upregulating several surface

Department of Microbiology and Immunology, University of California, San Francisco, CA 94143

molecules (Cyster et al., 2000). A key factor required for directing cells to lymphoid follicles is the CXC chemokine, CXCL13 (originally termed BLC (Gunn et al., 1998) or BCA1 (Legler et al., 1998)). CXCL13 is made by stromal cells within lymphoid follicles, and although CXCL13-producing stromal cells have not been examined in isolation, from an analysis of tissue sections hybridized with a CXCL13 probe and adjacent sections stained for the stromal marker BP3, it can be suggested that both molecules are made by the same cells (Ngo et al., 1999). This conclusion is also supported by the finding that the CXCL13-producing cells in the spleen are radiation resistant (Ansel et al., 2002). FDC correspond to a subset of the follicular stromal cells (Cyster et al., 2000) and they contribute to the production of CXCL13 (Ngo et al., 1999). The receptor for CXCL13, CXCR5 (originally termed BLR1 (Förster et al., 1996)) is expressed by all B cells and is upregulated during maturation in a similar manner to other maturation markers such as IgD (Cyster et al., 1999; Bowman et al., 2000). In gene-targeting experiments, both CXCL13 and CXCR5 are necessary for B cell homing to lymphoid follicles (Förster et al., 1996; Ansel et al., 2000). The impact of defective follicle formation on B cell function is still under investigation. One important observation is that follicles are not critical for survival of conventional B cells as B cell numbers in the spleen and blood of CXCL13-deficient mice are similar to wildtype controls (Ansel et al., 2000). The role of follicles in promoting B cell surveillance for antigen is under continued investigation but preliminary experiments indicate that CXCL13-deficient mice have diminished T-dependent antibody responses (K.M.Ansel and J.G.Cyster, unpublished observations).

3. B1 CELL HOMING TO BODY CAVITIES

While the conventional B cell compartment fills with recirculating B cells, a separate compartment of B cells develops and becomes enriched in the body cavities. These cells, termed B1 cells (Kantor and Herzenberg, 1993; Hardy and Hayakawa, 2001), are characterized by distinct surface features including expression of Mac1, high levels of IgM, low levels of IgD and CD23 and, for a fraction of the cells, high levels of CD5. Although they are the predominant B cell population within the body cavities, their total numbers in the body are about 50-fold lower than the number of conventional B cells. Despite their low number, B1 cells are characterized as major producers of natural antibodies, that is, antibodies that are detectable in the serum in the absence of intentional immunization (Hardy and Hayakawa, 2001). Natural antibodies typically have low affinities for multiple antigens and an increasing number of studies have highlighted their importance in helping to provide early protection against bacterial and viral infections (reviewed in Hardy and Hayakawa, 2001; Ansel et al., 2002). The natural antibodies produced by B1 cells are also thought to include antibodies against blood group antigens and carbohydrate xenoantigens.

The purpose of B1 cell lodgement in the body cavities has been difficult to firmly define, but studies in germ-free mice have shown that these mice make lower amounts of natural antibody. This has lead to the notion that body cavity comparmentalization provides B1 cells with low level exposure to gut flora, helping provoke production of 'natural antibodies'. As the homing properties of B1 and conventional B cells are quite different, it was initially assumed that the homing receptors that direct these cells to their subcompartments would be distinct. However, an early indication that this may not be the case was the finding that B1 cells express high levels of CXCR5 and can respond to CXCL13 (Cyster et al., 1999; Bowman et al., 2000). An examination of CXCL13-deficient mice revealed that these animals are deficient in body cavity B1 cells (Ansel et al., 2002). Bone marrow chimera experiments indicated that CXCL13 from both radiation resistant cells and bone marrow derived macrophages participated in promoting B1 cell accumulation in the body cavities. By Nothern blot analysis CXCL13 expression was detected in peritoneal macrophages and radiation resistant (presumably non-hematopoietic) cells within the omentum (Ansel et al., 2002). Therefore, B1 cell homing to body cavities and conventional B cell homing to lymphoid follicles both depend on CXCL13/CXCR5, raising the question of what provides the specificity for homing to one compartment or the other. The answer to this question is presently unresolved but it seems likely that other features of the body cavity homing address, such as the selectin or integrin requirements will be different from the requirements for entry to lymphoid follicles. Although B1 cells are enriched in the body cavities, they also represent about 1% of the B cells in spleen and blood. The presence of B1 cells in spleen and blood did not depend on CXCL13, establishing that B1 cell production is not CXCL13 dependent (Ansel et al., 2002). Using a short term transfer approach, B1 cells were found to home rapidly from blood to the peritoneum and the pleural cavity. In the case of the peritoneum, many B1 cells entered the omentum prior to the appearance of cells in the body cavities, suggesting that the omentum acts as a portal of B1 cell entry to the body cavities (Ansel et al., 2002).

The finding that B1 cells can home efficiently to the body cavities led us to test the extent to which B1 cells normally recirculate between blood and body cavities. For this purpose, pairs of adult Ly5.1 and Ly5.2 C57BL/6 mice were surgically joined in parabiosis (Harris et al., 1997). After 6-8 weeks of parabiosis, conventional B cells in spleen and lymph nodes had reached equilibrium between the two animals, consistent with the cells undergoing continual recirculation via the blood. Importantly, within the peritoneum, the B1 cells achieved partial equilibration between the Ly5.1 and Ly5.2 animals (Ansel et al., 2002). While this may in part reflect the appearance of new B1 cells in the body cavities that have derived from bone marrow precursors, the experiments also suggest that at least a fraction of the B1 cells are recirculating at a low rate between body cavity and blood. For B1 cells to recirculate from the peritoneum to the blood they would need to exit the cavity, most likely via lymphatic vessels. The principal lymphatic vessels leaving the body cavity drain to the parathymic lymph nodes. Therefore, it was predicted that if B1 cells are undergoing even a small degree of recirculation from the body cavities, B1 cells may be present within parathymic lymph nodes. Previous studies have shown that, in contrast to spleen, B1 cells are absent from lymph nodes (Kantor and Herzenberg, 1993). However, as can be seen in figure 1, parathymic lymph nodes contained a readily identifiable population of B1 cells. These observations support the conclusion that at least a fraction of body cavity B1 cells undergo recirculation between blood and lymph. The significance of such recirculation is presently unclear but it may help make the unique specifities of the B1 cell compartment available for antibody responses outside the body cavities.

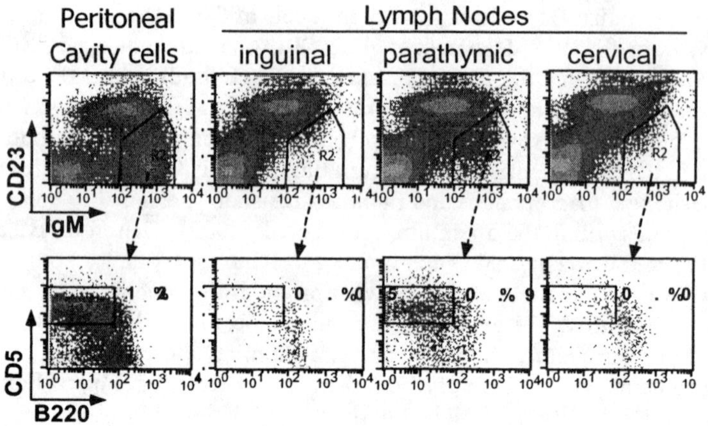

Figure 1. Detection of B1 cells in parathymic lymph nodes. Lymph node and peritoneal lavage cells from C57BL/6 mice were stained to detect IgM, B220, CD23 and CD5 as in (Ansel et al., 2002) and analyzed by 4-color flow cytometry. Upper panels are lymphocyte-size gated and lower panels are further gated on R2 (IgM^{hi} $CD23^{lo}$). The percentage value shown in each lower panel corresponds to the fraction of total cells that were IgM^{hi} $CD23^{lo}B220^{lo}CD5^{+}$ B1 cells.

4. PLASMA CELL HOMING

Early in the primary immune response to both T-independent and T-dependent antigens many IgM plasma cells are produced and these cells move away from the lymphoid zones into the highly vascular regions of the secondary lymphoid organs, the red pulp of spleen and medullary cords in lymph nodes. Plasma cells are believed to interact with supportive cells in these locations (Garcia De Vinuesa et al., 1999). After a period of a few days secreting large amounts of antibody, many of these plasma cells die. However, later in the primary response, as well as early in the secondary response to T-dependent antigens IgM and IgG plasma cells begin appearing in the bone marrow. A striking feature of plasma cells that home to the bone marrow is that they can survive for many months, contributing to a sustained period of antibody production (Slifka and Ahmed, 1998). In an effort to define the factors regulating plasma cell homing to these compartments, we examined the chemokine receptor profile and chemotactic responsiveness of plasma cells. Soon after their differentiation from activated B cells, at a time when the cells have begun to express the marker syndecan-1, plasma cells downregulate CXCR5 and CCR7 and lose responsiveness to CXCL13 and the CCR7 ligands, CCL19 (ELC) and CCL21 (SLC) (Hargreaves et al., 2001; Wehrli et al., 2001). These changes are likely to play an important role in promoting exclusion of the cells from the follicles and T zones. Concomittant with the decrease in CXCR5 and CCR7, plasma cells upregulate CXCR4 (Fig. 2) and exhibit increased responsiveness to CXCL12 (Hargreaves et al., 2001). CXCL12 is highly expressed by bone marrow stromal cells (Bleul et al., 1996). Analysis of CXCL12 expression in spleen and lymph nodes revealed strong expression in the red pulp and medullary cords, respectively (Hargreaves et al., 2001). To test the role of CXCR4 in plasma cell homing we reconstituted irradiated mice with fetal liver cells from CXCR4-deficient mice (Hargreaves et al., 2001). After six weeks reconstitution, the fetal liver chimeras were immunized and the distribution of plasma cells assessed. CXCR4-deficient plasma cells were found to be mislocalized in spleen, failing to move into their normal location within the red-pulp and instead accumulating in the marginal zone (Hargreaves et al., 2001). Within lymph nodes, CXCR4-deficient plasma cells failed to localize efficiently in the medullary cords, often appearing near the edges of this compartment or between B cell follicles and the subcapsular sinus (D.C. Hargreaves and J.G. Cyster, unpublished observations). Analysis later in the primary response and in secondary responses revealed a defect in accumulation of CXCR4-deficient plasma cells in the bone marrow and increased numbers of plasma cells were present in the blood (Hargreaves et al., 2001). Mixed fetal liver/bone marrow chimera experiments demonstrated that the requirement for CXCR4 in plasma cell homing was cell intrinsic (Hargreaves et al., 2001).

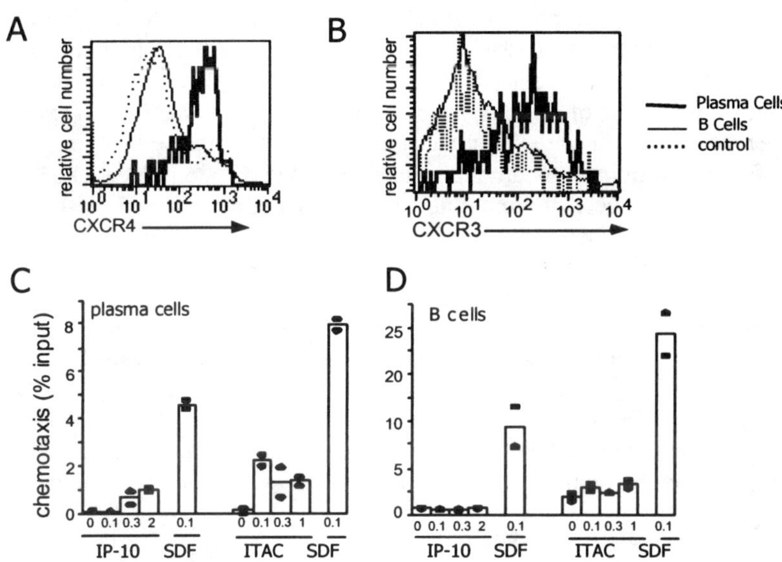

Figure 2. Plasma cells express CXCR3 and CXCR4 and respond to the corresponding ligands. (A, B) Spleen cells from mice immunized with NP-CGG/alum six days before were stained with B220, syndecan and rat anti-CXCR4 (Schabath et al., 1999) or polyclonal goat anti-CXCR3 (Santa Cruz Biotechnology), and analyzed by flow cytometry. B cells were identified as B220^{hi} syndecan^{-} and plasma cells were identified as B220^{lo} syndecan^{+}. In A, control refers to staining of plasma cells derived from CXCR4-deficient fetal liver chimers; in B, control refers to plasma cells stained with anti-CXCR3 in the presence of excess of the CXCR3 peptide that had served as the immunogen. (C, D) Summary of chemotaxis data in response to the indicated concentrations of CXCL10/IP10, CXCL11/ITAC and CXCL12/SDF1 (ng/ml). Cells were prepared and B cells and plasma cells identified as in A and B. Bars represent means of duplicate transwells. Data for CXCL10/IP10 and CXCL11/ITAC are from separate experiments and the CXCL12/SDF1 response for each experiment is shown for comparison.

In addition to upregulating CXCR4, we observed CXCR3 on many of the IgM- and IgG-expressing plasma cells detectable in the spleen following intraperitoneal immunization with antigen in alum (Fig. 2). Consistent with expression of CXCR3, plasma cells showed a small but reproducible chemotactic response to the CXCR3 ligands CXCL10/IP10, CXCL11/ITAC (Fig. 2) and CXCL9/MIG (not shown). The ability of plasma cells to respond to CXCL9 was recently reported (Bowman et al., 2002). The significance of CXCR3 expression on plasma cells is presently unclear, although it is important to note that CXCR3 ligands are all strongly induced by interferon and are expressed at many sites of inflammation (Kunkel and Butcher, 2002). CXCR3 may therefore contribute to plasma cell accumulation at sites of inflammation.

6. IMPLICATIONS

Chemokines serve as key guidance factors for lymphocyte homing during homeostatic trafficking and in the immune response. The limiting role of CXCL13/CXCR5 in directing B cells to lymphoid follicles and body cavities raises the issue of whether interfering with this pathway could be beneficial in treatment of follicle-associated or body cavity lymphomas. Importantly, inhibition of CXCL13 may not only limit B cell recruitment, but should also prevent CXCL13-mediated induction of lymphotoxin (LT) $\alpha 1\beta 2$ on B cells (Ansel et al., 2000). As LT$\alpha 1\beta 2$ is an important signal for the maintenance of follicular stromal cell function, reductions in expression of this cytokine may help reduce the stromal support for lymphoma cells. Perhaps as a consequence of the strong chemoattractant activity and ability to induce LT$\alpha 1\beta 2$, CXCL13 is a strong inducer of lymphoid tissue (Luther et al., 2000). It will be interesting to determine whether the induction of CXCL13 that occurs in inflammed tissue, such as in the rheumatoid synovium (Takemura et al., 2001), contributes to disease pathology. Similarly, the important function of CXCL12/CXCR4 in directing plasma cells to the bone marrow raises questions about the contribution this pathway plays to homing of chronic lymphocytic leukemia cells (CLL) and myeloma cells to this tissue. CLL cells have been observed to express both CXCR4 and CXCR3 (Burger et al., 1999; Trentin et al., 1999; Jones et al., 2000). Reciprocally, it is interesting to consider whether treatments that antagonize CXCR4 function could help displace and eliminate long-lived plasma cells and thereby be useful for the treatment of autoantibody mediated diseases.

REFERENCES

Ansel, K. M., Harris, R. B., and Cyster, J. G., 2002, CXCL13 Is Required for B1 Cell Homing, Natural Antibody Production, and Body Cavity Immunity, *Immunity*. **16**: 67-76.

Ansel, K. M., Ngo, V. N., Hyman, P. L., Luther, S. A., Förster, R., Sedgwick, J. D., Browning, J. L., Lipp, M., and Cyster, J. G., 2000, A chemokine driven positive feedback loop organizes lymphoid follicles, *Nature*. **406**: 309-314.

Bleul, C. C., Farzan, M., Choe, H., Parolin, C., Clark-Lewis, I., Sodroski, J., and Springer, T. A., 1996, The lymphocyte chemoattractant SDF-1 is a ligand for LESTR/fusin and blocks HIV-1 entry, *Nature*. **382**: 829-33.

Bowman, E. P., Campbell, J. J., Soler, D., Dong, Z., Manlongat, N., Picarella, D., Hardy, R. R., and Butcher, E. C., 2000, Developmental switches in chemokine response profiles during B cell differentiation and maturation, *J Exp Med*. **191**: 1303-18.

Bowman, E. P., Kuklin, N. A., Youngman, K. R., Lazarus, N. H., Kunkel, E. J., Pan, J., Greenberg, H. B., and Butcher, E. C., 2002, The Intestinal Chemokine Thymus-expressed Chemokine (CCL25) Attracts IgA Antibody-secreting Cells, *J Exp Med*. **195**: 269-275.

Burger, J. A., Burger, M., and Kipps, T. J., 1999, Chronic lymphocytic leukemia B cells express functional CXCR4 chemokine receptors that mediate spontaneous migration beneath bone marrow stromal cells, *Blood*. **94**: 3658-67.

Cyster, J. G., 1999, Chemokines and cell migration in secondary lymphoid organs, *Science*. **286**: 2098-2102.

Cyster, J. G., Ansel, K. M., Reif, K., Ekland, E. H., Hyman, P. L., Tang, H. L., Luther, S. A., and Ngo, V. N., 2000, Follicular stromal cells and lymphocyte homing to follicles, *Immunol Rev*. **176**: 181-93.

Cyster, J. G., Ngo, V. N., Ekland, E. H., Gunn, M. D., Sedgwick, J. D., and Ansel, K. M., 1999, Chemokines and B-cell homing to follicles, *Curr. Top. Micro. Immunol.* **246***:* 87-92.

Förster, R., Mattis, A. E., Kremmer, E., Wolf, E., Brem, G., and Lipp, M., 1996, A putative chemokine receptor, BLR1, directs B cell migration to defined lymphoid organs and specific anatomic compartments of the spleen, *Cell.* **87***:* 1037-47.

Garcia De Vinuesa, C., Gulbranson-Judge, A., Khan, M., O'Leary, P., Cascalho, M., Wabl, M., Klaus, G. G., Owen, M. J., and MacLennan, I. C., 1999, Dendritic cells associated with plasmablast survival, *Eur J Immunol.* **29***:* 3712-21.

Gunn, M. D., Ngo, V. N., Ansel, K. M., Ekland, E. H., Cyster, J. G., and Williams, L. T., 1998, A B-cell-homing chemokine made in lymphoid follicles activates Burkitt's lymphoma receptor-1, *Nature.* **391***:* 799-803.

Hardy, R. R., and Hayakawa, K., 2001, B cell development pathways, *Annu. Rev. Immunol.* **19***:* 595-621.

Hargreaves, D. C., Hyman, P. L., Lu, T. T., Ngo, V. N., Bidgol, A., Suzuki, G., Zou, Y. R., Littman, D. R., and Cyster, J. G., 2001, A coordinated change in chemokine responsiveness guides plasma cell movements, *J Exp Med.* **194***:* 45-56.

Harris, R. B., Zhou, J., Weigle, D. S., and Kuijper, J. L., 1997, Recombinant leptin exchanges between parabiosed mice but does not reach equilibrium, *Am J Physiol.* **272***:* R1800-8.

Jones, D., Benjamin, R. J., Shahsafaei, A., and Dorfman, D. M., 2000, The chemokine receptor CXCR3 is expressed in a subset of B-cell lymphomas and is a marker of B-cell chronic lymphocytic leukemia, *Blood.* **95***:* 627-32.

Kantor, A. B., and Herzenberg, L. A., 1993, Origin of murine B cell lineages, *Annual Review of Immunology.* **11***:* 501-38.

Kunkel, E. J., and Butcher, E. C., 2002, Chemokines and the tissue-specific migration of lymphocytes, *Immunity.* **16***:* 1-4.

Legler, D. F., Loetscher, M., Roos, R. S., Clark-Lewis, I., Baggiolini, M., and Moser, B., 1998, B cell-attracting chemokine 1, a human CXC chemokine expressed in lymphoid tissues, selectively attracts B lymphocytes via BLR1/CXCR5, *J. Exp. Med.* **187***:* 655-60.

Luther, S. A., Lopez, T., Bai, W., Hanahan, D., and Cyster, J. G., 2000, BLC expression in pancreatic islets causes B cell recruitment and lymphotoxin-dependent lymphoid neogenesis, *Immunity.* **12***:* 471-481.

Ngo, V. N., Korner, H., Gunn, M. D., Schmidt, K. N., Riminton, D. S., Cooper, M. D., Browning, J. L., Sedgwick, J. D., and Cyster, J. G., 1999, Lymphotoxin alpha/beta and tumor necrosis factor are required for stromal cell expression of homing chemokines in B and T cell areas of the spleen, *J Exp Med.* **189***:* 403-12.

Schabath, R., Muller, G., Schubel, A., Kremmer, E., Lipp, M., and Forster, R., 1999, The murine chemokine receptor CXCR4 is tightly regulated during T cell development and activation, *J Leukoc Biol.* **66***:* 996-1004.

Slifka, M. K., and Ahmed, R., 1998, Long-lived plasma cells: a mechanism for maintaining persistent antibody production, *Curr Opin Immunol.* **10***:* 252-8.

Takemura, S., Braun, A., Crowson, C., Kurtin, P.J., Cofield, R.H., O'Fallon, W.M., Goronzy, J.J., Weyand, C.M., 2001, Lymphoid neogenesis in rheumatoid synovitis, *J. Immunol.* **167**: 1072-80

Trentin, L., Agostini, C., Facco, M., Piazza, F., Perin, A., Siviero, M., Gurrieri, C., Galvan, S., Adami, F., Zambello, R., and Semenzato, G., 1999, The chemokine receptor CXCR3 is expressed on malignant B cells and mediates chemotaxis, *J Clin Invest.* **104***:* 115-21.

Wehrli, N., Legler, D. F., Finke, D., Toellner, K.-M., Loetscher, P., Baggiolini, M., MacLennan, I. C. M., and Acha-Orbea, H., 2001, Changing responsiveness to chemokines allows medullary plasmablasts to leave lymph nodes, *Eur. J. Immunol.* **31***:* 609-616.

Zlotnik, A., and Yoshie, O., 2000, Chemokines: A new classification system and their role in immunity, *Immunity.* **12***:* 121-127.

LYMPHOCYTE TRAFFIC IN LYMPHOID ORGAN NEOGENESIS

Differential Roles of Ltα and LTαβ

Danielle L. Drayton, Kee Chan, Werner Lesslauer, Jason Lee, Xiao Yan Ying, and Nancy H. Ruddle*

1. INTRODUCTION

The three members of the immediate lymphotoxin (LT) tumor necrosis factor (TNF) family play important roles in inflammation and lymphoid organ development. Here their differential roles and mechanisms in lymphoid organ genesis and neogenesis are considered with particular emphasis on LTα and the LTαβ complex. Elucidation of their different roles will provide insight into the processes that regulate cell trafficking in inflammation and development. A central theme is that the mechanisms by which these cytokines induce inflammation and lymphoid organs have much in common and depend in part on regulation of chemokines and endothelial adhesion molecules.

The genes for immediate LT/TNF family members are found in a tightly linked cluster in the major histocompatibility complex (MHC). $LT\alpha_3$ is secreted as a homotrimer that can bind to the $TNFR_1$ and $TNFR_2$ receptors. The LTαβ complex is exists most frequently as $LT\alpha_2\beta_1$ and is a membrane associated molecule that depends on LTα for transport to the cell surface (Androlewicz et al. 1992). This complex binds to the LTβR receptor. TNFα is both membrane bound, and after interaction with TNFα converting enzyme (TACE), is secreted and interacts with TNFR1 and TNFRII. Members of the LT/TNF family are made by a wide variety of cells. LTα and LTαβ are made by antigen activated CD4 Th1 cells, CD8 cells, NK cells, and early B cells (Sacca et al. 1997). TNFα expression is more widespread and includes those cells noted above, activated macrophages, and many other cell types. Those cells that express LTα and LTαβ in development are less well studied but include a $CD4^+$ $CD3^-$ $IL7R^+$ positive cell (Mebius et al. 1997).

2. DIFFERENTIAL ROLES OF LTα AND LTαβ IN LYMPHOID ORGAN DEVELOPMENT

The role of LTα in lymphoid organ development was first established by DeTogni et al (DeTogni et al. 1994) by elimination of its expression by homologous recombination.

*Yale University School of Medicine, New Haven, CT 06520

LTα-/- mice have no lymph nodes or Peyer's patches and exhibit a disorganized splenic architecture. At the time of the initial studies, it was unclear whether LTα was acting in its $LT\alpha_3$ form or through the LTαβ complex. Elimination of LTβ expression suggested that both were true. LTβ-/- lack PP and exhibit splenic disorganization, and have no peripheral LN, though they maintain mesenteric (MLN) and cervical LN (CLN) (Koni et al. 1997). These data suggested that LTα alone could induce MLN and CLN, whereas the LTαβ complex was necessary for PLN. These results were confirmed by studies in which LTbRFcIg fusion protein was administered to pregnant mice. Depending on the time of administration during gestation, different PLN were eliminated (Rennert et al. 1998). However, MLN were not eliminated, supporting the contention that functions of the LTα and LTαβ complex differ in some situations. This is supported by our recent data (Chan, K., Shenoy, P. & Ruddle, N.H. unpublished) that indicate that their roles also differ in development of nasal associated lymphoid tissue (NALT). Though both LTα-/- and LTβ-/- mice retain the NALT organ, cellular composition and expression of chemokines and endothelial adhesion molecules differs in the two knockout strains. Experiments are underway to determine whether these differences are qualitative or quantitative, particularly with regard to L-selectin and L-selectin ligand (peripheral node addressin, MECA 79) expression.

3. ROLES OF LTα AND LTαβ IN TERTIARY LYMPHOID ORGANS

3.1. Definition of Tertiary Lymphoid Organs

Tertiary lymphoid organs are accumulations of cells with the characteristics of lymphoid tissues that arise in the course of chronic inflammation in autoimmunity or microbial infection. These lymphoid accumulations, also called ectopic lymphoid organs, arise through a process that has been termed lymphoid organ neogenesis (Kratz et al. 1996; Ruddle 1999) These lymphoid accumulations variously exhibit characteristics of secondary lymphoid organs (e.g. lymph nodes, spleen) with regard to the presence of germinal centers, high endothelial venules, adhesion molecules including PNAd and MAdCAM, lymphoid organ chemokines, and the presence of dendritic cells and follicular dendritic cells. A few examples of tertiary lymphoid organs that have been described in chronic inflammation in autoimmune and microbial diseases are indicated in Table 1.

3.2. The Role of LTα in Lymphoid Organ Neogenesis

In our laboratory we have studied lymphoid organ neogenesis in transgenic mice in which LTα expression is driven by the rat insulin II promoter. This results in expression in the β cells in the islets of Langerhans in the pancreas as expected, and also in the kidney and skin because this promoter is not completely tissue specific. Cellular infiltration at the sites of RIPLTα transgene expression has characteristics of classical inflammation (VCAM, ICAM, RANTES, IP-10, MIP1-α, L selectin^{lo} cells) but also lymphoid organs (MAdCAM, PNAd, SLC, BLC, L selectin^{hi} cells, FDC, DC, B and T cell compartmentalization) (Cuff et al. 1999; Hjelmstrom et al. 2000; Kratz et al. 1996), and thus can be considered as a tertiary organ.

Table 1. Lymphoid organ neogenesis in chronic inflammation in autoimmune and microbial diseases [a]

Hashimoto's disease (thyroid)
- HEVs, cellular organization, GCs
- "Tertiary lymphoid tissue"

Multiple sclerosis (CNS)
- HEVs, cellular organization

Myasthenia gravis (thymic hyperplasia)
- HEVs, cellular organization, GCs

Rheumatoid arthritis (joints, lung)
- HEVs, cellular organization, GCs
- B cell differentiation & somatic hypermutation

Diabetes (pancreas)
- HEV addressins PNAd and MAdCAM; SLC, BLC (NOD; RIPLTa)

Lyme synovitis (joints)
- HEVs, cellular organization, FDCs, GCs

Hepatitis C (liver)
- "lymphoid follicles"; some FDCs

[a] The information in this table in summarized in Kratz et al (1996)

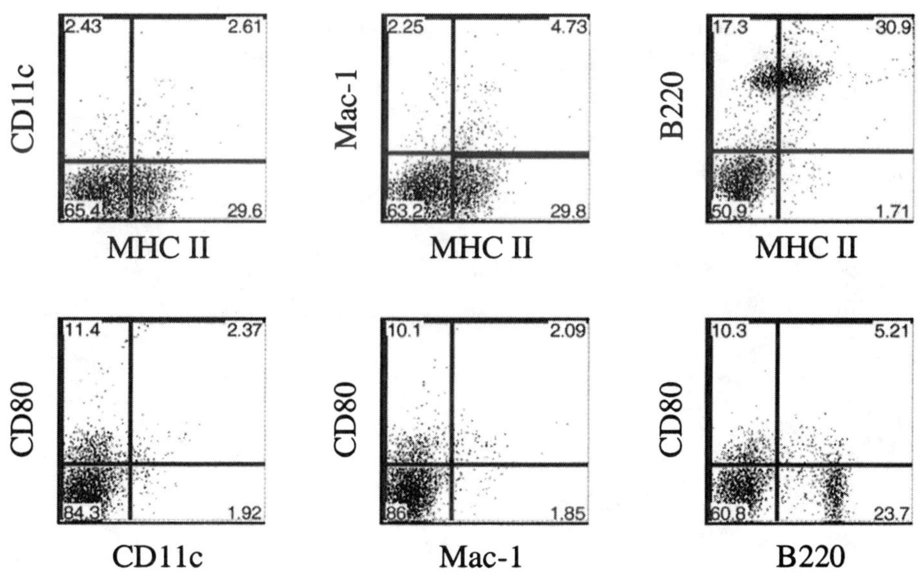

Figure 1. Potential APCs in RIPLTα kidney infiltrate

Furthermore, there are cells in the infiltrate that express high levels of MHC class II B7 (Figure 1), suggesting that these cells possess the potential to present antigen. We conclude that LTα in the absence of LTb directly induces many of these activities as it can induce MAdCAM, VCAM-1, ICAM, MCP-1, IP10, MIP-1α *in vitro* (Cuff et al. 1998).

3.3. LTβ is Necessary for PNAd, but not SLC and BLC in Tertiary Lymphoid Organs

In order to determine whether LTα alone was capable of lymphoid organ neogenesis or whether LTαβ complex was also required, RIPLTα mice were crossed to LTβ-/- or to TNFRI-/- or TNFRII-/- mice and then backcrossed to achieve homozygozity at the disrupted loci. These studies indicated that LTα alone through the TNFR1 receptor is responsible for the majority of the cellular infiltrate (Sacca et al. 1998). RIPLTα• LTβ-/- mice still exhibited inflammation, but the phenotype of cells in their infiltrates differed. The proportion of $CD44^{lo}$ L sel^{hi} cells was reduced form 20% to 8%. Furthermore, though MAdCAM expression was retained, PNAd was lost (Cuff et al. 1999) (Figure 2). These data indicate that LTαβ expression, most likely by cells in the infiltrate, influenced PNAd expression and attraction of L sel^{hi} cells. Nonetheless, even though SLC and BLC were reduced in the spleen of RIPLTα LTβ-/- mice, as expected from the studies of Ngo et. al (Ngo et al. 1999), their expression in the infiltrates was maintained (Hjelmstrom et al. 2000). Data from these studies and from the in vitro experiments noted above that indicate that LTα can induce MAdCAM from endothelial cells (Cuff et al. 1998) suggest that LTα can induce MAdCAM (and thus MLN) but the LTαβ complex is required for PNAd (and peripheral LN).

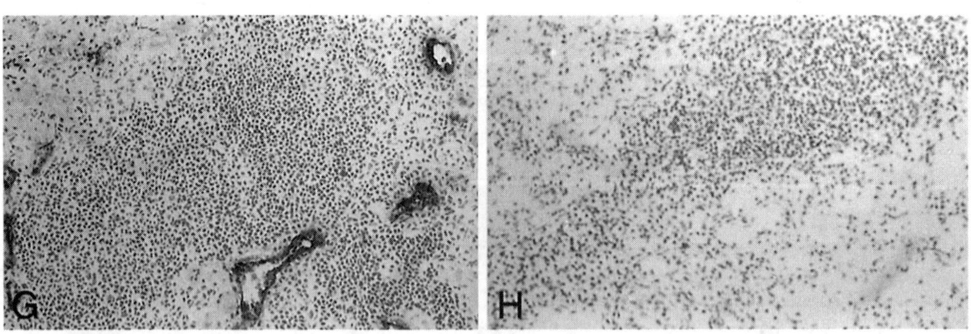

Figure 2. LTβ is required for PNAd expression in RIPLTα kidney infiltrates .
(G) RIPLTα kidney stained with MECA 79. (H). Absence of MECA 79 staining in infiltrate of RIPLTα.LTβ-/- mice. Reprinted by permission of the American Association of Immunologists (Cuff et al, 1999).

4. HOW DOES LTαβ INFLUENCE EXPRESSION OF PERIPHERAL NODE ADDRESSIN?

In order to determine how LTαβ expression induces expression of PNAd, we have prepared RIPLTβ mice that will be crossed to RIPLTα mice. The RIPII promoter was used to drive expression of the LTβ gene in transgenic mice. To date, 6 mice positive for the RIPLTβ transgene have been identified by southern blot hybridization. Copy numbers range from two to approximately 20. Four transgenic lines transmit through the germ line and express the transgene mRNA in pancreas or pancreas and kidney. In contrast to RIPLTα mice, expression of RIPLTβ did not result in infiltrates in those organs in which it was expressed. These mice will be crossed to RIPLTα mice. We expect their infiltrates will have a phenotype that is more similar to a peripheral node than a mesenteric node with more PNAd, and a higher proportion of naïve T cells than that found in infiltrates from RIPLTα mice.

One mechanism by which the LTαβ complex may influence PNAd expression is through regulation of an HEV sulfotransferase. MECA 79 is an epitope that is defined by the modification of a number of different proteins include GlyCAM-1 podocalyxin, CD34, and Sgp200. The modifications include sialyation, fucosylation and sulfation. A number of enzymes have been identified that participate in this process including Fuc-TIV, and Fuc-TVII. Of special interest is an enzyme called L-selectin ligand sulfotransferase (LSST) by Fukuda's group (Hiraoka et al. 1999) and HEC-GlcNAc6ST by Rosen's group (Bistrup et al. 1999). This enzyme is expressed by HEV in PLN. Mice deficient in HEC-GlcNAc6ST lack luminial expression of PNAd in PLN and MLN though they retain MAdCAM in MLN (Hemmerich et al. 2001). Our hypothesis is that the blood vessels in the infiltrates of tertiary lymphoid organs as defined in Table I will express HEC-GlcNAc6ST, and that the infiltrates of RIPLTα•RIPLTαβ mice will express even higher levels of this enzyme, PNAd, and L-selectin^{hi} cells.

5. SUMMARY AND CONCLUSIONS

In these studies the differential roles of LTα and LTαβ complex have been discussed with regard to development of lymphoid organs in ontogeny and in inflammation, LTα is necessary for PLNand MLN, most likely as both LTα and LTαβ complex, whereas only LTαβ is required for MLN. Both are involved in the cellularity of the NALT. When expressed as a transgene, LTα alone can induce cellular accumulation and MAdCAM, but not PNAd, an epitope associated with PLN HEV. These data suggest that LTαβ complex plays a crucial role in PNAd. One hypothesis is that LTαβ induces PNAd through modification via an HEV sulfotransferase. RIPLTα•RIPLTβ mice will provide an important tool to investigate this question.

6. ACKNOWLEDGEMENTS

Supported by NIH CA 16885, AI 44453, DK 57731 (NHR) and NIH F31 GM20919 (D.D.).

REFERENCES

Androlewicz, M. J., J. L. Browning and C. F. Ware (1992). "Lymphotoxin is expressed as a heteromeric complex with a distinct 33- kDa glycoprotein on the surface of an activated human T cell hybridoma." *J Biol Chem* **267**: 2542-7.

Bistrup, A., S. Bhakta, J. K. Lee, Y. Y. Belov, M. D. Gunn, F. R. Zuo, C. C. Huang, R. Kannagi, S. D. Rosen and S. Hemmerich (1999). "Sulfotransferases of two specificities function in the reconstitution of high endothelial cell ligands for L-selectin." *J Cell Biol* **145**: 899-910.

Cuff, C. A., R. Sacca and N. H. Ruddle (1999). "Differential induction of adhesion molecule and chemokine expression by LTa3 and LTab in inflammation and elucidates potential mechanisms of mesenteric and peripheral lymph node development." *J. Immunol.* **162**: 5965-5972.

Cuff, C. A., J. Schwartz, C. M. Bergman, K. S. Russell, J. R. Bender and N. H. Ruddle (1998). "Lymphotoxin alpha3 induces chemokines and adhesion molecules: insight into the role of LT alpha in inflammation and lymphoid organ development." *J Immunol* **161**: 6853-60.

DeTogni, P., J. Goellner, N. H. Ruddle, P. R. Streeter, A. Fick, S. Mariathasan, S. C. Smith, R. Carlson, L. P. Shornick, J. Strauss-Schoenberger, J. H. Russell, R. Karr and D. D. Chaplin (1994). "Abnormal development of peripheral lymphoid organs in mice deficient in lymphotoxin." *Science* **264**: 703-707.

Hemmerich, S., A. Bistrup, M. S. Singer, A. van Zante, J. K. Lee, D. Tsay, M. Peters, J. L. Carminati, T. J. Brennan, K. Carver-Moore, M. Leviten, M. E. Fuentes, N. H. Ruddle and S. D. Rosen (2001). "Sulfation of L-selectin ligands by an HEV-restricted sulfotransferase regulates lymphocyte homing to lymph nodes." *Immunity* **15**: 237-47.

Hiraoka, N., B. Petryniak, J. Nakayama, S. Tsuboi, M. Suzuki, J. C. Yeh, D. Izawa, T. Tanaka, M. Miyasaka, J. B. Lowe and M. Fukuda (1999). "A novel, high endothelial venule-specific sulfotransferase expresses 6-sulfo sialyl Lewis(x), an L-selectin ligand displayed by CD34." *Immunity* **11**: 79-89.

Hjelmstrom, P., J. Fjell, T. Nakagawa, R. Sacca, C. A. Cuff and N. H. Ruddle (2000). "Lymphoid tissue homing chemokines are expressed in chronic inflammation." *Am J Pathol* **156**: 1133-8.

Koni, P. A., R. Sacca, P. Lawton, J. L. Browning, N. H. Ruddle and R. A. Flavell (1997). "Distinct roles in lymphoid organogenesis for lymphotoxins alpha and beta revealed in lymphotoxin beta-deficient mice." *Immunity* **6**: 491-500.

Kratz, A., A. Campos-Neto, M. S. Hanson and N. H. Ruddle (1996). "Chronic inflammation caused by lymphotoxin is lymphoid neogenesis." *J.Exp.Med.* **183**: 1461-1472.

Mebius, R. E., P. Rennert and I. L. Weissman (1997). "Developing lymph nodes collect CD4+CD3- LTbeta+ cells that can differentiate to APC, NK cells, and follicular cells but not T or B cells." *Immunity* **7**: 493-504.

Ngo, V. N., H. Korner, M. D. Gunn, K. N. Schmidt, D. S. Riminton, M. D. Cooper, J. L. Browning, J. D. Sedgwick and J. G. Cyster (1999). "Lymphotoxin alpha/beta and tumor necrosis factor are required for stromal cell expression of homing chemokines in B and T cell areas of the spleen." *J Exp Med* **189**: 403-12.

Rennert, P. D., D. James, F. Mackay, J. L. Browning and P. S. Hochman (1998). "Lymph node genesis is induced by signaling through the lymphotoxin beta receptor." *Immunity* **9**: 71-9.

Ruddle, N. H. (1999). "Lymphoid neo-organogenesis: lymphotoxin's role in inflammation and development." *Immunol Res* **19**: 119-25.

Sacca, R., C. A. Cuff, W. Lesslauer and N. H. Ruddle (1998). "Differential activities of secreted lymphotoxin-alpha3 and membrane lymphotoxin-alpha1beta2 in lymphotoxin-induced inflammation: critical role of TNF receptor 1 signaling." *J Immunol* **160**: 485-91.

Sacca, R., C. A. Cuff and N. H. Ruddle (1997). "Mediators of inflammation." *Curr Opin Immunol* **9**: 851-7.

REGULATION OF SPLEEN WHITE PULP STRUCTURE AND FUNCTION BY LYMPHOTOXIN

David D. Chaplin*

1. INTRODUCTION

The secondary lymphoid tissues are thought to provide sites for efficient interactions between the cells of the immune system, permitting rare antigen-specific B lymphocytes and T lymphocytes to interact with each other and with antigen non-specific dendritic cells to interact and generate a properly regulated immune response. Few experiments, however, directly test the role of the secondary lymphoid organs in the immune response.

We initially observed that targeted deletion of the gene encoding lymphotoxin-α (LTα, also designated TNFβ) in mice resulted in congenital absence of peripheral lymph nodes and Peyer's patches, and disruption of normal spleen white pulp architecture.[1] The alterations in spleen structure broadly affected all aspects of white pulp organization. There was loss of distinct white pulp T cell and B cell zones and loss of clustered follicular dendritic cells and detectable germinal centers.[2] These studies demonstrated that lymphotoxin has a major role as a signal supporting the induction or maintenance of normal secondary lymphoid organ structures. Subsequent studies by several groups demonstrated that targeted ablation of the genes encoding lymphotoxin-β (LTβ) and the LTβ receptor (LTβR) showed a similar phenotype.[3, 4] This established that the actions of lymphotoxin controlling lymphoid organogenesis are mediated by the membrane lymphotoxin heterotrimer ($LT\alpha_1\beta_2$) interacting with the LTβR.

The availability of mice with congenital disturbances of lymphoid tissue organogenesis provides an opportunity to investigate the role of normal lymphoid organ structure in various aspects of the immune response. Perhaps not surprisingly, the broad disturbances of LN and spleen white pulp structure observed in $LT\alpha^{-/-}$ mice were associated with impaired production of high affinity isotype switched antibody following immunization with T cell-dependent antigens such as sheep red blood cells (SRBC)

* David D. Chaplin, Department of Microbiology, Univ. of Alabama at Birmingham, Birmingham, AL 35294.

(Figure 1) or keyhole limpet hemocyanin (KLH) when they were administered without adjuvant.[2, 5]

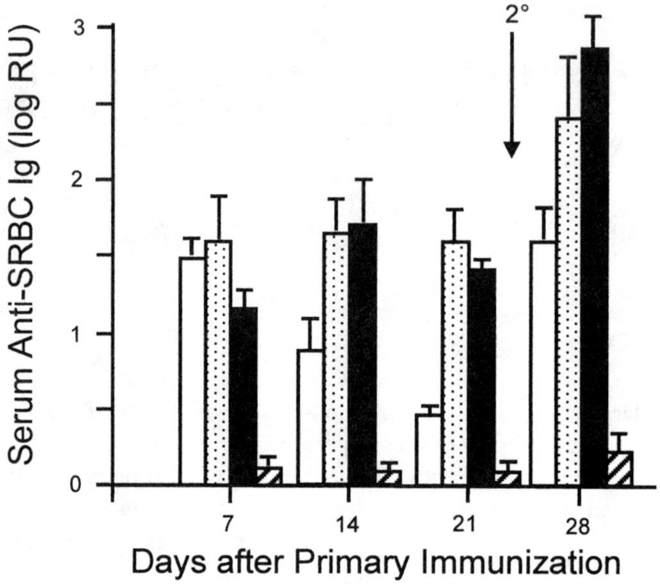

Figure 1. LTα$^{-/-}$ mice fail to generate high affinity serum IgG antibody responses after immunization with a T cell-dependent antigen. Mice were immunized i.v. with 10^8 SRBC suspended in PBS on day 0. A booster immunization was administered on day 21 (2°). Serum samples collected at days 7, 14, 21, and 28 were analyzed for the presence of IgM anti-SRBC antibodies (open and stippled bars) and for IgG anti-SRBC antibodies (solid and cross hatched bars) using an ELISA in which SRBC were bound to the plate. This ELISA detects high affinity somatically mutated anti-SRBC IgG antibodies, but not low-affinity germ-line encoded anti-SRBC IgG antibodies. Open and solid bars represent samples from wild type mice. Stippled and cross hatched bars represent samples from LTα$^{-/-}$ mice. Modified from Fu, et al.[5]

When LTα$^{-/-}$ mice are lethally irradiated and reconstituted with bone marrow from a wild type animal, repopulation of the spleen tissue with wild type hematopoietic cells is associated with restoration of follicular dendritic cell (FDC) clusters, and the ability to form germinal centers (GC) and the adjuvant-independent IgG responses.[5] This indicates that some aspects of the spleen white pulp microenvironment are plastic and dependent on the presence of LT-expressing bone marrow-derived cells. In these experiments, the ability to mount a high affinity isotype-switched antibody response was correlated with the presence of clusters of FDC. The experiments described here were designed to test the requirements for normal white pulp structure in the formation of memory cells and for the expression of the memory response.

2. DEFECTIVE MEMORY IGG RESPONSES IN LT$\alpha^{-/-}$ MICE

Many studies suggest that memory B cells form largely in GC in association with FDC.[6, 7] Under normal circumstances, memory B-lymphocytes manifest antigen receptors carrying substantial numbers of somatic mutations and showing high affinity for their antigen, supporting a model in which the GC is critical in memory B cell formation. Antigen-specific T cells also undergo phenotypic alterations in GC.[8, 9] The relationship between these GC-dependent changes in T cell surface phenotype and maturation into a T memory cell remains less well defined. Furthermore, it is not clear whether the changes in T and B cell phenotype that occur in the GC are essential elements required for memory differentiation, or are simply associated with the development of memory.

2.1 Impaired memory IgG responses in LT$\alpha^{-/-}$ mice

The finding that LT$\alpha^{-/-}$ mice immunized with SRBC showed reduced high affinity IgG responses (Fig. 1) suggested that aspects of the maturation of the B cell response that have been associated with the induction of B cell memory did not occur in the defective LTα-deficient microenvironment. In order to test directly whether immunized LT$\alpha^{-/-}$ mice could manifest a memory response, mice that had been primed i.p. with SRBC were rested for three months and then challenged by i.p. immunization with 10^8 SRBC (Fig. 2). Wild type mice primed and challenged in this fashion showed the robust high affinity IgG anti-SRBC response characteristic of B cell memory. In contrast, LT$\alpha^{-/-}$ mice showed a recall response nearly a thousand-fold reduced compared to the wild type response.

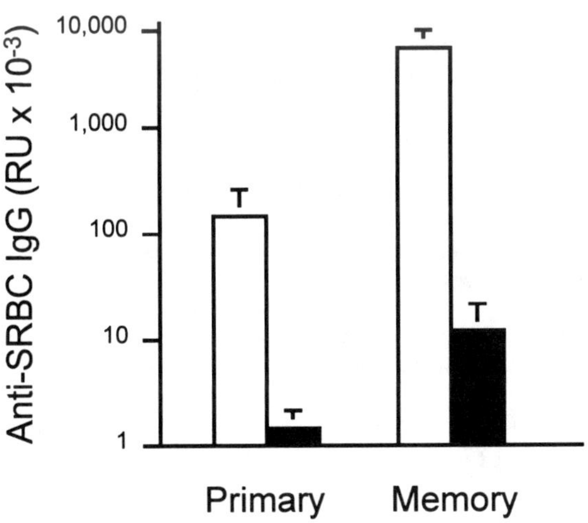

Figure 2. Lymphotoxin-deficient mice show impaired IgG memory responses. Wild type (open bars) and LTα-deficient (filled bars) were primed by immunization i.p. with 10^8 SRBC, then three months later were challenged with a similar immunization. Serum anti-SRBC IgG was measured 5 days after the boost using an ELISA. Modified from Fu, et al.[10]

2.2 Impaired IgG memory in LTα$^{-/-}$ mice is not restored by immunization with adjuvant

Previous studies had shown that use of adjuvants such as incomplete Freund's when immunizing LTα$^{-/-}$ mice with T dependent antigens largely restored the development of high affinity somatically mutated IgG responses.[5] Interestingly, this induction of somatically mutated antibodies was achieved in the absence of morphologically defined follicular dendritic cells and GC, suggesting that neither the follicular dendritic cells nor the GC structure are absolutely required for this important aspect of the mature B cell response. Given the association of somatic mutation with the induction of B cell memory, we tested whether immunization with adjuvant at the time of priming permitted the expression of a memory response. Both wild type and LTα$^{-/-}$ mice were immunized i.p. with 10^8 SRBC emulsified in incomplete Freund's adjuvant, then challenged 3 months later by i.p. immunization with 10^8 SRBC without adjuvant (Fig. 3). Although immunization with adjuvant restored the high affinity IgG response measured 10 days after the priming immunization, it did not lead to the expression of a memory response after challenge without antigen.

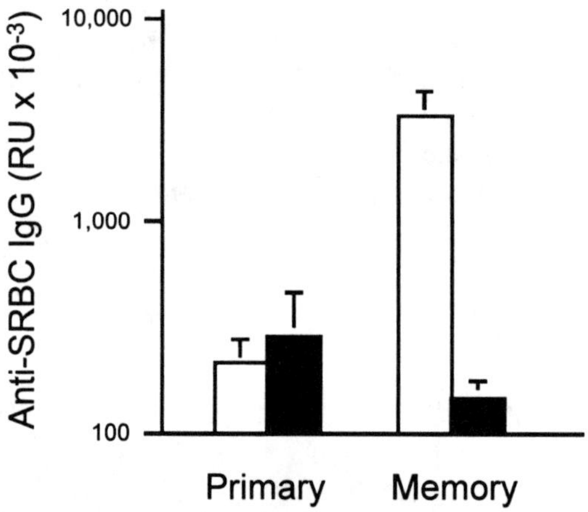

Figure 3. Immunization with adjuvant fails to restore the memory response in LTα$^{-/-}$ mice. Wild type (open bars) and LTα$^{-/-}$ (filled bars) mice were primed with SRBC in incomplete Freund's adjuvant and serum was collected 10 days later (Primary). Three months later, all mice were challenged by immunization i.p. with SRBC without adjuvant and 5 days later serum was collected (Memory). We analyzed all sera for the presence of high affinity IgG anti-SRBC antibodies using an ELISA. Modified from Fu, et al.[10]

2.3 Induction of B cell memory is blocked in LTα$^{-/-}$ mice, whereas induction of T cell memory is independent of LTα

Normal memory responses require primed B cells and primed T cells.[7, 8] In order to test whether both T cell and B cells require lymphotoxin, we used a hapten/carrier system to distinguish the roles of the T cell and B cell arms of the response. To test whether memory B cells were formed in LTα$^{-/-}$ mice, we attempted to prime memory B cells by immunizing wild type and LTα$^{-/-}$ mice with nitrophenyl-haptenated keyhole limpet hemocyanin (NP-KLH). Two weeks after priming, B cells were purified from the spleens of these immunized mice and transferred adoptively to sublethally irradiated wild type recipient mice together with memory T cells from a lymphotoxin-sufficient donor. In order to assure that the memory T cell preparation did not introduce functional B cells, we used memory T cells that had been generated by immunizing a B cell-deficient mouse (Ig heavy chain targeted strain[11]) with SRBC. T cells were purified from the spleens of these immunized mice and adoptively transferred together with purified B cells from NP-KLH immunized mice. The mice were then challenged with 10^8 NP-haptenated SRBC and serum and NP responses were measured 5 days later. Mice that received NP-specific B cells derived from wild type donors together with wild type SRBC-specific T cells showed a robust NP-specific memory response (Fig. 4), whereas mice that received B cells from an NP-sensitized LTα-deficient donor showed a nearly undetectable memory NP response.

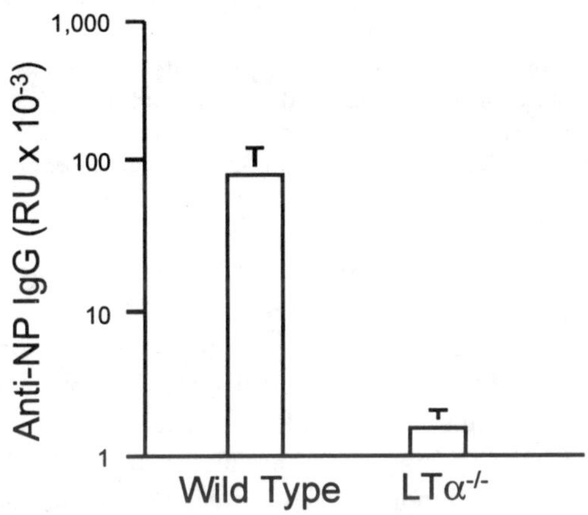

Figure 4. B cell memory does not form in LTα$^{-/-}$ mice. Partially purified B cells from the spleens of wild type and LTα$^{-/-}$ mice that had been immunized with NP-KLH were adoptively transferred together with T cells from SRBC-sensitized B cell-deficient mice. Recipient mice were sublethally irradiated wild type mice. Five days after immunization of the recipient mice with NP-haptenated SRBC, we measured serum anti-NP IgG using an ELISA. Modified from Fu, et al.[10]

In order to test whether a lymphotoxin-dependent microenvironment is required for the formation of memory T cells, we attempted to prime T cells by immunizing both wild type and LTα$^{-/-}$ mice with SRBC and to prime B cells by immunizing a second set of mice with NP-KLH. After adoptive transfer of these cells to sublethally irradiated wild type recipient mice, we challenged the recipients by i.v. immunization with NP-SRBC. Five days after adoptive transfer and challenge, there was a robust serum anti-NP IgG response in mice that had received wild type sensitized NP-specific B cells and SRBC-specific T cells that had been generated in either a wild type or an LTα$^{-/-}$ environment (Fig. 5). Thus, unlike the LTα-dependent development of memory B cells, development of memory T cells is independent of LTα.

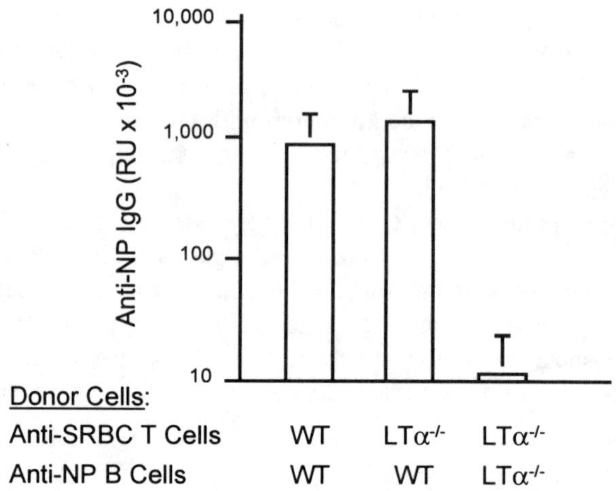

Figure 5. LTα$^{-/-}$ T cells show memory activity equal to wild type T cells. Five days after i.v. immunization of wild type mice that had received either adoptively transferred NP-sensitized B cells from either wild type (WT) or LTα-deficient (LTα$^{-/-}$) donors and SRBC-sensitized T cells from either wild type or LTα-deficient donors, serum anti-NP-IgG responses showed effective memory T cell function independent of the expression of LTα in the T cell donor. Modified from Fu, et al.[10]

2.4 Impaired expression of memory Ig response in LTα-deficient mice

In order to begin to define the elements required for the expression of a memory IgG response, we tested whether wild type memory cells formed in a normal environment could manifest their memory phenotype after adoptive transfer and challenge in a LTα-deficient recipient. When SRBC-primed splenocytes are transferred into a sublethally irradiated normal recipient, they support a more than 230-fold greater anti-SRBC IgG response than naïve splenocytes (Fig. 6). In contrast, primed splenocytes support a less than 10-fold greater anti-SRBC IgG response than naïve splenocytes when transferred into a lymphotoxin-α-deficient recipient. This identifies a previously unrecognized requirement for either LTα-expressing cells or an LTα-specified microenvironment for normal expression of memory antibody responses.

3. CONCLUSION

The induction and expression of a well-coordinated memory antibody response requires many steps, from the initial sensitization of both B and T cells to a T-dependent antigen, to maturation of the activated B cells into a memory phenotype, and finally to antigen-induced expression of the memory response. The studies described here demonstrate a differential requirement for lymphotoxin-α in the generation of antigen-specific memory B and memory T cells. Memory T cells are induced effectively in the absence of either lymphotoxin or the complex lymphoid tissue microenvironment that is specified by lymphotoxin. In contrast, memory B cells are profoundly dependent on the

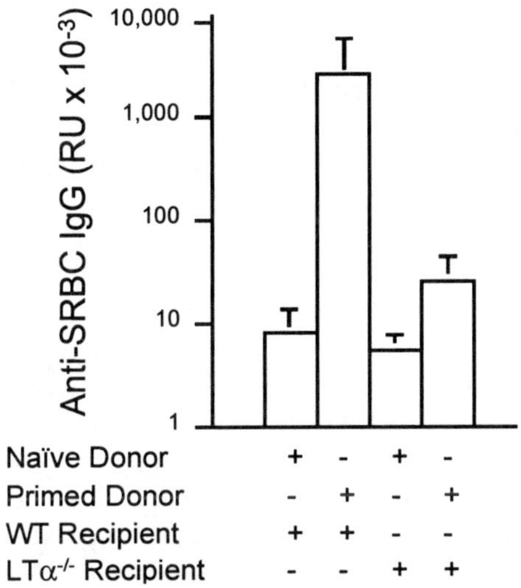

Figure 6. Normal memory cells cannot express their memory phenotype in the disturbed microenvironment of lymphotoxin-deficient mice. Total splenocytes from either naïve or SRBC-sensitized donors were adoptively transferred into either wild type (WT) or lymphotoxin-α-deficient ($LT\alpha^{-/-}$) sublethally irradiated recipients. These chimeric mice then received an antigen challenge by i.v. infusion of SRBC. Five days later, we measured the serum anti-SRBC IgG antibody response using an ELISA. Modified from Fu, et al.[10]

expression either of lymphotoxin itself, or of some aspect of the lymphoid tissue microenvironment specified by lymphotoxin. We suspect that major determinants of the ability to support the formation of memory B cells are the structures required for the formation of the germinal center response. Clusters of follicular dendritic cells appear central in this process, providing an environment supportive of the processes underlying the somatic hypermutation that is characteristic of B cell memory.

Of particular interest, in addition to being required for the formation of memory cells, some aspect of the lymphotoxin-dependent microenvironment is required for the expression of the memory response by normally formed memory B and T cells. The current experiments have not addressed what features of the LT-dependent structural phenotype support the memory response. Prior studies have not excluded a role for the LT-dependent clusters of follicular dendritic cells; however, other LT-dependent aspects of normal lymphoid tissue architecture (including normally segregated T and B cell zones, LT-dependent bone marrow derived dendritic cells, and expression of LT-dependent chemokines) should all be considered. Experiments using selective reconstitution of individual aspects of the normal spleen microenvironment should permit dissection of the key elements supporting both the induction and the expression of memory B responses. An understanding of the structures essential for the normal memory response may provide new targets for immunomodulatory therapies designed to enhance or inhibit normal and pathologic persistent immune responses.

REFERENCES

1. P. De Togni, J. Goellner, N. H. Ruddle, P. R. Streeter, A. Fick, S. Mariathasan, S. C. Smith, R. Carlson, L. P. Shornick, J. Strauss-Schoenberger, J. H. Russell, R. W. Karr, and D. D. Chaplin, Abnormal development of peripheral lymphoid organs in mice deficient in lymphotoxin, *Science* **264,** 703-707 (1994).
2. M. Matsumoto, S. F. Lo, C. J. Carruthers, J. Min, S. Mariathasan, G. Huang, D. R. Plas, S. M. Martin, R. S. Geha, M. H. Nahm, and D. D. Chaplin, Affinity maturation without germinal centers in lymphotoxin-α (LTα) deficient mice, *Nature* **382,** 462-466 (1996).
3. A. Futterer, K. Mink, A. Luz, M. H. Kosco-Vilbois, and K. Pfeffer, The lymphotoxin β receptor controls organogenesis and affinity maturation in peripheral lymphoid tissues, *Immunity* **9,** 59-70 (1998).
4. P. A. Koni, R. Sacca, P. Lawton, J. L. Browning, N. H. Ruddle, and R. A. Flavell, Distinct roles in lymphoid organogenesis for lymphotoxin α and β revealed in lymphotoxin β-deficient mice, *Immunity* **6,** 491-500 (1997).
5. Y.-X. Fu, H. Molina, M. Matsumoto, G. Huang, J. Min, and D. D. Chaplin, Lymphotoxin-α supports development of splenic follicular structure that is required for IgG responses, *J. Exp. Med.* **185,** 2111-2120 (1997).
6. Y. J. Liu, Sites of B lymphocyte selection, activation, and tolerance in spleen, *J. Exp. Med.* **186,** 625-629 (1997).
7. K. Rajewsky, Clonal selection and learning in the antibody system, *Nature* **381,** 751-758 (1996).
8. B. Zheng, S. Han, and G. Kelsoe, T helper cells in murine germinal centers are antigen-specific emigrants that downregulate Thy-1, *J. Exp. Med.* **184,** 1083-1091 (1996).
9. B. Zheng, S. Han, Q. Zhu, R. Goldsby, and G. Kelsoe, Alternative pathways for the selection of antigen-specific peripheral T cells, *Nature* **384,** 263-266 (1996).
10. Y.-X. Fu, G. Huang, Y. Wang, and D. D. Chaplin, Lymphotoxin-α-dependent spleen microenvironment supports the generation of memory B cells and is required for their subsequent antigen-induced activation, *J. Immunol.* **164,** 2508-2514 (2000).
11. D. Kitamura, J. Roes, R. Kuhn, and K. Rajewsky, A B cell-deficient mouse by targeted disruption of the membrane exon of the immunoglobulin mu chain gene, *Nature* **350,** 423-426 (1991).

ENZYMATIC CONTROL OF LEUKOCYTE TRAFFICKING: ROLE OF VAP-1

Marko Salmi and Sirpa Jalkanen•

1. INTRODUCTION

Lymphocytes need to extravasate from the blood into secondary lymphoid tissues, such as lymph nodes, when patrolling the body in search of harmful antigens. Moreover, both lymphocytes and granulocytes must leave the blood to execute their effector functions in peripheral tissues during the defence reaction. Leukocytes normally emigrate from the blood using a multistep adhesion cascade[1-3]. Selectins and their carbohydrate ligands mediate the intial tethering of blood-borne leukocytes and are also critical in the ensuing rolling phase. Chemoattractants and their serpentine receptors appear crucial for subsequent triggering events, which lead to stable shear-resistant adhesion. This firm adhesion and the final diapedesis through the vascular wall heavily relies on leukocyte integrins and endothelial members of the immunoglobulin superfamily. Nevertheless, the currently known molecules cannot explain the full diversity, specificity and mechanics of immune cell trafficking, and newly discovered molecules are constantly refining the picture of extravasation.

2. EARLY HISTORY OF VAP-1

In the late 80's we were interested in analyzing the endothelial determinants directing lymphocyte homing to inflamed joints. Since it had been shown that lymphocyte binding to synovial vessels displays marked differences in respect of binding to either gut or PLN vessels and since healthy synovial tissue lining the joint cavity is practically devoid of lymphocytes, we hypothesized that a unique endothelial addressin might be

• MediCity Research Laboratory, University of Turku and National Public Health Institute, Department in Turku, Turku, Finland

expressed in inflamed synovial vessels[4]. A panel of mAbs was produced against purified synovial vessels from a joint of a patient suffering from rheumatoid arthritis. We identified one mAb, 1B2, which reacted with the synovial vessels[5]. Most importantly, the same antibody also diminished binding of human peripheral blood lymphocytes to postcapillary venules in inflamed synovial membranes by 50%. Since the molecular mass and the tissue and cell distribution of this antigen were different from those of other endothelial adhesion molecules known at the time, this novel antigen was named as vascular adhesion protein-1 or VAP-1[5].

3. VAP-1 IS UPREGULATED AT SITES OF INFLAMMATION

Anti-VAP-1 mAbs stain high endothelial cells of lymphatic organs, sinusoidal endothelial cells in liver and small-caliber venules in many other tissues[6]. Moreover, VAP-1 is expressed in dendritic cells of germinal centers. VAP-1 is also abundantly present in pericytes, smooth muscle cells, and fat cells. In contrast, most capillaries and endothelial cells in large vessels are devoid of VAP-1 reactivity, as are epithelial and fibroblastoid cells. VAP-1 is also absent from all leukocytes, and from *in vitro* growing cell lines.

VAP-1 is up-regulated on vasculature at many sites of inflammation, such as in synovitis, allergic and other skin inflammations, and inflammatory bowel diseases[6]. However, the expression of functional VAP-1 appears to be more drastically controlled by another mechanism. In endothelial cells, VAP-1 is stored in intracellular granules that are distinct from Weibel-Palade bodies and luminal VAP-1 is only induced upon elicition of inflammation. In human trials, anti-VAP-1 mAbs given intravenously only binds to endothelial cells at sites where an inflammation has been induced (eg. in skin) and spares the normal tissue. The maximal luminal synthesis is found in 6-8 h old lesions.

4. VAP-1 HAS ENZYMATIC ACTIVITY

Biochemical analyses show that VAP-1 is a heavily sialylated glycoprotein consisting of two identical monomeric subunits of 90 kDa[7]. The oligosaccharide modifications of VAP-1 differ between the endothelial and smooth muscle cells suggesting possibly divergent functions[8]. In support of this, binding assays show that although smooth muscle cells bind lymphocytes, VAP-1 appears not to be involved in this interaction.

Cloning and sequencing of VAP-1 revealed that it has significant identity with semicarbazide sensitive mono-amine oxidases (SSAO). The closest homology (about 85%) is to a molecule called bovine serum amine oxidase[9]. What are SSAO? They are mostly soluble enzymes which catalyze oxidative deamination of primary amines[10-12]. The reaction results in formation of the corresponding aldehyde and release of hydrogen peroxide and ammonium. These enzymes are notably different from better characterized monoamine oxidases A and B in terms of their substrates, inhibitors, cofactors, subcellular localization and function. Despite 40 years of intense research, no physiological function has been firmly alluded to SSAOs, and even the nature of physiological substrates is under dispute. In different cell types SSAO have been implicated in metabolism of exo- and endogenous amines and in regulation of glucose transport[11, 13].

VAP-1 cDNA encodes a transmembrane molecule. It is type 2 membrane protein with a very short cytoplasmic N-terminus, a single transmembrane domain and a large extracellular domain. Based on limited sequence homology with distantly related bacterial SSAO crystals, VAP-1 appears to be a mushroom shaped dimer[14]. Each monomer contains 6 putative attachment sites for N-linked sugars, which all seem to be in use based on site-directed mutagenesis, and at least 3 extend from the upper surface of VAP-1. The catalytic center of the enzyme is buried within the globular head of the molecule, but is accessible from the surface via a substrate channel.

VAP-1 displays the SSAO activity[15]. VAP-1 but not control transfectants and affinity-purified VAP-1 readily oxidize benzylamine, which is a widely used artificial substrate in SSAO studies. Human VAP-1 also displays activity against methylamine, which can be formed during intermediary metabolism in man. The enzymatic activity of VAP-1 is completely abrogated by semicarbazide and hydroxylamine, which have been used extensively to define SSAOs.. Thus, VAP-1 is structurally very different from classical endothelial molecules involved in adhesion.

5. VAP-1 IS INVOLVED IN LEUKOCYTE ROLLING

The adhesive function of VAP-1 was initially deduced from the *in vitro* binding assays, in which anti-VAP-1 mAbs block lymphocyte binding to venules in synovial, peripheral lymph node and, to some extent, in appendix tissue sections[5]. Using the same assay VAP-1 proved to be important for lymphocyte adherence to vessels in inflamed skin, inflamed gut and in heart during ischemia-reperfusion in man[6, 16, 17]. Moreover, in liver lymphocyte adherence to sinusoidal endothelium is to a large extent VAP-1 dependent[18]. VAP-1 is also important for lymphocyte adherence to isolated sinusoidal endothelial cells of liver, which were the first non-transfected in vitro growing cell type that retain VAP-1 on the surface[19]. When isolated populations of mononuclear cells were tested for their VAP-1 dependent adhesion to high endothelial venules, only CD8-positive T killer cells and CD16-positive natural killer cells, but not monocytes, B-cells or T helper cells, bound to VAP-1. Thus, VAP-1 appears to contribute to lymphocyte-endothelial interactions at multiple tissues as a part of the adhesion cascade under non-static conditions.

Intravital videomicroscopy is an extremely useful technique in determining the adhesive role of various molecules under physiologically relevant conditions of blood shear. Using anti-VAP-1 mAbs we have shown that VAP-1 blockade diminishes granulocyte rolling, firm adhesion and transmigration in a model of acute peritoneal inflammation[20]. During rolling, VAP-1 blockade increases the rolling velocity, which results in diminished opportunities for the cells to adhere firmly. However, since the 70 % reduction of granulocyte extravasation into inflamed peritoneum was much more dramatic than the effect on rolling, these data hint to the possibility that VAP-1 may contribute to other steps of extravasation in addition to the initial rolling. In particular, due to the VAP-1 expression in pericytes, its role in diapedesis warrants further investigation.

6. VAP-1 IS ALSO PRESENT IN SERUM

In serum of healthy adults a soluble form of VAP-1 (sVAP-1) is found at concentration of 80 ng/ml[21]. Interestingly, sVAP-1 levels increase quite specifically in certain liver diseases and in diabetes, but remain normal at many other serious inflammations. Hence, in contrast to the soluble forms of most other endothelial molecules, which seem to be elevated in a multitude of different diseases[22], sVAP-1 is not merely a marker of general inflammation.

SSAO activity has been defined in enzymatic terms for 40 yrs in serum[23], and it has been a subject to an intense research for clinical applications[11]. With the advent of molecular characterization of VAP-1, we now know that VAP-1 is the major, if not the only, soluble SSAO species in man[24]. Hence, the availability of a sandwich-ELISA for sVAP-1 should bring significant advances over the radiochemical enzymatic determinations, if sVAP-1 determinations prove to be useful for monitoring or predicting certain specific disease processes.

sVAP-1 has an N-terminal amino acid sequence identical to the proximal extracellular sequence of membrane bound form of VAP-1. There is also compelling evidence that sVAP-1 is produced in the liver[24]. Hence, the current view holds that at least a significant portion of sVAP-1 is formed by a proteolytic cleavage of sinusoidal VAP-1 in liver.

Like other soluble adhesion molecules, sVAP-1 modulates lymphocyte adherence. In fact, presence of sVAP-1 appears to augment rather than to inhibit lymphocyte binding to endothelial cells[21]. These data indicate that the soluble VAP-1 may trigger biological responses in ligand bearing cells, and possibly contribute to the pathological accumulation of lymphocytes in diabetes and liver diseases.

7. ADHESIVE AND ENZYMATIC FUNCTIONS OF VAP-1 ARE INTERCONNECTED

The discovery of the oxidative activity of VAP-1 immediately raises the question, whether the adhesive and enzymatic properties are interconnected or distinct functions of the same molecule. Establishment of VAP-1 positive primary endothelial cultures has allowed approaching of this enigma using flow chamber assays. In this technique lymphocytes are perfused with a defined shear stress over a monolayer of endothelial cells in vitro. In this assay, anti-VAP-1 mAbs block about 50% of lymphocyte rolling. Consequently, the firm adhesion is also reduced by about half. Most notably, chemical inhibitors of SSAO activity prevent lymphocyte rolling and firm adhesion to the same extent. The end products (aldehydes or hydrogen peroxide) of the enzyme reaction do not appear to be important in this form of lymphocyte-endothelial cell interaction [25].

Since the physiological substrates of SSAO remain unknown, a provocative prediction has been made that a free amino group of an amino acid might serve as the substrate for VAP-1. Molecular modelling is compatible with this possibility, since an ε-amino group of a lysine side chain, for example, is long enough to become in contact with the catalytic center of VAP-1. Actually, there is a groove on the surface of VAP-1, which easily accommodates peptides so that their side chains have no restrains in penetrating to the cavity leading to the catalytic center. In fact, a synthetic peptide can bind to the catalytic site of VAP-1. Thus, it seems plausible that a lymphocyte surface protein could present an amino group to VAP-1. This enzyme-substrate reaction would then effectively lead to a covalent, but transient, linkage between the two cell types (Fig. 1). This kind of regulation can be envisioned to be compatible with the rapid time constants needed for the molecular interactions during rolling. Moreover, they reveal a completely new way how enzymes may regulate leukocyte extravasation.

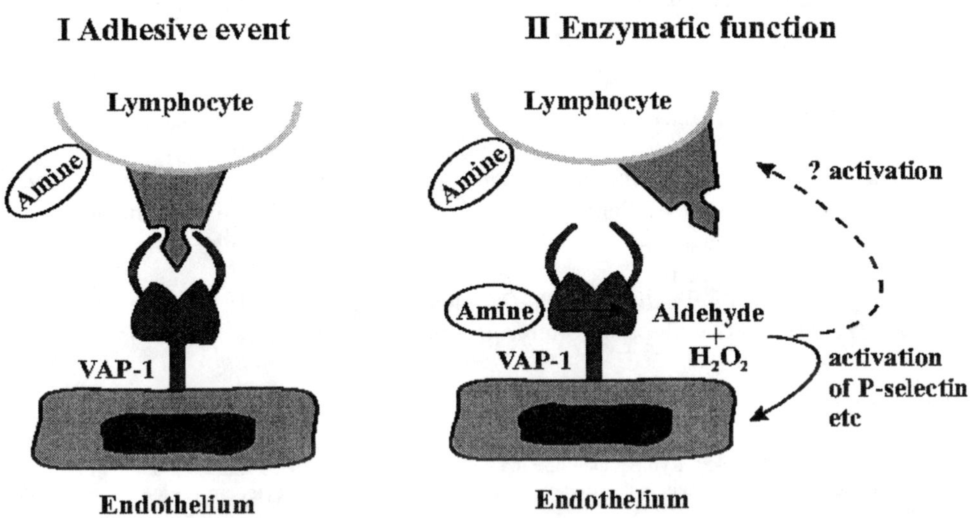

Figure 1. Dual functions of VAP-1. VAP-1 is an endothelial adhesion molecule and SSAO. The adhesive function involves binding of an uncharacterized lymphocyte counter-receptor to the sialic acid containing oligosaccharide modifications of VAP-1. As an enzyme VAP-1 can convert soluble primary amines into corresponding aldehydes in the reaction which results in a formation of biologically active products. VAP-1 can also bind to primary amines presented by lymphocyte surface proteins (eg. N-terminus of a protein, aminosugars etc). In this reaction the enzyme-substrate interaction results in a transient, but covalent cross-linking of the lymphocyte and the endothelial cell during the rolling step. The adhesive and enzymatic functions of VAP-1 can be separate or, if the adhesive counter-receptor is the amine containing substrate, the same molecule may be used at both steps.

8. CONCLUSIONS

VAP-1 is an inducible endothelial enzyme, which regulates leukocyte subtype specific adhesion. The function of VAP-1 at non-endothelial cell types remains to be determined, and they may prove to serve processes unrelated to leukocyte adhesion. The hydrogen peroxide and aldehydes produced during the VAP-1 catalyzed reaction are biologically potent substances. At high concentrations these compounds are cytotoxic and may contribute to the pathogenesis of different vasculopathies, since SSAO inhibition ameliorates the development of atherosclerotic lesions in diabetes models[26]. Equally important, however, is the signalling function of hydrogen peroxide. Rapidly increasing literature points to the importance of low concentration of this reactive oxygen species as a key signalling mechanism controlling the proliferation and gene expression of vascular endothelial and smooth muscle cells[27, 28]. In fact, also a direct link between hydrogen peroxide formation and leukocyte adhesion has already been established, since H_2O_2 induces P-selectin expression and leukocyte rolling[29]. Hence, characterization of VAP-1 in vascular wall provides a potential, local and natural source for the generation of this reactive oxygen species. The transgenic and knock-out animals currently under investigation will eventually help to elucidate these questions, as well as to help to establish the role of enzymatic activity in adhesion *in vivo*. Nevertheless, the proven efficacy of anti-VAP-1 mAbs in abrogating inflammatory infiltrate *in vivo*, and the existence of small molecular inhibitors of SSAO opens exciting avenues for developing new anti-adhesive therapeutics in future.

9. ACKNOWLEDGEMENTS

The original work reviewed in this paper was supported by the Finnish Academy, the Technology Development Centre of Finland, the Sigrid Juselius Foundation, the Finnish Cancer Union and EU (QLG7-CT-1999-00295). We thank Anne Sovikoski-Georgieva for secretarial help.

Adapted from Publication: Salmi M, Jalkanen S. 'VAP-1: an adhesin and an enzyme' 211-216, 2001 with permission from Elsevier Science".

REFERENCES

1. E. C. Butcher and L. J. Picker, Lymphocyte homing and homeostasis, *Science* **272**(5258), 60-66 (1996).
2. T. A. Springer, Traffic signals for lymphocyte recirculation and leukocyte emigration: the multistep paradigm, *Cell* **76**, 301-314 (1994).
3. M. Salmi and S. Jalkanen, How do lymphocytes know where to go: current concepts and enigmas of lymphocyte homing, *Adv. Immunol.* **64**, 139-218 (1997).
4. S. Jalkanen, A. C. Steere, R. I. Fox and E. C. Butcher, A distinct endothelial cell recognition system that controls lymphocyte traffic into inflamed synovium, *Science* **233**(4763), 556-558 (1986).
5. M. Salmi and S. Jalkanen, A 90-kilodalton endothelial cell molecule mediating lymphocyte binding in humans, *Science* **257**(5075), 1407-1409 (1992).
6. M. Salmi, K. Kalimo and S. Jalkanen, Induction and function of vascular adhesion protein-1 at sites of inflammation, *J. Exp. Med.* **178**(6), 2255-2260 (1993).
7. M. Salmi and S. Jalkanen, Human vascular adhesion protein-1 (VAP-1) is a unique sialoglycoprotein that mediates carbohydrate-dependent binding of lymphocytes to endothelial cells, *J. Exp. Med.* **183**(2), 569-579 (1996).
8. K. 'aakkola, K. Kaunismäki, S. Tohka, G. Yegutkin, E. Vänttinen, T. Havia, L. J. Pelliniemi, M. Virolainen, S. Jalkanen and M. Salmi, Human vascular adhesion protein-1 in smooth muscle cells, *Am. J. Pathol.* **155**(6), 1953-1965 (1999).
9. D. Mu, K. F. Medzihradszky, G. W. Adams, P. Mayer, W. M. Hines, A. L. Burlingame, A. J. Smith, D. Cai and J. P. Klinman, Primary structures for a mammalian cellular and serum copper amine oxidase, *J. Biol. Chem.* **269**(13), 9926-9932 (1994).
10. J. P. Klinman and D. Mu, Quinoenzymes in biology, *Annu. Rev. Biochem.* **63**, 299-344 (1994).
11. G. A. Lyles, Mammalian plasma and tissue-bound semicarbazide-sensitive amine oxidases: biochemical, pharmacological and toxicological aspects, *Int. J. Biochem. Cell Biol.* **28**(3), 259-274 (1996).
12. C. M. Wilmot, J. Hajdu, M. J. McPherson, P. F. Knowles and S. E. Phillips, Visualization of dioxygen bound to copper during enzyme catalysis, *Science* **286**(5445), 1724-1728 (1999).
13. G. Enrique-Tarancón, L. Marti, N. Morin, J. M. Lizcano, M. Unzeta, L. Sevilla, M. Camps, M. Palacín, X. Testar, C. Carpéné and A. Zorzano, Role of semicarbazide-sensitive amine oxidase on glucose transport and GLUT4 recruitment to the cell surface in adipose cells, *J. Biol. Chem.* **273**(14), 8025-8032 (1998).
14. T. A. Salminen, D. J. Smith, S. Jalkanen and M. S. Johnson, Structural model of the catalytic domain of an enzyme with cell adhesion activity: human vascular adhesion protein-1 (HVAP-1) D4 domain is an amine oxidase, *Protein Eng.* **11**(12), 1195-1204 (1998).

15. D. J. Smith, M. Salmi, P. Bono, J. Hellman, T. Leu and S. Jalkanen, Cloning of vascular adhesion protein-1 reveals a novel multifunctional adhesion molecule, *J. Exp. Med.* **188**(1), 17-27 (1998).

16. A.-M. Arvilommi, M. Salmi, K. Kalimo and S. Jalkanen, Lymphocyte binding to vascular endothelium in inflamed skin revisited: a central role for vascular adhesion protein-1 (VAP-1), *Eur. J. Immunol.* **26**(4), 825-833 (1996).

17. K. Jaakkola, S. Jalkanen, K. Kaunismäki, E. Vänttinen, P. Saukko, K. Alanen, M. Kallajoki, L.-M. Voipio-Pulkki and M. Salmi, Vascular adhesion protein-1, intercellular adhesion molecule-1 and P- selectin mediate leukocyte binding to ischemic heart in humans, *J. Am. Coll. Cardiol.* **36**(1), 122-129 (2000).

18. G. McNab, J. L. Reeves, M. Salmi, S. Hubscher, S. Jalkanen and D. H. Adams, Vascular adhesion protein 1 mediates binding of T cells to human hepatic endothelium, *Gastroenterology* **110**(2), 522-528 (1996).

19. K. F. Yoong, G. McNab, S. G. Hübscher and D. H. Adams, Vascular adhesion protein-1 and ICAM-1 support the adhesion of tumor- infiltrating lymphocytes to tumor endothelium in human hepatocellular carcinoma, *J. Immunol.* **160**(8), 3978-3988 (1998).

20. S. Tohka, M.-L. Laukkanen, S. Jalkanen and M. Salmi, Vascular adhesion protein 1 (VAP-1) functions as a molecular brake during granulocyte rolling and mediates their recruitment *in vivo*, *Faseb J.* **15**(2), 373-382 (2001).

21. R. Kurkijärvi, D. H. Adams, R. Leino, T. Möttönen, S. Jalkanen and M. Salmi, Circulating form of human vascular adhesion protein-1 (VAP-1): increased serum levels in inflammatory liver diseases, *J. Immunol.* **161**(3), 1549-1557 (1998).

22. A. J. H. Gearing and W. Newman, Circulating adhesion molecules in disease, *Immunol. Today* **14**(10), 506-512 (1993).

23. B. Bergeret, H. Blaschko and R. Hawes, Occurrence of an amine oxidase in horse serum, *Nature* **180**, 1127-1128 (1957).

24. R. Kurkijärvi, G. G. Yegutkin, B. K. Gunson, S. Jalkanen, M. Salmi and D. H. Adams, Circulating soluble vascular adhesion protein 1 accounts for the increased serum monoamine oxidase activity in chronic liver disease, *Gastroenterology* **119**(4), 1096-1103 (2000).

25. M. Salmi, G. Yegutkin, R. Lehvonen, K. Koskinen, T. Salminen and S. Jalkanen, A cell surface amine oxidase directly controls lymphocyte migration, *Immunity* **14**(3), 265-276 (2001).

26. P. H. Yu and D. M. Zuo, Aminoguanidine inhibits semicarbazide-sensitive amine oxidase activity: implications for advanced glycation and diabetic complications, *Diabetologia* **40**(11), 1243-1250 (1997).

27. T. Finkel, Oxygen radicals and signaling, *Curr. Opin. Cell Biol.* **10**(2), 248-253 (1998).

28. C. Bogdan, M. Röllinghoff and A. Diefenbach, Reactive oxygen and reactive nitrogen intermediates in innate and specific immunity, *Curr. Opin. Immunol.* **12**(1), 64-76 (2000).

29. B. Johnston, S. Kanwar and P. Kubes, Hydrogen peroxide induces leukocyte rolling: modulation by endogenous antioxidant mechanisms including NO, *Am. J. Physiol.* **271**(2 Pt 2), H614-621 (1996).

HOMEOSTATIC CHEMOKINES AND THE TARGETING OF REGIONAL IMMUNITY

Eric J. Kunkel and Eugene C. Butcher*

1. ABSTRACT

Tissue-selective trafficking of memory and effector T and B lymphocytes is mediated by unique combinations of adhesion molecules and chemokines. The discovery of several related epithelial-expressed chemokines (TECK/CCL25 in small intestine, CTACK/CCL27 in skin, and MEC/CCL28 in diverse mucosal sites) now highlights an important role for epithelial cells in controlling homeostatic lymphocyte trafficking, including the localization of cutaneous and intestinal memory T cells, and of IgA plasma cells. Constitutively expressed epithelial chemokines may help determine the character of local immune responses and contribute to the systemic organization of the immune system.

2. INTRODUCTION

Nearly 40 years ago, James Gowans demonstrated that circulating lymphocytes enter secondary lymphoid tissues and subsequently return to the circulation through the efferent lymph (Gowans and Knight, 1964). Later data showed that memory T lymphocytes selectively recirculated back through tissues from which they came (including the skin and intestines), and B immunoblasts expressing IgA specifically migrated to mucosal tissues (Cahill et al., 1977; McDermott and Bienenstock, 1979). Together, these findings laid the groundwork for the concept of tissue-specific lymphocyte recirculation, and the idea of specific lymphocyte trafficking has become a fundamental concept in the study of acquired immunity.

Physiologically, tissue-restricted recirculation of memory and effector lymphocytes may serve to 1) increase the efficiency and robustness of regional immune responses while decreasing the possibility of tissue antigen cross-reactivity; and 2) allow

* Laboratory of Immunology and Vascular Biology, Department of Pathology, Stanford University School of Medicine, Stanford, CA 94305 and Center for Molecular Biology and Medicine, Veterans Affairs Palo Alto Health Care System, Palo Alto, CA 94304.

functional immune specialization of particular tissues (*e.g.* skin versus intestines) or tissue systems (*e.g.* mucosal epithelial surfaces). Lymphocyte recirculation is tightly regulated by the expression of particular adhesion molecules and chemoattractant receptors on lymphocytes, combined with the spatial and temporal expression of ligands for these receptors by a variety of tissue cells. The present review will examine some fundamental concepts underlying tissue-specific lymphocyte recirculation, focusing primarily on epithelial chemokines with restricted tissue expression as newly appreciated regulators of lymphocyte trafficking through tissues.

3. MULTIPLE STEPS PROMOTE SPECIFICITY IN LYMPHOCYTE TRAFFICKING

Lymphocyte recirculation into lymphoid and non-lymphoid tissues was described ten years ago as a series of adhesive and activation processes that impart specificity while retaining robustness in the face of mutation and regulatory variability (Butcher, 1991). The requirement for multiple protein-protein interactions allows a genetically limited receptor-ligand repertoire to be used combinatorially to control the recirculation of different lymphocyte (and other leukocyte) subsets. Specificity in one step alone can lead to enrichment of the final recruited population, but often, several steps in the process are selective for a particular tissue site, resulting in a high degree of specificity.

Recirculation begins with blood lymphocytes interacting transiently and reversibly with the vascular endothelium through villous-expressed adhesion receptors (usually selectins and selectin ligands, but also α_4 integrins) in a process called rolling. Rolling brings lymphocytes into contact with the endothelium where they can sample the surface for activating factors. Activating factors (often chemokines for lymphocytes) bind to pertussis toxin-sensitive G protein-coupled receptors on the rolling lymphocytes, triggering rapid activation of integrins that results in reversible arrest and firm adhesion of the lymphocyte. Additional adhesion molecules, endothelium-associated chemokines, and potentially other signals then lead the adherent lymphocyte to migrate across the endothelium where tissue-associated chemokine gradients may also direct localization. Because chemokines contribute to both lymphocyte exit from the circulation and localization and retention within tissues, they are important regulators of the lymphocyte recruitment process.

4. SECONDARY LYMPHOID ORGANS AND TISSUE-SPECIFIC LYMPHOCYTE DEVELOPMENT

Naive T and B cells are preprogrammed to migrate to and recirculate through secondary lymphoid organs (*e.g.* spleen, lymph nodes, and Peyer's patches) in part through expression of the chemokine receptor CCR7, a ligand for the lymphoid tissue chemokines SLC/CCL21 and MIP-3β/CCL19 (reviewed in (von Andrian and Mackay, 2000). Cells encountering antigen in the lymphoid environment undergo reprogramming of homing properties during antigen-driven proliferation and differentiation, allowing their progeny to traffic to extra-lymphoid tissue sites, although early arising immunoblasts can display substantial promiscuity for different inflamed tissues, perhaps reflecting their return to the blood while still expressing activated integrins. Imprinting, or selection for differential homing properties, is determined by the local lymphoid organ microenvironment (perhaps by dendritic cells and chemokines), and begins almost immediately during the naive-to-memory/effector T cell transition. For example, naïve T cells responding to antigen begin to upregulate the intestinal homing receptor $\alpha_4\beta_7$ and responses to the intestinal chemokine TECK/CCL25 (see below) within one to two cell divisions in intestine-associated lymphoid tissues, while conversely, homing properties targeting cells to non-intestinal tissues are upregulated during the initial proliferative response in peripheral lymph nodes (Campbell and Butcher, 2002).

5. TISSUE-SPECIFIC LYMPHOCYTE POPULATIONS AT EPITHELIAL SURFACES

Arguably, the main function of the mammalian immune system is to protect the host from foreign pathogens, and thus, evolution would favor the development of lymphocyte trafficking patterns that protect sites of pathogen entry, particularly epithelial tissues. The skin and the gastrointestinal tract are clearly two of the largest epithelial organ systems in contact with the outside world, and accordingly, the circulating memory lymphocyte pool contains two clearly separable subsets of lymphocytes with skin versus gut homing potential. Memory for skin-associated antigens is found in a population of skin-homing circulating lymphocytes expressing the cutaneous lymphocyte antigen (CLA), a carbohydrate-defined cutaneous "homing receptor". CLA^+ lymphocytes are highly enriched in cutaneous inflammatory sites and the oral mucosa (which like the skin, has a stratified squamous epithelium), but not in other tissues. Conversely, memory for rotavirus (an intestinal epithelial-restricted pathogen) has been localized to a population of circulating lymphocytes expressing high levels of the $\alpha_4\beta_7$ integrin. Lymphocytes that are not in the intestinal $\alpha_4\beta_7^{hi}$ population generally express high levels of the VCAM-1 receptor $\alpha_4\beta_1$ (including many skin-homing lymphocytes), and these $\alpha_4\beta_1^{hi}$ lymphocytes make up a pool of lymphocytes with homing potential to tissues outside the gastrointestinal tract but including non-intestinal mucosal sites such as the bronchi (reviewed in (Shaw and Brenner, 1995; Butcher et al., 1999; Robert and Kupper, 1999).

Recent data reveal that chemokines and their receptors also help control the specificity of memory lymphocyte subsets for skin and gut. Two homeostatically expressed chemokines appear to participate in selective T lymphocyte recruitment into the skin. One of these chemokines, TARC/CCL17, is expressed by cutaneous, but not intestinal, endothelium and binds to the chemokine receptor CCR4 expressed at high levels on CLA^+ skin memory lymphocytes (and at lower levels on lung lymphocytes (Campbell et al., 2001). TARC is hypothesized to trigger the adhesive arrest of rolling lymphocytes in cutaneous venules, especially during chronic inflammation (Campbell et al., 1999). More recently, keratinocytes, the epithelial cells of the skin, have been shown to express the chemokine CTACK/CCL27, that binds to the receptor CCR10 also specifically expressed on circulating skin-homing CLA^+ T lymphocytes (Morales et al., 1999). At least in some settings of inflammation, CCR4 and CTACK/CCR10 have overlapping, redundant roles in cutaneous lymphocyte recruitment since blockade of lymphocyte homing from the blood in a delayed-type hypersensitivity (DTH) model requires simultaneous deficiency or inhibition of CCR4 and CTACK/CCR10 (Reiss et al., 2001). In addition to mediating vascular arrest, CCR4 may be able to support diapedesis and chemotaxis through a gradient of MDC/CCL22 (another CCR4 ligand) secreted by resident activated macrophages. Similarly, in addition to providing an epithelial-derived gradient for diapedesis and recruitment to the epithelium, CTACK may be transcytosed and presented on the endothelium (Middleton et al., 1997) to support skin-homing lymphocyte adhesion triggering. Importantly, these results demonstrate that inflammatory chemokines cannot substitute for these two skin-selective chemokines in cutaneous lymphocyte recruitment in DTH.

In the intestines, part of the selectivity of lymphocyte homing is due to the expression of the $\alpha_4\beta_7$ integrin ligand MAdCAM-1 on normal and inflamed intestinal endothelium (reviewed in (Butcher et al., 1999). Recent results indicate that a specific subset of circulating $\alpha_4\beta_7^{hi}$ lymphocytes is in fact targeted to the small intestine by virtue of expression of the chemokine receptor CCR9 (Zabel et al., 1999). The CCR9 ligand, TECK, is expressed by epithelial cells in the small intestine, especially those in the crypt region most closely associated with the MAdCAM-1$^+$ vessels involved in lymphocyte recruitment (Kunkel et al., 2000; Wurbel et al., 2000). Indeed, TECK is detected on small intestinal endothelium (Papadakis et al., 2000), and may be produced by endothelial cells or simply transcytosed after being secreted by intestinal epithelium. Virtually all T lymphocytes in the small intestine express CCR9 (Zabel et al., 1999; Kunkel et al., 2000; Papadakis et al., 2000) suggesting that CCR9 and TECK play a critical role in lymphocyte biology in this tissue. Even more interesting is the finding that TECK is absent or only weakly expressed in other segments of the gastrointestinal tract (*e.g.* colon and stomach) and only a small fraction of colon lymphocytes express CCR9 (Kunkel et al., 2000; Wurbel et al., 2000; Papadakis et al., 2000). This finding suggests that even organs thought to be part of a common mucosal immune system may have different lymphocyte homing pathways distinguished by chemokine usage, providing a potential mechanism for further specialization of mucosal immune responses in various intestinal and other mucosal sites.

Mucosal epithelial tissues are also the major site of secretory IgA production by resident plasma cells, and IgA-dependent pathogen neutralization at mucosal sites is critical to host protection. B cell immunoblasts secreting IgA (IgA ASC) also migrate preferentially to the small intestines (and other mucosal sites) (McDermott and Bienenstock, 1979) and accordingly, express the mucosal homing receptor $\alpha_4\beta_7$ (Butcher et al., 1999). As during T lymphocyte recirculation, these antibody secreting cells likely require a chemoattractant signal for extravasation or tissue localization, and indeed, IgA ASC respond chemotactically to unknown factors present in mucosal epithelial tissues such as the mammary gland (Czinn and Lamm, 1986). Interestingly, splenic, mesenteric lymph node, and Peyer's patch IgA ASC migrate efficiently to TECK, and express CCR9 (Bowman et al., 2002). Thus, epithelial chemokines such as TECK may also participate in ASC localization to epithelial surfaces, in this case the small intestine. Interestingly, while IgG ASC do not respond to TECK (Bowman et al., 2002), they are responsive to other chemokines including SDF-1α/CXCL12 and Mig/CXCL9 (ligands for CXCR4 and CXCR3, respectively) (Hargreaves et al., 2001; Bowman et al., 2002) and most express the integrin $\alpha_4\beta_1$ (Finke et al., 2001). IgG ASC may use these receptors to localize to sites of chronic inflammation. Thus, selectivity in isotype-specific ASC chemokine receptor expression is likely a major determinant of IgA ASC localization to mucosal epithelial surfaces and of IgG ASC localization to systemic sites of chronic inflammation.

Interestingly, the epithelial chemokines TECK and CTACK are very closely related, suggesting that these homeostatic tissue-selective chemokines may have evolved from a common (potentially epithelial) precursor. Indeed, based on sequence homology, a third closely related chemokine was recently discovered (Wang et al., 2000; Pan et al., 2000). MEC (mucosal epithelial chemokine, CCL28) is abundantly expressed by epithelia in the bronchi, colon, salivary gland, and mammary gland, and at lower levels in the small intestine. Importantly, even though MEC is apparently absent from skin, it, like CTACK, is also a ligand for CCR10. Indeed, MEC can attract circulating CLA^+ skin-homing lymphocytes, even though these lymphocytes are essentially undetectable in the colon or small intestine. (This can be explained in the context of the combinatorial control of vascular interaction: CLA^+ T cells lack the $\alpha_4\beta_7$ integrin expression required for tethering and rolling on intestinal lamina propria venules, and thus are expected to be unable to sample the intestinal endothelium for CCR10 ligand activity.) Nevertheless, CCR10 mRNA is abundantly expressed in the colon and small intestine where MEC is also expressed (Jarmin et al., 2000). Therefore, MEC may serve to recruit, or retain, populations of mucosal lymphocytes (*e.g.* B or T immunoblasts) that are not well-represented in the circulation. Teleologically, the striking overlap of bacterial species in the oral cavity and colon (Kroes et al., 1999), and the obvious relationship between the antigen exposure of the oral cavity and airways, suggests that the unique tissue pattern of MEC expression may serve to unify trafficking of specialized immune cell populations to these physically dispersed but immunologically related mucosal sites.

The evolutionary pressures driving the specialization of lymphocyte subset recruitment to different epithelial organs (manifested in part by differential chemokine expression) is still unclear, but likely relates to the type and intensity of antigen exposure in various organs. The specialization between segments of the intestines itself is surprising and may be related to the differing physiological functions and immune requirements of each intestinal segment. For instance, the small intestine is exposed to food-derived antigens to which an immune response is unnecessary and potentially debilitating, as in the case of celiac disease. In contrast, the colon (and oral cavity) are sites of heavy commensal bacterial population. Thus, specialized trafficking mechanisms may allow distinctive immune responses in the small intestine, where immune tolerance to food antigens is important, versus highly colonized sites where constitutive antibacterial immunity or symbiosis is essential. Differences in lymphocyte homing character may also be related to the characteristic segmental involvement of the distal small intestine (ileum) and colon in Crohn's disease and of the colon and rectum in ulcerative colitis (Fiocchi, 1998).

The recent findings reviewed here thus highlight an important role for epithelial cells as active participants in defining the specificity of lymphocyte trafficking and the nature of local immune specialization. Because the role of epithelial cells in the induction and resolution of inflammation, including secretion of inflammatory chemokines, is well-appreciated (reviewed in (Pitman and Blumberg, 2000), a role for these cells in tissue specific homeostatic lymphocyte recirculation is not surprising. It will be important to define the developmental and regulatory mechanisms underlying the tissue selectivity of endothelial and epithelial expression of homeostatic chemokines (*e.g.* TARC, TECK, CTACK, and MEC), and to determine if these mechanisms are shared with those controlling expression of vascular addressins (*e.g.* MAdCAM-1).

While this review has focused primarily on a newly emerging model of epithelial chemokine control of the specificity of lymphocyte homing, significant experimental effort has gone into identifying lymphocyte subsets that specifically traffic through many internal organs, such as the liver, kidneys, and synovial joints (far too much work to adequately cite here). While tissue-specific subsets for these organs cannot be ruled out, the idea of an evolutionary bias towards specific protection of pathogen entry sites may suggest that trafficking of lymphocytes through such "protected" internal organs (in the sense that they are not directly in contact with the outside world) may occur principally under inflammatory conditions, and data from many studies support the hypothesis that lymphocytes infiltrating these organs may rely on a similar handful of widely utilized "inflammatory" adhesion and chemokine receptors (*e.g.* $\alpha_4\beta_1$ and VCAM-1, P-selectin and PSGL-1, CXCR3, CCR5 *etc.*) which can likely also complement tissue selective mechanisms in external epithelia-associated tissues in the setting of strong inflammatory insult or tissue damage. A common mechanism of trafficking control in internal organs would not imply that such trafficking is less important or poorly regulated, only that no evolutionary pressure existed to drive the development of internal organ immune response specialization, which would require additional tissue-specific homing mechanisms.

6. CONCLUSIONS

The identification of novel chemokines and chemokine receptors has revolutionized our understanding of tissue-specific lymphocyte recirculation and localization. The differential expression of particular chemokines within epithelial tissues further suggests that organ systems previously thought to be relatively immunologically uniform may have important differences in terms of their immune character (*e.g.* the small and large intestines) while those thought to be more diverse may be linked in a previously unrecognized way (*e.g.* the colon, salivary glands and oral cavity, and airways). At the same time, although the discovery of new chemokines and receptors appears to offer answers to long-standing questions in lymphocyte homing, many fundamental questions still remain, including the mechanisms behind the development of tissue-homing subsets, the mechanisms controlling tissue-specific endothelial and epithelial expression of particular chemokines and adhesion molecules, and in some instances, the physiological rationale behind the evolutionary separation or unification of particular tissues through lymphocyte recirculation pathways.

7. ACKNOWLEDGEMENTS

Because of the extent and complexity of the lymphocyte homing field, we could not discuss many interesting studies, and we apologize to those whose excellent work could not be cited due to space limitations.

Reprinted from Immunity, Vol 16, No 1, 2002, pp. 1-4, Kunkel and Butcher: "Chemokines and the tissue specific migration of lymphocytes", with permission from Elsevier Science.

REFERENCES

Bowman, E.P., Kuklin, N., Youngman, K.R., Lazarus, N., Kunkel, E.J., Pan, J., Greenberg, H.B., and Butcher, E.C., 2002, The intestinal chemokine TECK (CCL25) attracts IgA antibody secreting cells, *J.Exp.Med.*, **195**:269.

Butcher, E.C., 1991, Leukocyte-endothelial cell recognition: three (or more) steps to specificity and diversity, *Cell*, **67**:1033.

Butcher, E.C., Williams, M., Youngman, K., Rott, L., and Briskin, M., 1999, Lymphocyte trafficking and regional immunity, *Adv.Immunol.*, **72**:209.

Cahill, R.N., Poskitt, D.C., Frost, D.C., and Trnka, Z., 1977, Two distinct pools of recirculating T lymphocytes: migratory characteristics of nodal and intestinal T lymphocytes, *J.Exp.Med.*, **145**:420.

Campbell, D.J. and Butcher, E.C., 2002, Rapid acquisition of tissue-specific homing phenotypes by CD4+ T cells activated in cutaneous or mucosal lymphoid tissues, *J.Exp.Med.*, **195**:135.

Campbell, J.J., Brightling, C.E., Symon, F.A., Qin, S., Murphy, K.E., Hodge, M., Andrew, D.P., Wu, L., Butcher, E.C., and Wardlaw, A.J., 2001, Expression of chemokine receptors by lung T cells from normal and asthmatic subjects, *J.Immunol.*, **166**:2842.

Campbell, J.J., Haraldsen, G., Pan, J., Rottman, J., Qin, S., Ponath, P., Andrew, D.P., Warnke, R., Ruffing, N., Kassam, N., Wu, L., and Butcher, E.C., 1999, The chemokine receptor CCR4 in vascular recognition by cutaneous but not intestinal memory T cells, *Nature*, **400**:776.

Czinn, S.J. and Lamm, M.E., 1986, Selective chemotaxis of subsets of B lymphocytes from gut-associated lymphoid tissue and its implications for the recruitment of mucosal plasma cells, *J.Immunol.*, **136**:3607.

Finke, D., Baribaud, F., Diggelmann, H., and Acha-Orbea, H., 2001, Extrafollicular plasmablast B cells play a key role in carrying retroviral infection to peripheral organs, *J.Immunol.*, **166**:6266.

Fiocchi, C., 1998, Inflammatory bowel disease: etiology and pathogenesis, *Gastroenterology*, **115**:182.

Gowans, J.L. and Knight, E.J., 1964, The route of recirculation of lymphocytes in the rat., *Proc.R.Soc.London Ser.B.*, **159**:257

Hargreaves, D.C., Hyman, P.L., Lu, T.T., Ngo, V.N., Bidgol, A., Suzuki, G., Zou, Y.R., Littman, D.R., and Cyster, J.G., 2001, A coordinated change in chemokine responsiveness guides plasma cell movements, *J.Exp.Med.*, **194**:45.

Jarmin, D.I., Rits, M., Bota, D., Gerard, N.P., Graham, G.J., Clark-Lewis, I., and Gerard, C., 2000, Cutting edge: identification of the orphan receptor G-protein-coupled receptor 2 as CCR10, a specific receptor for the chemokine ESkine, *J.Immunol.*, **164**:3460.

Kroes, I., Lepp, P.W., and Relman, D.A., 1999, Bacterial diversity within the human subgingival crevice, *Proc.Natl.Acad.Sci.U.S.A.* , **96**:14547.

Kunkel, E.J., Campbell, J.J., Haraldsen, G., Pan, J., Boisvert, J., Roberts, A.I., Ebert, E.C., Vierra, M.A., Goodman, S.C., Genovese, M.C., Wardlaw, A.J. , Greenberg, H.B., Parker, C.M., Butcher, E.C., Andrew, D.P., and Agace, W.W., 2000, Lymphocyte CCR9 and epithelial TECK expression distinguish the small intestinal immune compartment: epithelial expression of tissue-specific chemokines as an organizing principle in regional immunity, *J.Exp.Med.*, **192**:761.

McDermott, M.R. and Bienenstock, J., 1979, Evidence for a common mucosal immunologic system. I. Migration of B immunoblasts into intestinal, respiratory, and genital tissues, *J.Immunol.*, **122**:1892.

Middleton, J., Neil, S., Wintle, J., Clark-Lewis, I., Moore, H., Lam, C., Auer, M., Hub, E., and Rot, A., 1997, Transcytosis and surface presentation of IL-8 by venular endothelial cells, *Cell*, **91**:385.

Morales, J., Homey, B., Vicari, A.P., Hudak, S., Oldham, E., Hedrick, J., Orozco, R., Copeland, N.G., Jenkins, N.A., McEvoy, L.M., and Zlotnik, A., 1999, CTACK, a skin-associated chemokine that preferentially attracts skin-homing memory T cells, *Proc.Natl.Acad.Sci.U.S.A.* , **96**:14470.

Pan, J., Kunkel, E.J., Gosslar, U., Lazarus, N., Langdon, P., Broadwell, K., Vierra, M.A., Genovese, M.C., Butcher, E.C. , and Soler, D., 2000, A novel chemokine ligand for CCR10 and CCR3 expressed by epithelial cells in mucosal tissues, *J.Immunol.*, **165**:2943.

Papadakis, K.A., Prehn, J., Nelson, V., Cheng, L., Binder, S.W., Ponath, P.D., Andrew, D.P., and Targan, S.R., 2000, The role of thymus-expressed chemokine and its receptor CCR9 on lymphocytes in the regional specialization of the mucosal immune system, *J.Immunol.*, **165**:5069.

Pitman, R.S. and Blumberg, R.S., 2000, First line of defense: the role of the intestinal epithelium as an active component of the mucosal immune system, *J.Gastroenterol.*, **35**:805.

Reiss, Y., Proudfoot, A.E., Campbell, J.J., and Butcher, E.C., 2001, CCR4 and the CCR10 ligand CTACK in lymphocyte trafficking to inflamed skin, *J.Exp.Med.*, **194** :1541.

Robert, C. and Kupper, T.S., 1999, Inflammatory skin diseases, T cells, and immune surveillance, *N.Engl.J.Med.*, **341**:1817.

Shaw, S.K. and Brenner, M.B., 1995, The beta 7 integrins in mucosal homing and retention, *Semin.Immunol.*, **7**:335.

von Andrian, U.H. and Mackay, C.R., 2000, T-cell function and migration. Two sides of the same coin, *N.Engl.J.Med.*, **343**:1020.

Wang, W., Soto, H., Oldham, E.R., Buchanan, M.E., Homey, B., Catron, D., Jenkins, N., Copeland, N.G., Gilbert, D.J., Nguyen, N., Abrams, J., Kershenovich, D., Smith, K., McClanahan, T., Vicari, A.P., and Zlotnick, A., 2000, Identification of a Novel CC Chemokine (CCL28) which binds CCR10 (GPR2), *J.Biol.Chem.*, **275**:22313.

Wurbel, M.-A., Philippe, J.-M., Nguyen, C., Victorero, G., Freeman, T., Wooding, P., Miazek, A., Mattel, M.-G., Malissen, M., Jordan, B.R., Malissen, B., Carrie, A., and Naquet, P., 2000, The chemokine TECK is expressed by thymic and intestinal epithelial cells and attracts double- and single-positive thymocytes expressing the TECK receptor CCR9, *Eur.J.Immunol.*, **30**:262.

Zabel, B.A., Agace, W.W., Campbell, J.J., Heath, H.M., Parent, D., Roberts, A.I., Ebert, E.C., Kassam, N., Qin, S., Zovko, M., LaRosa, G.J., Yang, L.L., Soler, D., Butcher, E.C., Ponath, P.D., Parker, C.M., and Andrew, D.P., 1999, Human G protein-coupled receptor GPR-9-6/CC chemokine receptor 9 is selectively expressed on intestinal homing T lymphocytes, mucosal lymphocytes, and thymocytes and is required for thymus-expressed chemokine-mediated chemotaxis, *J.Exp.Med.*, **190**:1241.

FACTORS REGULATING NAÏVE T CELL HOMEOSTASIS

Charles D. Surh, Joyce Tan, William C. Kieper, and Bettina Ernst

Department of Immunology, The Scripps Research Institute, La Jolla, CA 92037

INTRODUCTION

The mature T cells in the peripheral secondary lymphoid tissues are controlled by homeostatic mechanisms that regulate their survival and the overall cell counts. Such regulation applies to populations of naïve and memory T cells. The prevailing view in the past was that the signals essential for regulating T cell homeostasis were thought to be largely derived from contact with foreign antigens. However, recent work from several groups has strongly indicated that homeostasis of naïve T cells is crucially dependent on interaction with two self-ligands — self-MHC/peptide complexes and the cytokine IL-7. This article will present our recent work supporting the revised view on the regulation of naïve T cell homeostasis.

T CELL DEVELOPMENT

Prior to considering homeostasis of mature T cells in the secondary lymphoid organs, it is beneficial to briefly review the process of T cell development in the thymus. Typical αβ TCR^+ T cells develop in the thymus through a combined process of positive and negative selection (Sprent et al., 1988; Goldrath and Bevan, 1999). Positive selection rescues a small fraction of immature $CD4^+8^+$ cells with a minimal affinity to self-MHC/peptide ligands expressed on cortical epithelial cells from default death. Positively selected $CD4^+8^+$ cells become either mature $CD4^+$ or $CD8^+$ cells depending on whether the cells' interaction was made with MHC class II or class I molecules, respectively. A proportion of the positively selected cells, which has overtly high affinity to self-MHC/peptide ligands, subsequently undergoes negative selection in the thymic medulla. The resulting repertoire of T cells has a low, but significant, affinity to self-MHC/peptide ligands.

Elegant work with TCR transgenic mice and mice engineered to express MHC molecules loaded with a single species of self peptides over the past several years has shown that both MHC-bound self peptides play a major role in shaping the T cell repertoire (Hogquist et al., 1994; Fung-Leung et al., 1996; Ignatowicz et al., 1996; Martin et al., 1996; Miyazaki et al., 1996; Bevan, 1997; Surh et al., 1997). Although the exact role of peptide in T cell selection is yet to be firmly established, the general consensus from these investigations is that thymic selection is peptide-dependent with a certain level of degeneracy. Thus, the data are consistent with the view that a given self-peptide is capable of selecting a narrow repertoire of T cells, and a given clone can be selected by only a limited set of self peptides that presumably have a weak homology to the agonist peptide (Bevan, 1997; Marrack and Kappler, 1997). The point to stress is that thymic selection of the majority of T cells appears to require recognition of specific self-MHC/peptide complexes.

SURVIVAL OF NAÏVE T CELLS

Despite the controversy in the past, it is now widely believed that naïve T cells in the secondary lymphoid tissues have a long life span (von Boehmer and Hafen, 1993; Sprent and Tough, 1994). Most of the naïve T cells survive at an interphase, but despite being in a quiescent state, it is now well established that prolonged survival of naïve T cells requires signaling through the TCR from contact with self-MHC/peptide ligands (Goldrath and Bevan, 1999). Thus, it has been found that naïve T cells will die off slowly if they are transferred into MHC-deficient conditions (Takeda et al., 1996; Brocker, 1997; Kirberg et al., 1997; Rooke et al., 1997; Tanchot et al., 1997; Nesic and Vukmanovic, 1998; Murali-Krishna et al., 1999). This applies generally to both $CD4^+$ and $CD8^+$ cells, although not all investigators agree that survival of naïve $CD4^+$ cells is MHC-dependent (Clarke and Rudensky, 2000; Dorfman et al., 2000). More recently, the requirement for TCR signaling for naïve T cell survival was addressed through the generation of engineered mice in which the TCR expression can be abolished at the level of mature T cells. In support of the idea for MHC-dependency, TCR-deficient mature T cells slowly disappeared with an average half-life of 4-8 weeks for naïve $CD4^+$ cells and 2-3 weeks for naïve $CD8^+$ cells (Labrecque et al., 2001; Polic et al., 2001). These findings collectively lend strong support for the idea that continuous contact with self-MHC/peptide ligands is required to maintain survival of naïve T cells.

HOMEOSTATIC PROLIFERATION OF NAÏVE T CELLS

In addition to regulating survival of mature T cells, homeostatic mechanisms also regulate the overall size of the mature T cell pool to remain at a constant level (Freitas and Rocha, 1993; Bell and Sparshott, 1997; Mackall et al., 1997). A clear manifestation of such mechanisms is the well-known finding that mature T cells undergo a slow form of "homeostatic" proliferation when transferred into syngeneic T cell (T)-depleted irradiated, nude, SCID or RAG-deficient hosts (Bell et al., 1987; Rocha et al., 1989). Until recently, homeostatic proliferation was thought to be driven by foreign antigens augmented by inflammatory cytokines (Mackall et al., 1997; Tanchot et al., 1997). This view was largely based on the finding that anti-HY TCR transgenic $CD8^+$ cells failed to undergo homeostatic expansion in syngeneic female lymphopenic hosts (Rocha and von Boehmer, 1991).

Past studies on homeostatic proliferation relied on measuring the relative expansion of the donor T cells in lymphopenic hosts by counting the number of donor T cells that are recovered from the hosts at an arbitrary later time. The waiting time was often long and the counting process was cumbersome and prone to mistakes. To facilitate accurate analysis of homeostatic proliferation in a relatively short period of time, recent experiments involved adoptive transfer of donor T cells labeled with intracellular fluorescent dye 5,6-carboxyfluorescein diacetate succinimyl ester (CFSE) (Lyons and Parish, 1994). The intensity of the dye decreased in a two-fold step-wise manner with each cell division and the homeostatic proliferation could be observed within the period of a week.

Role of Self-MHC/Peptide Ligands in Homeostatic Proliferation

Through the adoptive transfer of CFSE-labeled T cells into mice deficient in MHC class I or II molecules, it was realized that efficient homeostatic proliferation of polyclonal $CD4^+$ and $CD8^+$ cells requires contact with self-MHC class II and I molecules, respectively (Bender et al., 1999; Ernst et al., 1999; Goldrath and Bevan, 1999; Kieper and Jameson, 1999; Murali-Krishna et al., 1999). Using purified naïve phenotype cells also revealed that homeostatic proliferation applied to naïve cells (Bender et al., 1999; Ernst et al., 1999; Goldrath and Bevan, 1999; Kieper and Jameson, 1999; Murali-Krishna et al., 1999). In addition, testing various lines of TCR transgenic cells showed that many naïve transgenic T cell lines are capable of undergoing homeostatic proliferation. Thus, transgenic lines such as $CD8^+$ OT-I, 2C, P14 and $CD4^+$ DO11, 1H3.1 are able to undergo homeostatic proliferation (Ernst et al., 1999; Goldrath and Bevan, 1999; Kieper and Jameson, 1999; Murali-Krishna et al., 1999; Viret et al., 1999). As previously found with the $CD8^+$ anti-HY clone (Rocha and von Boehmer, 1991), however, this was not an invariant finding with all clones: transgenic $CD4^+$ OT-II and AND clones also failed to undergo homeostatic proliferation (Ernst et al., 1999; Surh and Sprent, 2000). Why some clones cannot proliferate is currently unknown. The possible explanations are that the non-responsive clones express TCR that are inherently of low affinity to self-MHC/peptide ligands or that the specific self-MHC/peptide ligands required for homeostatic proliferation is present at a sub-optimal level in the periphery (see below).

Since many of the naïve polyclonal and TCR transgenic cells can undergo homeostatic proliferation, it is unlikely that homeostatic proliferation is driven by MHC molecules loaded with foreign peptides. In support of this view, compelling evidence was found that homeostatic proliferation is mediated by MHC molecules loaded with specific self-peptides, namely, the self-peptides that initially induced positive selection of the T cells in the thymus. Two different systems were used. The first involved use of $H2\text{-}M^-$ mice which express MHC class II (A^b) molecules loaded with virtually only a single species of self-peptides, class II-associated invariant chain peptides (CLIP) (Fung-Leung et al., 1996; Martin et al., 1996; Miyazaki et al., 1996). Janeway's group and ours found that $CD4^+$ cells from wild-type B6 mice, i.e. cells which were positively selected by A^b molecules loaded with a diverse spectrum of self-peptides, were unable to undergo efficient homeostatic proliferation in MHC-matched T-depleted H2-M-deficient mice (Ernst et al., 1999; Viret et al., 1999). In contrast, $CD4^+$ cells from H2-M-deficient mice, which were positively selected by CLIP, underwent efficient homeostatic proliferation in T-depleted H2-M-deficient mice (Ernst et al., 1999; Viret et al., 1999). Similarly, with the use of peptide transporter (TAP) deficient mice that were engineered to express MHC class I (K^b) molecules loaded with specific peptides, it was shown that K^b-restricted OT-I transgenic $CD8^+$ cells were able to undergo homeostatic proliferation only in mice that expressed peptides that are known to mediate positive selection of OT-I cells (Goldrath and Bevan, 1999). Minimal proliferation of OT-I cells was observed in the presence of irrelevant peptides (Goldrath and Bevan, 1999).

Role of Cytokine in Homeostatic Proliferation of Naïve T Cells

With the elucidation that homeostatic proliferation is driven by self-MHC/peptide ligands, the question remained as to how naïve T cells sense the severe reduction in the number of neighboring T cells and begin to undergo proliferation. One hint as to the possible mechanism arose from the finding that survival of naïve T cells is greatly enhanced by a few cytokines. Specifically, it has been shown that IL-4 or IL-7 can prolong survival of naïve T cells *in vitro* (Boise et al., 1995; Vella et al., 1997; Kishimoto and Sprent, 1999) and blocking these two cytokines has been reported to shorten the life span of naïve T cells (Boursalian and Bottomly, 1999). Indeed, Marrack and others have proposed IL-7 as the main regulator of naïve T cell homeostasis (Webb et al., 1999; Marrack et al., 2000).

To determine the role of cytokines in homeostatic proliferation, we devised a novel approach to study homeostatic proliferation *in vitro*. The new system involved culturing whole LN organs or spleen fragments from T-depleted mice injected with CFSE-labeled OT-I $CD8^+$ cells under high oxygen conditions (Tan et al., 2001). Using this approach we were able to evaluate the ability of various cytokines to augment homeostatic proliferation of naïve T cells. Of all the cytokines we tested, only IL-4, IL-7 or IL-15, which are all members of the common γ chain (γc) family of cytokines (Sugamura et al., 1996), were able to augment homeostatic proliferation of OT-I cells (Tan et al., 2001). In terms of the receptor expression, all naïve T cells expressed receptors for IL-4 and IL-7, whereas receptor for IL-15 is expressed on naïve $CD8^+$ cells but not on naïve $CD4^+$ cells (Ohara and Paul, 1987; Zhang et al., 1998; Zhu et al., 2000; Tan et al., 2001).

In order to determine whether IL-4, IL-7 or IL-15 is required for homeostatic proliferation of naïve T cells under *in vivo* conditions, experiments were performed using cytokine-deficient mice. Strikingly, whereas naïve T cells proliferated efficiently in T-depleted IL-4- or IL-15-deficient mice, homeostatic proliferation was nearly completely abrogated in IL-7-deficient mice (Tan et al., 2001). The inability to proliferate in IL-7-deficient hosts was observed for both polyclonal and TCR transgenic $CD4^+$ and $CD8^+$ cells. In addition, lack of proliferation in IL-7-deficient hosts was not due to a defect in the microenvironment of the acellular secondary lymphoid tissues; this is a distinct possibility because IL-7 is essential during early stages of T and B cell development. Thus, infusing exogenous IL-7 restored the ability of IL-7-deficient mice to support homeostatic proliferation of naïve T cells (Tan et al., 2001).

Besides supporting homeostatic proliferation, IL-7 is also required for naïve T cells to survive at a resting state. Thus, naïve T cells transferred into IL-7-deficient mice gradually disappeared over the next few weeks and the donor cells were not detectable in the hosts after one month (Tan et al., 2001). In contrast, naïve T cells transferred into unirradiated wild-type B6 mice were detectable for many months, albeit with a gradual reduction in the numbers. IL-7, therefore, is also required for prolonged survival of naïve T cells.

Similar evidence for the obligatory role of IL-7 for survival and homeostatic proliferation of naïve T cells was also reported by Lefrancois' group with the use of IL-7-deficient mice (Schluns et al., 2000). The requirement for IL-7, but not IL-4 or IL-6, for long-term survival of naïve T cells was also reported by Mathis and Benoist's group

(Vivien et al., 2001). Collectively, these reports provide compelling evidence for IL-7 as the primary regulator of the naïve T cell homeostasis and suggest the following model on how IL-7 regulates naïve T cell homeostasis. Under normal T-sufficient conditions, IL-7, in conjunction with low level TCR signals from contact with self-MHC/peptide ligands, keeps naïve T cells alive in a resting state. The overall size of the naïve T cell pool is thus likely to be determined by the basal level of IL-7, the production of IL-7 presumably being tightly regulated. In T-deficient conditions, the available concentration of IL-7 rises from lack of consumption, and the increased concentration of IL-7 presumably serves as a costimulatory signal to augment TCR signaling from contact with self-MHC/peptide ligands and drives naïve T cells to undergo homeostatic proliferation.

Effect of IL-7 Over-Expression

IL-7 was originally identified as a B cell growth factor produced by bone marrow stromal cells (Namen et al., 1988), but subsequently shown to be produced by other tissues such as thymic and intestinal epithelial cells (Gutierrez and Palacios, 1991; Watanabe et al., 1995). IL-7 was later found to be a non-redundant cytokine for B and T cell development as demonstrated by the absence of lymphopoeisis in IL-7-deficient and IL-7R-deficient mice (Peschon et al., 1994; von Freeden-Jeffry et al., 1995). Four independent lines of transgenic mice over-expressing IL-7 were also generated several years ago and were found to possess increased numbers of B and T cells (Samaridis et al., 1991; Rich et al., 1993; Uehira et al., 1993; Mertsching et al., 1995). However, careful examination of the T cells in these mice was not performed.

To examine the effect of over-producing IL-7 on the homeostasis of naïve T cells, an IL-7 transgenic line expressing a high copy number of IL-7 cDNA under the control of the MHC class II promoter (Mertsching et al., 1995) was analyzed. As predicted from the newly discovered role of IL-7, these mice possess 10-20-fold higher number of mature T cells in the peripheral lymphoid tissues even though the size and cellularity of the thymus is comparable to normal mice (Mertsching et al., 1995). The transgenic mice are otherwise normal, with no elevated infiltration of T cells into non-lymphoid tissues. Remarkably, although most of the $CD4^+$ cells in the transgenic mice are comparable to wild-type $CD4^+$ cells in terms of diversity and the expression of various markers, most of the $CD8^+$ cells are of memory-phenotype with upregulated expression of CD44 and CD122 markers. The following crucial observations were made through the analysis of the IL-7 transgenic mice (manuscript submitted).

First, the $CD44^{hi}$ $CD122^{hi}$ $CD8^+$ cells in IL-7 transgenic mice have characteristics of memory-phenotype from wild-type mice in their ability to synthesize cytokines upon stimulation directly *ex vivo* and in their lack of requirement for MHC class I molecules to undergo homeostatic proliferation. Second, by breeding two different $CD8^+$ TCR transgenic mice, OT-I and HY, it was apparent that generation of $CD44^{hi}$ $CD122^{hi}$ $CD8^+$ cells in IL-7 transgenic mice could arise from response to self-MHC/peptide ligands. Thus, OT-I cells converted to $CD44^{hi}$ $CD122^{hi}$ phenotype in an IL-7 transgenic background whereas HY cells remained $CD44^{lo}$. Since it is known that OT-I cells are able to undergo homeostatic proliferation and upregulate CD44, whereas HY cells are not capable of undergoing such a response (Ernst et al., 1999; Goldrath and Bevan, 1999), these findings indicate a homeostasis-driven mechanism for generation of memory-phenotype $CD8^+$ cells in IL-7 transgenic mice. Finally, the IL-7 transgenic mice were crossed into an IL-15-deficient background to determine whether the generation or maintenance of $CD44^{hi}$ $CD122^{hi}$ $CD8^+$ cells is dependent on IL-15. This is because IL-15 has been recently shown to be essential for survival of $CD44^{hi}$ $CD122^{hi}$ $CD8^+$ cells under normal conditions. Thus, IL-15 can induce survival of $CD44^{hi}$ $CD122^{hi}$ $CD8^+$ cells *in vitro* and mice deficient in either IL-15 or IL-15R are depleted of $CD44^{hi}$ $CD122^{hi}$ $CD8^+$ cells (Lodolce et al., 1998; Zhang et al., 1998; Kennedy et al., 2000; Ku et al., 2000). Strikingly, IL-7 transgenic mice deficient in IL-15 were found to be comparable to regular IL-7 transgenic mice, including possession of massive numbers of $CD44^{hi}$ $CD122^{hi}$ $CD8^+$ cells. The implication of this finding is that high concentration of IL-7 may compensate for IL-15 in supporting survival of $CD44^{hi}$ $CD122^{hi}$ $CD8^+$ cells.

Conclusions

In addition to its established crucial role in development of B and T cells, recent work indicates that IL-7 is the primary agent in regulating homeostasis of naïve T cells. Thus, prolonged survival of naïve T cells under normal physiological conditions requires signals from contact with IL-7 and self-MHC/peptide ligands. Under T-depleted conditions, the increased availability of IL-7, presumably from lack of consumption, induces naïve T cells to undergo homeostatic proliferation. The basal level of IL-7 in the body presumably is tightly regulated and probably determines the overall size of the naïve T cell pool. Hence, over-expression of IL-7 in IL-7 transgenic mice results in proportional enlargement in the size of the mature T cell pool. Strikingly, most of the $CD8^+$ cells in IL-7 transgenic mice are of memory-phenotype, even in the absence of IL-15. In addition to naïve T cells, therefore, IL-7 may also participate in regulating homeostasis of memory $CD8^+$ cells.

Acknowledgements

This work was supported by U.S. Public Health Service Grants. J.T.T. and W.C.K are supported by U.S. Public Health Service Institute National Research Service Awards. C.D.S. is a Scholar of the Leukemia and Lymphoma Society.

References

Bell, E. B. and Sparshott, S. M., 1997, The peripheral T-cell pool: regulation by non-antigen induced proliferation?, *Sem. Immunol.* **9**:347.

Bell, E. B., Sparshott, S. M., Drayson, M. T. and Ford, W. L., 1987, The stable and permanent expansion of functional T lymphocytes in athymic nude rats after a single injection of mature T cells., *J. Immunol.* **139**:1379.

Bender, J., Mitchell, T., Kappler, J. and Marrack, P., 1999, CD4+ T cell division in irradiated mice requires peptides distinct from those responsible for thymic selection., *J. Exp. Med.* **190**:367.

Bevan, M. J., 1997, In thymic selection, peptide diversity gives and takes away., *Immunity* **7**:175.

Boise, L. H., Minn, A. J., June, C. H., Lindsten, T. and Thompson, C. B., 1995, Growth factors can enhance lymphocyte survival without committing the cell to undergo cell division., *Proc. Natl. Acad. Sci. U S A* **92**:5491.

Boursalian, T. E. and Bottomly, K., 1999, Survival of naive CD4 T cells: roles of restricting versus selecting MHC class II and cytokine milieu., *J. Immunol.* **162**:3795.

Brocker, T., 1997, Survival of mature CD4 T lymphocytes is dependent on major histocompatibility complex class II-expressing dendritic cells., *J. Exp. Med.* **186**:1223.

Clarke, S. R. and Rudensky, A. Y., 2000, Survival and homeostatic proliferation of naive peripheral CD4+ T cells in the absence of self peptide:MHC complexes., *J. Immunol.* **165**:2458.

Dorfman, J. R., Stefanova, I., Yasutomo, K. and Germain, R. N., 2000, CD4+ T cells survival is not directly linked to self-MHC-induced TCR signaling., *Nature Immunol.* **1**:329.

Ernst, B., Lee, D.-S., Chang, J. M., Sprent, J. and Surh, C. D., 1999, The peptide ligands mediating positive selection in the thymus control T cell survival and homeostatic proliferation in the periphery., *Immunity* **11**:173.

Freitas, A. A. and Rocha, B. B., 1993, Lymphocyte lifespan: homeostasis, selection and competition., *Immunol. Today* **14**:25.

Fung-Leung, W.-P., Surh, C. D., Liljedahl, M., Pang, J., Leturcq, D., Peterson, P. A., R., W. S. and Karlsson, L., 1996, Antigen presentation and T cell development in H2-M-deficient mice., *Science* **271**:1278.

Goldrath, A. W. and Bevan, M. J., 1999, Antagonist ligands for the TCR drive proliferation of mature CD8+ T cells in lymphopenic hosts., *Immunity* **11**:183.

Goldrath, A. W. and Bevan, M. J., 1999, Selecting and maintaining a diverse T-cell repertoire., *Nature* **402**:255.

Gutierrez, J. C. and Palacios, R., 1991, Heterogeneity of thymic epithelial cells in promoting T-lymphocyte differentiation in vivo, *Proc Natl Acad Sci U S A* **88**:642.

Hogquist, K. A., Jameson, S. C., Heath, W. R., Howard, J. L., Bevan, M. J. and Carbone, F. R., 1994, T cell receptor antagonist peptides induce positive selection., *Cell* **76**:17.

Ignatowicz, L., Kappler, J. and Marrack, P., 1996, The repertoire of T cells shaped by a single MHC/peptide ligand., *Cell* **84**:521.

Kennedy, M. K., Glaccum, M., Brown, S. N., Butz, E. A., Viney, J. L., Embers, M., Matsuki, N., Charrier, K., Sedger, L., Willis, C. R., Brasel, K., Morrissey, P. J., Stocking, K., Schuh, J. C., Joyce, S. and Peschon, J. J., 2000, Reversible defects in natural killer and memory CD8 T cell lineages in interleukin 15-deficient mice., *J. Exp. Med.* **191**:771.

Kieper, W. C. and Jameson, S. C., 1999, Homeostatic expansion and phenotypic conversion of naive T cells in response to self peptide/MHC ligands, *Proc Natl Acad Sci U S A* **96**:13306.

Kirberg, J., Berns, A. and von Boehmer, H., 1997, Peripheral T cell survival requires continual ligation of hte T cell receptor to major histocompatibility complex-encoded molecules., *J. Exp. Med.* **186**:1269.

Kishimoto, H. and Sprent, J., 1999, Strong TCR ligation without costimulation causes rapid onset of Fas-dependent apoptosis of naive murine $CD4^+$ T cells., *J. Immunol.* **163**:1817.

Ku, C. C., Murakami, M., Sakamoto, A., Kappler, J. and Marrack, P., 2000, Control of homeostasis of CD8+ memory T cells by opposing cytokines., *Science* **288**:675.

Labrecque, N., Whitfield, L. S., Obst, R., Waltzinger, C., Benoist, C. and Mathis, D., 2001, How much TCR does a T cell need?, *Immunity* **15**:71.

Lodolce, J. P., Boone, D. L., Chai, S., Swain, R. E., Dassopoulos, T., Trettin, S. and Ma, A., 1998, IL-15 receptor maintains lymphoid homeostasis by supporting lymphocyte homing and proliferation., *Immunity* **9**:669.

Lyons, A. B. and Parish, C. R., 1994, Determination of lymphocyte division by flow cytometry., *J. Immunol. Methods* **171**:131.

Mackall, C. L., Hakim, F. T. and Gress, R. E., 1997, Restoration of T-cell homeostasis after T-cell depletion., *Sem. Immunol.* **9**:339.

Marrack, P., Bender, J., Hildeman, D., Jordan, M., Mitchell, T., Murakami, M., Sakamoto, A., Schaefer, B. C., Swanson, B. and Kappler, J., 2000, Homeostasis of ab TCR+ T cells., *Nature Immunol.* **1**:107.

Marrack, P. and Kappler, J., 1997, Positive selection of thymocytes bearing ab T cell receptors., *Curr. Opin. Immunol.* **9**:250.

Martin, W. D., Hicks, G. G., Mendiratta, S. K., Leva, H. I., Ruley, H. E. and Van Kaer, L., 1996, H2-M mutant mice are defective in the peptide loading of class II molecules, antigen presentation, and T cell repertoire selection., *Cell* **84**:543.

Mertsching, E., Burdet, C. and Ceredig, R., 1995, IL-7 transgenic mice: analysis of the role of IL-7 in the differentiation of thymocytes in vivo and in vitro., *Int. Immunol.* **7**:401.

Miyazaki, T., Wolf, P., Tourne, S., Waltzinger, C., Dierich, A., Barois, N., Ploegh, H., Benoist, C. and Mathis, D., 1996, Mice lacking H2-M complexes, enigmatic elements of the MHC class II peptide-loading pathway., *Cell* **84**:531.

Murali-Krishna, K., Lau, L. L., Sambhara, S., Lemonnier, F., Altman, J. and Ahmed, R., 1999, Persistence of memory CD8 T cells in MHC class I-deficient mice., *Science* **286**:1377.

Namen, A. E., Lupton, S., Hjerrild, K., Wignall, J., Mochizuki, D. Y., Schmierer, A., Mosley, B., March, C. J., Urdal, D. and Gillis, S., 1988, Stimulation of B-cell progenitors by cloned murine interleukin-7, *Nature* **333**:571.

Nesic, D. and Vukmanovic, S., 1998, MHC class I is required for peripheral accumulation of CD8+ thymic emigrants., *J. Immunol.* **160**:3705.

Ohara, J. and Paul, W. E., 1987, Receptors for B-cell stimulatory factor-1 expressed on cells of haematopoietic lineage., *Nature* **325**:537.

Peschon, J. J., Morrissey, P. J., Grabstein, K. H., Ramsdell, F. J., Maraskovsky, E., Gliniak, B. C., Park, L. S., Ziegler, S. F., Williams, D. E., Ware, C. B., Meyer, J. D. and Davison, B. L., 1994, Early lymphocyte expansion is severly impaired in interleukin 7 receptor-deficient mice., *J. Exp. Med.* **180**:1955.

Polic, B., Kunkel, D., Scheffold, A. and Rajewsky, K., 2001, How alpha beta T cells deal with induced TCR alpha ablation, *Proc Natl Acad Sci U S A* **98**:8744.

Rich, B. E., Campos-Torres, J., Tepper, R. I., Moreadith, R. W. and Leder, P., 1993, Cutaneous lymphoproliferation and lymphomas in interleukin 7 transgenic mice., *J. Exp. Med.* **177**:305.

Rocha, B., Dautigny, N. and Pereira, P., 1989, Peripheral T lymphocytes: expansion potential and homeostatic regulation of pool sizes and CD4/8 ratios *in vivo*., *Eur. J. Immunol.* **19**:905.

Rocha, B. and von Boehmer, H., 1991, Peripheral selection of the T cell repertoire., *Science* **251**:1225.

Rooke, R., Waltzinger, C., Benoist, C. and Mathis, D., 1997, Targeted complemetation of MHC class II deficiency by intrathymic delivery of recombinant adenoviruses., *Immunity* **7**:123.

Samaridis, J., Casorati, G., Traunecker, A., Iglesias, A., Gutierrez, J. C., Muller, U. and Palacios, R., 1991, Development of lymphocytes in interleukin 7-transgenic mice., *Eur. J. Immunol.* **21**:453.

Schluns, K. S., Kieper, W. C., Jameson, S. C. and Lefrancois, L., 2000, Interleukin-7 mediates the homeostasis of naive and memory CD8 T cells in vivo., *Nature Immunol.* **1**:426.

Sprent, J., Lo, D., Gao, E. K. and Ron, Y., 1988, T cell selection in the thymus., *Immunol. Rev.* **101**:173.

Sprent, J. and Tough, D. F., 1994, Lymphocyte life-span and memory., *Science* **265**:1395.

Sugamura, K., Asao, H., Kondo, M., Tanaka, N., Ishii, N., Ohbo, K., Nakamura, M. and Takeshita, T., 1996, The interleukin-2 receptor gamma chain: its role in the multiple cytokine receptor complexes and T cell development in XSCID., *Annu. Rev. Immunol.* **14**:179.

Surh, C. D., Lee, D.-S., Fung-Leung, W.-P., Karlsson, L. and Sprent, J., 1997, Thymic selection by a single MHC/peptide ligand produces a semidiverse repertoire of $CD4^+$ T cells., *Immunity* **7**:209.

Surh, C. D. and Sprent, J., 2000, Homeostatic T Cell Proliferation. How far can T cells be activated to self-ligands?, *J. Exp. Med.* **192**:F9.

Takeda, S., Rodewald, H.-R., Arakawa, H., Bluethmann, H. and Shimizu, T., 1996, MHC class II molecules are not required for survival of newly generated $CD4^+$ T cells but affect their long-term life span., *Immunity* **5**:217.

Tan, J. T., Dudl, E., LeRoy, E., Murray, R., Sprent, J., Weinberg, K. I. and Surh, C. D., 2001, IL-7 is critical for homeostatic proliferation and survival of naive T cells., *Proc. Natl. Acad. Sci. U S A* **98**:8732.

Tanchot, C., Lemonnier, F. A., Pérarnau, B., Freitas, A. A. and Rocha, B., 1997, Differential requirements for survival and proliferation of CD8 naïve or memory T cells., *Science* **276**:2057.

Tanchot, C., Rosado, M. M., Agenes, F., Freitas, A. A. and Rocha, B., 1997, Lymphocyte homeostasis., *Sem. Immunol.* **9**:331.

Uehira, M., Matsuda, H., Hikita, I., Sakata, T., Fujiwara, H. and Nishimoto, H., 1993, The development of dermatitis infiltrated by gamma delta T cells in IL-7 transgenic mice., *Int. Immunol.* **5**:1619.

Vella, A., Teague, T. K., Ihle, J., Kappler, J. and Marrack, P., 1997, Interleukin 4 (IL-4) or IL-7 prevents the death of resting T cells: stat6 is probably not required for the effect of IL-4., *J. Exp. Med.* **186**:325.

Viret, C., Wong, F. S. and Janeway, C. A., Jr., 1999, Designing and maintaining the mature TCR repertoire: the continuum of self-peptide:self-MHC complex recognition., *Immunity* **10**:559.

Vivien, L., Benoist, C. and Mathis, D., 2001, T lymphocytes need IL-7 but not IL-4 or IL-6 to survive in vivo, *Int Immunol* **13**:763.

von Boehmer, H. and Hafen, K., 1993, The life span of naive a/b T cells in secondary lymphoid organs., *J. Exp. Med.* **177**:891.

von Freeden-Jeffry, U., Vieira, P., Lucian, L. A., McNeil, T., Burdach, S. E. G. and Murray, R., 1995, Lymphopenia in interleukin (IL)-7 gene-deleted mice identifies IL-7 as a nonredundant cytokine., *J. Exp. Med.* **181**:1519.

Watanabe, M., Ueno, Y., Yajima, T., Iwao, Y., Tsuchiya, M., Ishikawa, H., Aiso, S., Hibi, T. and Ishii, H., 1995, Interleukin 7 is produced by human intestinal epithelial cells and regulates the proliferation of intestinal mucosal lymphocytes, *J Clin Invest* **95**:2945.

Webb, L. M., Foxwell, B. M. and Feldmann, M., 1999, Putative role for interleukin-7 in the maintenance of the recirculating naive CD4+ T-cell pool., *Immunology* **98**:400.

Zhang, X., Sun, S., Hwang, I., Tough, D. F. and Sprent, J., 1998, Potent and selective stimulation of memory-phenotype CD8+ T cells in vivo by IL-15., *Immunity* **8**:591.

Zhu, J., Huang, H., Guo, L., Stonehouse, T., Watson, C. J., Hu-Li, J. and Paul, W. E., 2000, Transient inhibition of interleukin 4 signaling by T cell receptor ligation., *J. Exp. Med.* **192**:1125.

IL-7, THE THYMUS, AND NAÏVE T CELLS

Yukari Okamoto, Daniel C. Douek, Richard D. McFarland, and Richard A. Koup*

1. INTRODUCTION

It has been reported that T cell numbers are maintained in adults predominantly through the expansion of post-thymic, memory T cells, whereas in infants, T cells are predominantly maintained through the production of new naïve T cells by the thymus [1]. However, we and others have recently demonstrated that the adult thymus is still capable of thymopoiesis and can contribute to T cell reconstitution in adults [2, 3]. Several methods have been used to measure thymopoietic capacity. Thymic size as measured by radiographic imaging [1] or volumetric computerized tomograph measurements [4, 5] has been correlated with numbers of $CD4^+CD45RA^+$ naïve T cells and the number of phenotypically naïve T cells after transplantation has been shown to correlate with antigen-specific function [6]. However, there are concerns about limitations of estimating thymic function based on naïve T cell phenotype alone. T cells expressing a naïve phenotype are not necessarily accurate surrogate markers of thymic function. Following thymic emigration, $CD45RA^+$ naïve T cells can have a long quiescent lifespan [7], may proliferate in an antigen-independent manner [8], or may rapidly convert to $CD45RO^+$ memory/effector phenotype T cells [9]. Furthermore, naïve T cell markers may be acquired by memory T cells (especially $CD8^+$ T cells) [9], further compounding the difficulty in accurate enumeration of naïve T cells [10, 11].

In order to measure thymic function more directly in humans, we recently described an assay that quantifies an episomal DNA by-product of the T cell receptor (TCR) rearrangement process [2]. These TCR rearrangement excision circles (TREC) contain the signal joint sequences from the *TCRAD* locus δRec to ψJα recombination event, which is common to approximately 70% of thymocytes destined to become mature TCRαβ T cells. TREC are stable and do not replicate with cellular proliferation [2, 12],

* Yukari Okamoto, Daniel C. Douek, and Richard A. Koup, Vaccine Research Center, NIAID, NIH, Bethesda, MD 20892, Richard D. McFarland, Center for Biologics Evaluation and Research, Food and Drug Administration, Rockville, MD 20850.

Reprinted in part from: Okamoto Y, DC Douek, RD McFarland, and RA Koup. Effects of interleukin-7 on human thymus function. *Blood* 99:2851-2858,2002. Copyright American Society of Hematology, used by permission

therefore the TREC content in a peripheral cell population is proportional to the frequency of recent thymic émigrants (RTE). This frequency, however, is affected not only by changes in thymic output but also by the proliferative history of the cells. Because RTE have TREC at levels which would not have been affected by peripheral expansion and cellular replication, quantification of TREC levels in peripheral blood represents a sensitive measurement of thymopoietic capacity.

While thymic function declines with age, substantial output is maintained into late adulthood [2, 3]. Furthermore, the adult thymus can contribute to immune reconstitution in individuals following antiretroviral therapy [2, 13, 14], and following myeloablative chemotherapy and autologous haematopoietic stem-cell transplantation [15]. T cells generated *de novo* from thymopoiesis have a broad TCR repertoire and are theoretically more capable of responding to neoantigens effectively [15, 16]. In contrast, peripheral expansion of existing T cell pools may lead to limited T cell repertoires and antigen-responsiveness [1, 16-21]. Therefore in patients with HIV infection or who have received chemotherapy, the ability of the thymus to generate naïve T cells with a broad TCR repertoire should allow for recovery of T cell mediated immunity that is qualitatively better than if the recovery were only through expansion of pre-existing naïve and memory T cells.

Interleukin-7 (IL-7), which was originally reported as a pre-B cell growth factor [22], is produced by stromal cells in the thymus and bone marrow and appears to play a role at multiple stages of T and B lymphocyte development [23, 24]. In mice, the IL-7 receptor (IL-7R) is first expressed in lymphoid lineage-restricted progenitors in bone marrow [25] and later can be detected in various tissues including thymus [26]. IL-7R is composed of the IL-7Rα [26] chain and the common cytokine receptor γ chain (γc) [27], the latter of which is the indispensable subunit receptor for several lymphoid-related cytokines such as IL-2, IL-4, IL-9 and IL-15 [28]. Several studies have shown the essential role of this cytokine in the survival and normal differentiation of thymocytes into mature naïve T cells in mice. It has been suggested that IL-7 is a cofactor for V(D)J rearrangement of the *TcRb* gene locus [29], and that it is also required for *TcRb* D-J rearrangement [30]. In mice with targeted deletions of the IL-7 gene (IL-7$^{-/-}$), there was a 20-fold reduction in the number of pro-T cells [31], yet the percentage of TcR$\alpha\beta$ thymocytes was relatively normal [24, 31], while the number of TcR$\gamma\delta$ thymocytes was substantially reduced [31]. In mice with deletions of the IL-7R gene (IL-7R$\alpha^{-/-}$ and γc$^{-/-}$), the number of TcR$\alpha\beta$ thymocytes was markedly decreased and TcR$\gamma\delta$ thymocytes were completely absent [32, 33]. On the other hand, in humans, it has been reported that mutations that prevent the expression of IL-7Rα chain and γc result in severe combined immunodeficiency (SCID) [34, 35]. Thus, it appears that IL-7 is necessary for the survival and proliferation of early thymic progenitor cells and production of TcR$\gamma\delta$ T cells, although it may not be unconditionally required for rearrangement of *TcRa* and *b* genes in mice. However, the regulatory mechanism of TCR$\alpha\beta$ or $\gamma\delta$ rearrangement in human thymus and the roles of IL-7 in thymopoiesis are as yet not fully understood. Importantly, it has been recently shown that IL-7 levels are increased during T lymphopenia in HIV infection, strongly suggesting that IL-7 may play a major role in both peripheral and central T cell homeostasis [36, 37].

2. TREC LEVELS IN FACS-SORTED THYMOCYTE SUBSETS IN THE FRESH HUMAN THYMUS

To investigate the timing of TREC generation and dilution during thymic development *in vivo*, we measured TREC levels in sorted thymocyte subsets from a newborn infant. The cells were separated into $CD3^-$ and $CD3^+$ fractions by magnetic cell sorting (MACS), and further purified by flow cytometry based on surface phenotype. Commitment of progenitor cells to the T cell lineage occurs at or around the transition of $CD1a^-CD34^+$ thymocytes into $CD1a^+CD34^+$ thymocytes [38, 39]. Cells were sorted by flow cytometry for $CD1a^{+/++}CD3^-$ and $CD3^+$ subsets so as to exclude the $CD1a^-CD3^-$ subset which contains precursors for NK cells and thymic dendritic cells [40-42]. Quantitative PCR demonstrated that the TREC level in unfractionated thymocytes was 12,194 per 10,000 cells. No TREC were detected at the $CD3^-CD4^-CD8^-$ triple negative (TN), $CD3^-CD4^{low}CD8^-$, or $CD3^-CD4^+CD8^-$ stages of thymocyte development. However, the highest level of TREC were detected within $CD1a^+$ cells that are $CD3^-CD4^+CD8^+$ double positive ($CD3^-$ DP) (14,692 per 10,000 cells), confirming previous experiments that address the timing of TCRD gene rearrengement [43]. This indicates that the excision leading to TREC which commits a cell to the TCRαβ lineage occurs concomitant with the expansion of $CD3^-$ DP thymocytes. Assuming that TCRA recombination occurs in both alleles [44], a TREC level of 14,692 per 10,000 cells indicates that ~75% of the cells have undergone TREC generation in this subpopulation. Since cellular divisions after TREC generation will dilute TREC, it should be that the initiation of cell expansion precedes TREC generation in $CD3^-$DP cells. As far as the authors know, this is the first report to directly demonstrate the timing of cell expansion and TREC generation in this subset. This information would be useful for understanding the underlying mechanism for IL-7 effects on TCR rearrangement.

High levels of TREC were still detected in the $CD3^+$ DP subset which have undergone expansion (6900 per 10,000 cells), and also in the more mature $CD4^+$ and $CD8^+$ single positive (SP) subsets (6064 and 4099 per 10,000 cells, respectively). The decrease in TREC in mature cells is likely to represent dilution secondary to cellular proliferation that occurs during the processes of positive and negative selection. There was no evidence to suggest that TREC generation still occurs after the $CD3^+$ stage.

3. IL-7Rα EXPRESSION IN THYMOCYTE SUBSETS

As a basis for studying the role of IL-7 in TREC generation in human thymus, IL-7Rα expression was examined by flow cytometry on thymocytes from a fetus, a 2 month-old infant and a 14 year-old youth. IL-7Rα expression was highest in immature $CD3^-CD4^{low}CD8^-$ cells (89.0 - 95.5%), just before TREC generation. Since cells with this phenotype are thought to be initiating the process of TCRαβ or γδ lineage commitment and are known to undergo proliferation [45], these results suggest the involvement of IL-7Rα in regulating the proliferation of thymocytes and their commitment to TCRαβ or γδ lineage. Although IL-7Rα expression tend to decrease after the $CD3^-CD4^{low}CD8^-$ stage, it was still fairly high (52.2 - 85.9%) in $CD3^-$ DP cells regardless of the age of the thymus. IL-7Rα expression was lowest in $CD3^+$ DP cells (31.4 - 57.7%) and then increased again in $CD3^+$ SP cells (43.6 - 69.2%).

4. THE EFFECTS OF IL-7 ON PROLIFERATION AND APOPTOSIS OF THYMOCYTES

Several studies have suggested that IL-7 is a physiological survival factor for early lymphoid progenitor cells [23, 46, 47, 48]. The evaluation of TREC levels in each thymic subpopulation will be affected by proliferation and apoptosis of thymocytes. In order to evaluate the *in vitro* effects of exogenous IL-7 on TREC generation in our TOC system, we first assessed the effect of IL-7 on thymocyte proliferation by measuring the incorporation of BrdU. The rate of BrdU incorporation was highest in $CD3^{-/low}CD4^{+}CD8^{-}$ cells and low in mature $CD3^{+}$ cells. We observed a marked increase in BrdU uptake in most thymocyte populations with the administration of 10 ng/ml of IL-7. Increased BrdU uptake with IL-7 was observed consistently in five different TOC, not only in immature cells (4.0 ± 3.3-fold and 2.3 ± 0.8-fold in TN and $CD3^{-}4^{low}8^{-}$, respectively) but also in more mature $CD3^{+}4^{+}$ SP (6.6 ± 3.3-fold) and $CD3^{+}8^{+}$ SP cells (18.8 ± 8.6-fold). This effect was dependent on IL-7 dose ranging from 1 to 50 ng/ml.

We next assessed the effects of IL-7 on apoptosis in TOC by measuring expression of annexin V after 4 days of TOC in the presence or absence of IL-7. IL-7 significantly lowered Annexin V expression by about 50% in $CD3^{-}$ $CD8^{-}$ immature thymocytes, which were pre-TREC generating cells. We then examined BCL-2, a well-established suppressor of apoptosis, for potential regulation of this apoptosis. It has been reported that BCL-2 expression during T cell differentiation is multiphasic; it is high in $CD3^{-}$ immature thymocytes, down-regulated in $CD4^{+}CD8^{+}$ thymocytes to facilitate selection process, and up-regulated in mature $CD3^{+}$ SP thymocytes thus enhancing prolonged survival [49]. As expected, IL-7 enhanced BCL-2 expression in $CD3^{-}$ immature thymocytes in dose dependent way. These results were consistent with previous studies in mice demonstrating that short-term culture with IL-7 of immature thymocytes from $IL\text{-}7^{-/-}$ mice caused up-regulation of BCL-2 and increased cell survival [47]. Thus in human thymus, IL-7 led to an increase in proliferation and decrease in apoptosis of immature $CD3^{-}$ to $CD3^{low}$ cells, which correlated with the increased expression of BCL-2.

5. THE IN VITRO EFFECTS OF IL-7 ON TREC GENERATION IN TOC

The effects of IL-7 on TREC generation were examined in TOC. After culturing newborn thymus in the presence (10 and 50 ng/ml) or absence of IL-7 for 4 days, thymocytes were separated into $CD3^{-}$ and $CD3^{+}$ fractions by MACS. The majority of $CD3^{low}$ cells were contained in $CD3^{-}$ fraction, while most of $CD3^{+}$ DP cells were collected into $CD3^{+}$ fraction. IL-7 significantly increased TREC levels in both $CD3^{-}$ and $CD3^{+}$ subsets in a dose-dependent manner.

If IL-7 were to be used as a therapeutic agent to improve immune reconstitution, its effect would have to be maintained in post-natal thymuses. The effects of IL-7 on TREC levels were further evaluated in TOC using thymuses from subjects of differing ages. Fetal (18, 19 and 22 week gestation), newborn (2 and 15 day, and 3, 4, 7 and 11 month) and infant (3 and 5.6 year) thymuses were cultured in TOC in the presence of IL-7 and TREC were measured. In all TOC, TREC levels increased in the presence of IL-7 (1.3 ± 0.1-fold for whole cells, 2.0 ± 0.8-fold for $CD3^{-}$ fractions, and 1.5 ± 0.2 -fold for $CD3^{+}$ fractions).

6. THE *IN VIVO* EFFECTS OF IL-7 ON TREC GENERATION IN NOD-SCID-HU MICE

Several studies have demonstrated the *in vivo* effects of IL-7 on mouse thymopoiesis. Aspinall et al. showed that the weight and cell density of thymus, and the number of mature T cells in the peripheral blood increased after administration of 500 ng IL-7 twice daily for 5 days in adult mice [50]. Furthermore, Abdul-Hai et al. showed that transferred mouse bone marrow CD34$^+$ cells proliferated with normal TCR Vβ8 usage in thymus and spleen of irradiated recipient mice administered with 100 ng of IL-7 twice daily for 10 days [51]. In the present study we used NOD-SCID-hu chimeric mice to measure the *in vivo* effects of IL-7 on human thymopoiesis. Twenty-eight days after engraftment of human fetal thymus and liver under the mouse kidney capsule, the growth of thymic grafts was readily apparent; typically a 2 mm^3 fragment grew up to a diameter of 5 to 7 mm. These grafts contained comparable levels of TREC per thymocytes to those of native human thymuses, indicating the generation of new thymocytes in the grafts. More importantly, we were able to detect human CD45$^+$ cells in the peripheral blood and spleen. Most of these human cells were positive for CD3 and represented 1.0 and 1.3% of the total lymphocytes in PBMC and spleen, respectively. These CD3$^+$ cells comprised mature CD4$^+$ and CD8$^+$ SP cells, the majority of which were CD45RA$^+$, suggesting that they were naïve T cells. The frequency of human CD3$^+$CD45RA$^+$CD4$^+$ or CD3$^+$CD45RA$^+$CD8$^+$ T cell correlated positively with the size of the graft.

We treated NOD-SCID-hu mice with 100 ng of IL-7 (n=11) or saline (n=8) twice daily for 10 days beginning 4 days after checking the size of the grafts. TREC levels in thymocytes from the grafts were significantly higher in the IL-7 treated group (14276 $\pm$ 1290 and 14344 $\pm$ 2449 for day 7 and 21, respectively) than the control group (8135 $\pm$ 1070 and 9877 $\pm$ 1869 for day 7 and 21, respectively) at day 7 ($p<0.002$) and day 21 ($p<0.007$) after the last administration of IL-7. Although the average graft weight was higher in IL-7-treated mice after 7 days, the difference between the two groups was not statistically significant either after 7 or 21 days . Moreover, there was no change by IL-7 treatment for the percentage of Ki67 positive cells in any of the subset except for CD3+ DP and CD3+CD8 SP at day 7. Also, there was no significant increase in the percentage of human CD3$^+$ cells in the peripheral blood after IL-7 treatment.

7. DISCUSSION

It is well established that IL-7 exerts potent effects on T-cell progenitors and required for T-cell development. It is critical for homeostatic proliferation and survival of not only thymocytes but also naïve T-cells in peripheral blood and lymph node [52]. In the present study we employed the quantitative TREC assay to assess more definitively the effects of exogenous IL-7 on TCR rearrangement in thymocytes using both *in vitro* TOC and thymic engraftments in NOD-SCID mice.

To understand the timing of TREC generation and its dilution during thymic development we quantified TREC level in each thymic subpopulation. We found the highest level of TREC in the CD1a$^+$CD3$^-$ DP stage. Recently, TCRβ selection in human was reported to be initiated at the transition of CD3$^-$CD4$^+$CD8$^-$ into the CD4$^+$CD8α$^+$β$^-$ stage [53], while TCRα gene expression has been known to be restricted to CD3$^+$ DP and later stage of T cell development. Together with these, our data suggest that TCRβ gene rearrangement precedes and induces the initiation of TREC generation, followed by TCRα rearrangement. TREC generation may represent a decisive step in the αβ versus γδ lineage commitment of differentiating thymocytes because this is an intermediate rearrangement between *TCRD* and *TCRA* gene rearrangement which excises most of the *TCRD* locus [54]. The only apparent reason for this rearrangement is to delete the *TCRD* locus to prepare the allele for subsequent *TCRA* gene rearrangement. Thus, our finding that IL-7 increases TREC levels of thymocytes most likely indicates that IL-7 eventually leads to the enhancement of induction of *TCRA* and *B* gene rearrangement.

Although IL-7 increased TREC in our TOC system, it remains unclear whether this is due to the promotion of TREC generation in the individual thymocytes, because TREC levels can be affected by the possible effect of IL-7 to regulate the apoptosis and proliferation of cells at pre-TREC generating stage [45, 46, 47, 55]. Indeed, BrdU, annexin V and BCL-2 staining in our TOC system clearly demonstrated that exogenous IL-7 can enhance the proliferation and prevent apoptosis of immature $CD3^{-/low}$ subsets which, according to our results, precede TREC generation. It is highly possible, therefore, that IL-7 indirectly increases the TREC frequency in thymus by promoting the proliferation and by lengthening survival of the pre-TREC generating thymocytes in *in vitro* TOC system.

On the other hand, our *in vivo* experiments suggested another pathway of IL-7 effects. The exogenous IL-7 were administered to NOD-SCID-hu, where the growth of human thymic grafts appeared to be fairly active. At 7 and 21 days after the last administration, TREC levels in thymocytes from grafts were significantly higher in the IL-7 treated group than the control group. However, the graft weights and also Ki67 expression in any of the subsets did not show significant change by IL-7 treatment. Considering that IL-7 can exhibit distinct effects on T cells at different concentration[56], one possible interpretation of our data would be that the endogenous human IL-7 in these grafts was sufficient to maintain the normal development of thymocytes, whereas the addition of exogenous IL-7 further promoted TCR gene rearrangement and therefore TREC generation, but was not enough to promote further proliferation of immature cells. In this expriment we did not measure the rate of apoptosis in the thymic grafts. However, it is unlikely that this increase of TREC by IL-7 was caused solely by lengthening the survival of immature cells because the percentage of AnnexinV positive cells in the fresh human thymus were typically as low as 0.5-4% in each subpopulation. Together, it is possible that IL-7 can directly stimulate TREC generation. Based upon transgenic mouse studies, it has been suggested that there is a regulatory sequence near the human δRec gene segment, which is recognized by putative regulatory proteins that induce δRec rearrangement [57]. Moreover, transcription of the T-early α (TEA) element is assumed to be responsible for opening the *TCRAJ* locus to the V(D)J recombinase complex [58, 59]. Further identification of regulatory mechanism of δRec-ψJα recombination event will eventually allow a definition whether there is a direct effect of IL-7 upon intracellular processes that lead to enhanced TREC production.

In our mouse system, we observed no significant increase for the percentage of human $CD3^+$ cells in the peripheral blood by IL-7 treatment. It has been reported that the recipient's NK cells affect the efficiency of engraftment of human cells [60]. Although NOD/LtSz-scid mice have relatively low NK cell activity [61], it is possible that IL-7 caused an enhancement of this [62, 63]. Be that as it may, our data clearly show that IL-7 increased TCR gene rearrangement and TREC generation, which would suggest that thymopoietic function was augmented.

It has been suggested that complete recovery of broad T cell immunity, after insults such as HIV-infection or chemotherapy, requires the generation of new naïve T cells from the thymus [2, 15, 16, 20]. The administration of thymopoietic cytokines may be able to improve immune reconstitution by augmenting thymic function in these situations. IL-7 is a major factor for thymopoiesis, and systemic administration of recombinant IL-7 to murine bone marrow transplant (BMT) recipients has been shown to normalize thymopoiesis and improve post-BMT immune function [20, 64, 65]. Furthermore, endogenous IL-7 may play a major role in the homeostatic response to T lymphopenia [36]. Our data suggest that IL-7 can enhance the function of fetal and infant thymuses, although in therapeutic practice it may be necessary to antagonize the effects of thymic atrophic factors [66]. Since TREC levels in the peripheral blood reflects RTE [2], the effects of IL-7 on thymopiesis should be able to be evaluated by monitoring the change of blood TREC levels. In this regard, our results provide a basis for understanding some of the regulatory mechanisms of thymocyte development in the human thymus, and encourage the further evaluation of cytokine-based immune reconstitution strategies, through the stimulation of thymus-dependent T cell generation.

REFERENCES

1. C. L. Mackall, T. A. Fleisher, M. R. Brpwn, M. P. Andrich, C. C. Chen, I. M. Feuerstein, M. E. Horowitz, I. T. Magrath, A. T. Shad, S. M. Steinberg, L. H. Wexler, and R. E. Gress, Age, thymopoiesis, and CD4+ T-lymphocyte regeneration after intensive chemotherapy. *N. Engl. J. Med.* **332**(3), 143-149 (1995).
2. C. D. Douek, D. R. McFarland, H. P. Keiser, A. E. Gage, M. J. Massey, F. B. Haynes, A. M. Polis, T. A. Haase, B. M. Feinberg, L. J. Sullivan, D. B. Jamieson, A. J. Zack, J. L. Picker, and A. R. Koup, Changes in thymic function with age and during the treatment of HIV infection. *Nature* **396,** 690-695 (1998).
3. B. D. Jamieson, D. C. Douek, S. Killian, L. E. Hultin, D. D. Scripture-Adams, J. V. Giorgi, D. Marelli, R. A. Koup, and J. A. Zack, Generation of functional thymocytes in the human adult. *Immunity* **10(5),** 569-575 (1999).
4. J. M. McCune, R. Loftus, D. K. Schmidt, P. Carroll, D. Webster, L. B. Swor-Yim, I. R. Francis, B. H. Gross, and R. M. Grant, High prevalence of thymic tissue in adults with human immunodeficiency virus-1 infection. *J. Clin. Invest.* **101**(4873), 2301-2308 (1998).
5. K. Y. Smith, H. Valdez, A. Landay, J. Spritzler, H. A. Kessler, E. Connick, D. Kuritzkes, B. Gross, I. Francis, J. M. McCune and M. M. Lederman, Thymic size and lymphocyte restoration in patients with human immunodeficiency virus infection after 48 weeks of zidovudine, lamivudine, and ritonavir therapy. *J. Infect. Dis.* **181**(1), 141-147 (2000).
6. G. Socie, J. V. Stone, J. R. Wingard, D. Weisdorf, P. J. Henslee-Downey, C. Bredeson, J. Y. Cahn, J. R. Passweg, P. A. Rowlings, H. C. Schouten, H. J. Kolb, and J. P. Klein, Long-term survival and late deaths after allogeneic bone marrow transplantation. Late Effects Working Committee of the International Bone Marrow Transplant Registry. *N. Engl. J. Med.* **341**(1), 14-21 (1999).
7. A. R. McLean, and C. A. Michie, In vivo estimates of division and death rates of human T lymphocytes. *Proc. Natl. Acad. Sci. U. S. A.* **92**(9), 3707-3711 (1995).
8. M. V. Soares, N. J. Borthwick, M. K. Maini, G. Janossy, M. Salmon, and A. N. Akbar, IL-7-dependent extrathymic expansion of CD45RA+ T cells enables preservation of a naive repertoire. *J. Immunol.* **161**(11), 5909-17 (1998).
9. L. J. Picker, J. R. Treer, B. Ferguson-Darnell, P. A. Collins, D. Buck, and L. W. Terstappen,, Control of lymphocyte recirculation in man. I. Differential regulation of the peripheral lymph node homing receptor L-selection on T cells during the virgin to memory cell transition. *J. Immunol.* **150**(3), 1105-1121 (1993).
10. D. Hamann, P. A. Baars, M. H. Rep, B. Hooibrink, R. S. Kerkhof-Garde, M. R. Klein, and van R. A. Lier,, Phenotypic and functional separation of memory and effector human CD8+ T cells. *J. Exp. Med.* **186**(9), 1407-18 (1997).
11. J. L. Young, J. M. Ramage, J. S. Gaston, and P. C. Beverley, In vitro responses of human CD45R0brightRA- and CD45R0-RAbright T cell subsets and their relationship to memory and naive T cells. *Eur. J. Immunol.* **27**(9), 2383-2390 (1997).
12. F. Kong, C. H. Chen, and M. D. Cooper, Thymic function can be accurately monitored by the level of recent T cell emigrants in the circulation. *Immunity* **8**(1), 97-104 (1998).

13. J. F. Poulin, and R. P. Sekaly, Function of the thymus in HIV-infected adults. *J.A.M.A.* **282**(3), 219 (1999).
14. L. Zhang, S. R. Lewin, M. Markowitz, H. H. Lin, E. Skulsky, R. Karanicolas, Y. He, X. Jin, S. Tuttleton, M. Vesanen, H. Spiegel, R. Kost, J. van Lunzen, H. J. Stellbrink, S. Wolinsky, W. Borkowsky, P. Palumbo, L. G. Kostrikis, and D. D. Ho, Measuring recent thymic emigrants in blood of normal and HIV-1-infected individuals before and after effective therapy. *J. Exp. Med.* **190**(5), 725-732 (1999).
15. D. C. Douek, R. A. Vescio, M. R. Betts, J. M. Brenchley, B. J. Hill, L. Zhang, J. R. Berenson, R. H. Collins, and R. A. Koup, Assessment of thymic output in adults after haematopoietic stem-cell transplantation and prediction of T-cell reconstitution. *Lancet* **355**(9218), 1875-1881(2000).
16. A. Heitger, N. Neu, H. Kern, E. R. Panzer-Grumayer, H. Greinix, D. Nachbaur, D. Niederwieser, and F. M. Fink, Essential role of the thymus to reconstitute naive (CD45RA+) T-helper cells after human allogeneic bone marrow transplantation. *Blood* **90**(2), 850-7(1997).
17. C. A. Keever, Small, T. N., Flomenberg, N., Heller, G., Pekle, K., Black, P., Pecora, A., Gillio, A., Kernan, N. A. and O'Reilly, R. J., Immune reconstitution following bone marrow transplantation: comparison of recipients of T-cell depleted marrow with recipients of conventional marrow grafts. *Blood* **73**(5),1340-1350 (1989).
18. C. L. Mackall, C. V. Bare, L. A. Granger, S. O. Sharrow, J. A. Titus, and R. E. Gress, Thymic-independent T cell regeneration occurs via antigen-driven expansion of peripheral T cells resulting in a repertoire that is limited in diversity and prone to skewing. *J. Immunol.* **156**(12), 4609-16 (1996).
19. J. Storek, R. P. Witherspoon, and R. Storb, T cell reconstitution after bone marrow transplantation into adult patients does not resemble T cell development in early life. *Bone Marrow Transplant.* **16**(3), 413-25 (1995).
20. K. Weinberg, G. Annett, A. Kashyap, C. Lenarsky, S. J. Forman, and R. Parkman, The effect of thymic function on immunocompetence following bone marrow transplantation. *Biol. Blood Marrow Transplant.* **1**(1), 18-23 (1995).
21. T. N. Small, E. B. Papadopoulos, F. Boulad, P. Black, H. Castro-Malaspina, B. H. Childs, N. Collins, A. Gillio, D. George, A. Jakubowski, G. Heller, M. Fazzari, N. Kernan, S. MacKinnon, P. Szabolcs, J. W. Young, and R. J.O'Reilly, Comparison of immune reconstitution after unrelated and related T-cell-depleted bone marrow transplantation: effect of patient age and donor leukocyte infusions. *Blood* **93**(2), 467-80 (1999).
22. A. E. Namen, A. E. Schmierer, C. J. March, R. W. Overell, L. S. Park, D. L. Urdal, and D. Y. Mochizuki, B cell precursor growth-promoting activity. Purification and characterization of a growth factor active on lymphocyte precursors. *J. Exp. Med.* **167**(3), 988-1002 (1988).
23. J. J. Peschon,, P. J. Morryssey, K. H. Grabstein, F. J. Ramsdell, E. Maraskovsky, B. C. Gliniak, L. S. Park, S. F. Ziegler, D. E. Williams, C. B. Ware, J. D. Meyer, and B. L. Davison, Early lymphcyte expansion is severely impaired in interleukin 7 receptor-deficient mice. *J. Exp. Med.* **180**, 1955-1960 (1994).
24. U. von Freeden-Jeffry, N. Solvason, M. Howard, and R. Murray, The earliest T lineage-committed cells depend on IL-7 for Bcl-2 expression and normal cell cycle progression. *Immunity* **7**,147-154 (1997).
25. M. Kondo, T. Takeshita, M. Higuchi, M. Nakamura, T. Sudo, S. Nishikawa, and K. Sugamura, Functional participation of the IL-2 receptor gamma chain in IL-7 receptor complexes. *Science* **263**(5152), 1453-1454 (1994).
26. R. G. Goodwin, D. Friend, S. F. Ziegler, R. Jerzy, B. A. Falk, S. Gimpel, D. Cosman, S. K. Dower, C. J. March, and A. E. Namen, Cloning of the human and murine interleukin-7 receptors: demonstration of a soluble form and homology to a new receptor superfamily. *Cell* **60**(6), 941-51 (1990).
27. M. Noguchi, Y. Nakamura, S. M. Russell, S. F. Ziegler, M. Tsang, X. Cao, and W. J. Leonard, Interleukin-2 receptor gamma chain: a functional component of the interleukin-7 receptor. *Science* **262**(5141), 1877-1880 (1993).
28. J. X. Lin, T. S. Migone, M. Tsang, M. Friedmann, J. A. Weatherbee, L. Zhou, A. Yamauchi, E. T. Bloom, J. Mietz, and S. John, The role of shared receptor motifs and common Stat proteins in the generation of cytokine pleiotropy and redundancy by IL-2, IL-4, IL-7, IL-13, and IL 15. *Immunity* **2**(4), 331-339 (1995).
29. K. Muegge, M. P. Vila, and S. K. Durum, Interleukin-7: a cofactor for V(D)J rearrangement of the T cell receptor beta gene. *Science* **261**(5117), 93-95 (1993).
30. S. Tsuda, S. Reike, Y. Hashimoto, H. Nakauchi, and Y. Takahama, IL-7 supports D-J but not V-DJ rearrangement of TCR-β gene in fetal liver progenitor cells. *J. Immunol.* **156**, 3233-3242 (1996).
31. T. A. Moore, U. von Freeden-Jeffry, R. Murray, and A. Zlotnik, Inhibition of gamma delta T cell development and early thymocyte maturation in IL-7 -/- mice. *J. Immunol.* **157**(6), 2366-2373 (1996).
32. K. Maki, S. Sunaga, Y. Komagata, A. Mabuchi, H. Karasuyama, K. Yokomuro, J-I. Miyazaki, and K. Ikuta, Interleukin 7 receptor-deficient mice lack γδ T cells. *Proc. Natl Acad. Sci.* USA. **93**, 7172-7177 (1996).

33. Y. W. He, and T. R. Malek, Interleukin-7 receptor alpha is essential for the development of gamma delta + T cells, but not natural killer cells. *J. Exp. Med.* **184**(1), 289-293 (1996).
34. A. Puel, S. F. Ziegler, R. H. Buckley, and W. J. Leonard, Defective IL7R expression in T(-)B(+)NK(+) severe combined immunodeficiency. *Nat. Genet.* **20**(4), 394-397 (1998).
35. M. Noguchi, H. Yi, H. M. Rosenblatt, A. H. Filipovich, S. Adelstein, W. S. Modi, O. W. McBride, and W. J. Leonard, Interleukin-2 receptor gamma chain mutation results in X-linked severe combined immunodeficiency in humans. *Cell* **73**(5141), 147-157 (1993).
36. L. A. Napolitano, R. M. Grant, S. G. Deeks, D. Schmidt, S. C. De Rosa, L. A. Herzenberg, B. G. Herndier, J. Andersson, and J. M. McCune, Increased production of IL-7 accompanies HIV-1-mediated T-cell depletion: implications for T-cell homeostasis. *Nat. Med.* **7**(1), 73-79 (2001).
37. T. J. Fry, B. L. Christensen, K. L. Komschlies, R. E. Gress, and C. L. Mackall, Interleukin-7 restores immunity in athymic T-cell-depleted hosts. *Blood* **97**(6), 1525-1533 (2001).
38. B. Blom, P. Res, E. Noteboom, K. Weijer, and H. Spits, Prethymic CD34+ progenitors capable of developing into T cells are not committed to the T cell lineage. *J. Immunol.* **158**(8), 3571-3577 (1997).
39. H. Spits, B. Blom, A. C. Jaleco, K. Weijer, M. C. Verschuren, J. J. van Dongen, M. H. Heemskerk, and P. C. Res, Early stages in the development of human T, natural killer and thymic dendritic cells. *Immunol. Rev.* **165**, 75-86 (1998).
40. C. Marquez, C. Trigueros, E. Fernandez, and M. L. Toribio, The development of T and non-T cell lineages from CD34+ human thymic precursors can be traced by the differential expression of CD44. *J. Exp. Med.* **181**(2), 475-83 (1995).
41. P. Res, E. Martinez-Caceres, A. Cristina Jaleco, F. Staal, E. Noteboom, K. Weijer, and H. Spits, CD34+CD38dim cells in the human thymus can differentiate into T, natural killer, and dendritic cells but are distinct from pluripotent stem cells. *Blood* **87**(12), 5196-206 (1996).
42. A. H. Dalloul, C. Patry, J. Salamero, B. Canque, F. Grassi, and C. Schmitt, Functional and phenotypic analysis of thymic CD34+CD1a- progenitor- derived dendritic cells: predominance of CD1a+ differentiation pathway. *J. Immunol.* **162**(10), 5821-5828 (1999).
43. M. C. Verschuren, B. Blom, A. J. Bogers, H. Spits, and J. J. van Dongen, PJA-BP expression and TCR delta deletion during human T cell differentiation. Int. Immunol. **10**(12), 1873-1880 (1998).
44. H, Petrie, F. Livak, D. Burtrum, and S. Mazel, T cell receptor gene reconbination patterns and mechanisms: cell death, rescue, and T cell production. *J Exp Med.* **182**(1), 121-127 (1995).
45. L. Wu, R. Scollay, M. Egerton, M. Pearse, G. J. Spangrude, and K. Shortman, CD4 expressed on earliest T-lineage precursor cells in the adult murine thymus. *Nature* **349**(6304), 71-74 (1991).
46. R. Murray, T. Suda, N. Wrighton, F. Lee, and A. Zlotnik, IL-7 is a growth and maintenance factor for mature and immature thymocyte subsets. Int. Immunol. **1**(5), 526-531 (1989).
47. T. A. Moore, and A. Zlotnik, T-cell lineage commitment and cytokine responses of thymic progenitors. *Blood* **86**(5), 1850-1860 (1995).
48. U. Von Freeden-jeffry, P. Vieira, L. A. Lucian, T. McNeil, S. Burdach, and R. Murry, Lymphopenia in interleukin (IL)-7 gene-depleted mice identifies IL-7 as a nonredundant cytokine. *J. Exp. Med.* **181**, 1519-1526 (1995).
49. J. Gratiot-Deans, R. Merino, G. Nunez, and L. A. Turka, Bcl-2 expression during T-cell development: Early loss and late return occur at specific stages of commitment to differentiation and survival. *Proc. Natl. Acad. Sci. USA.* **91**, 10685-10689 (1994).
50. R. Aspinall, and D. Andrew, Thymic atrophy in the mouse is a soluble problem of the thymic environment. *Vaccine* **18**(16), 1629-1637 (2000).
51. A. Abdul-Hai, R. Or, S. Slavin, G. Friedman, L. Weiss, D Matsa. and A. Ben-Yehuda, Stimulation of immune reconstitution by interleukin-7 after syngeneic bone marrow transplantation in mice. *Exp. Hematol.* **24**(12), 1416-1422 (1996).
52. J. T. Tan, E. Dudl, E. LeRoy, R. Murray, J. Sprent, K. I. Weinberg, and C. D. Surh, IL-7 is critical for homeostatic proliferation and survival of naive T cells. Proc Natl Acad Sci U S A **98**(15), 8732-7 (2001).
53. B. Blom, M. C. Verschuren, M. H. Heemskerk, A. Q. Bakker, E. J. van Gastel-Mol, I. L. Wolvers-Tettero, J. J. van Dongen, and H. Spits, TCR gene rearrangements and expression of the pre-T cell receptor complex during human T-cell differentiation. *Blood* **93**(9), 3033-3043 (1999).
54. R. D. Hockett, J. P. de Villartay, K. Pollock, D. G. Poplack, D. I. Cohen, and S. J. Korsmeyer, Human T-cell antigen receptor (TCR) delta-chain locus and elements responsible for its deletion are within the TCR alpha-chain locus. *Proc Natl. Acad. Sci. U. S. A.* **85**(24), 9694-9698 (1988).
55. J. Plum, M. De Smedt, G. Leclercq, B. Verhasselt, and B. Vandekerckhove, Interleukin-7 is a critical growth factor in early human T-cell development. *Blood.* **88**(11), 4239-45 (1996).
56. F. Offner, and J. Plum, The role of interleukin-7 in early T-cell development. *Leuk Lymphoma* **30**(1-2), 87-99 (1998).

57. J. Shutter, J. A. Cain, S. Ledbetter, M. D. Rogers, and R. D. Hockett, A d T-cell receptor deleting element transgenic reporter construct is rearranged in $\alpha\beta$ but not $\gamma\delta$ T-cell lineages. *Mol. Cell. Biol.* **15**, 7022-7031 (1995).
58. R. de Chasseval, and J. P. de Villartay, Functional characterization of the promoter for the human germline T cell receptor J alpha (TEA) transcript. *Eur J Immunol* **23**(6), 1294-1298 (1993).
59. B. P. Sleckman, C. G. Bardon, R. Ferrini, L. Davidson, and F. W. Alt, Function of the TCR alpha enhancer in alphabeta and gammadelta T cells. *Immunity* **7**(4), 505-515 (1997).
60. W. J. Murphy, V. Kumar, and M. Bennett, Rejection of bone marrow allografts by mice with severe combined immune deficiency (SCID). Evidence that natural killer cells can mediate the specificity of marrow graft rejection. *J. Exp. Med.* **165**(4), 1212-1217 (1987).
61. L. D. Shultz, P. A. Schweitzer, S. W. Christianson, B. Gott, I. B. Schweitzer, B. Tennent, S. McKenna, L. Mobraaten, T. V. Rajan, and D. L. Greiner, Multiple defects in innate and adaptive immunologic function in NOD/LtSz-scid mice. *J. Immunol.* **154**(1), 180-191 (1995).
62. R. Dadmarz, D. C. Bockstoce, and S. H. Golub, Interleukin-7 selectively enhances natural kill cytotoxicity mediated by the CD56bright natural killer subpopulation. *Lymphokine Cytokine Res.* **13**(6), 349-357 (1994).
63. E. K. Wong, C. Eaves, and H. G. Klingemann, Comparison of natural killer activity of human bone marrow and blood cells in cultures containing IL-2, IL-7 and IL-12. *Bone Marrow Transplant.* **18**(1), 63-71 (1996).
64. E. Bolotin, M. Smogorzewska, S. Smith, M. Widmer, and K. Weinberg, Enhancement of thymopoiesis after bone marrow transplant by in vivo interleukin-7. *Blood* **88**(5), 1887-1894 (1996).
65. E. Bolotin, G. Annett, R. Parkman, and K. Weinberg, Serum levels of IL-7 in bone marrow transplant recipients: relationship to clinical characteristics and lymphocyte count. *Bone Marrow Transplant.* **23**(8), 783-788 (1999).
66. G. D. Sempowski, L. P.Hale, J. S. Sundy, J. M. Massey, R. A. Koup, D. C. Douek, D. D. Patel, and B. F. Haynes, Leukemia inhibitory factor, oncostatin M, IL-6, and stem cell factor mRNA expression in human thymus increases with age and is associated with thymic atrophy. *J. Immunol.* **164**(4), 2180-2187 (2000).

NEONATES SUPPORT "HOMEOSTATIC" PROLIFERATION

Booki Min, Gregory D. Sempowski, and William E. Paul*

1. INTRODUCTION

The number of lymphocytes in the periphery is maintained at a constant level throughout life. Positive or negative disturbances of such balance by either antigenic challenge or lymphoablative therapies lead to activation of homeostatic regulatory processes including massive activation-induced cell death and homeostatic proliferation of peripheral T cells, respectively (Marrack et al., 2000). Experimentally, it has been well demonstrated that lymphopenic environments generated by irradiation or genetic manipulation drives the proliferation of transferred na ve T cells, which rarely divide in their normal environment (Goldrath and Bevan, 1999). It has been argued that this process is fundamentally different from antigen-driven proliferation, implying that different mechanisms are involved (Clarke and Rudensky, 2000). Both the physiologic significance and mechanisms of homeostatic proliferation are poorly understood.

Recent studies suggest that the recognition of self-MHC molecules loaded with the self-peptides involved in positive thymic selection plays a crucial role in initiating homeostatic proliferation (Ernst et al., 1999; Ge et al., 2001). Interleukin-7 has been recently demonstrated to be a critical cytokine for homeostatic regulation of CD4 and CD8 T cells (Schluns et al., 2000; Tan et al., 2001).

Studies reported thus far utilize lymphopenic recipients generated by irradiation or genetic manipulation. Since these are artificial environments, the physiologic significance of homeostatic proliferation remains to be examined. Indeed, it has been shown that the peptides other than selecting peptides in the thymus are responsible for irradiation-induced homeostatic proliferation (Bender et al., 1999).

T cells first appear in the periphery at the end of gestation. The secondary lymphoid organ of the early postnatal mouse contains relatively few lymphocytes (Garcia et al.,

* Booki Min and William E. Paul, Laboratory of Immunology, National Institute of Allergy and Infectious Diseases, National Institutes of Health, Bethesda, MD 20892-1892. Gregory D. Sempowski, Department of Medicine, Duke University, Durham, NC 27710.

2000). Therefore, we hypothesized that early cohorts of thymic emigrants may sense a lymphopenic (empty) environment and undergo homoeostatic proliferate. Indeed, this may well be the evolutionary pressure that led to the appearance of this mode of regulating numbers of CD4 T cells.

2. PROLIFERATION AND ACQUISITION OF MEMORY PHENOTYPE AFTER TRANSFER OF CD4 T CELLS INTO NEONATES

We used an adoptive transfer model system to study this problem. Lymph node CD4 T cells from B6.Ly5.1 adult mice were purified by negative selection using a MACS or by positive selection by cell sorting and labeled with CFSE. One million CD4 T cells were transferred into 1 day-old B6 (Ly5.2) mice. CFSE profiles and the phenotype of transferred cells were analyzed 16-18 days after transfer.

Of the surviving transferred cells, ~20% had undergone >7 cell divisions whereas very few cells had divided upon similar transfer into adult recipients (Table 1). Cells that had divided >7 times had acquired the phenotype of memory cells. They were CD44high, could enter the liver, and some could secrete IFNγ upon in vitro stimulation. To exclude the possibility that the proliferating cells derived from memory cells in the transferred cell population, we prepared purified CD44dull CD4 T cells; these cells behaved similarly upon transfer to neonatal recipients. Similarly, purified CD4 single positive thymocytes from 2 week old donors also underwent extensive proliferation upon transfer to neonates, indicating that immature cells can respond to the neonatal environment by proliferating.

Table 1. CD4 T cell proliferation in neonates[a]

Recipients	>7 divisions
Neonates	**18%**
Adults	**<5%**

[a] CFSE labeled CD4 T cells (Ly5.1) were transferred into 1 day old or adult mice. CFSE profiles of transferred cells were analyzed from the lymph nodes after 16-18 days of transfer.

3. REGULATION OF T CELL PROLIFERATION IN NEONATES

The extent of T cell proliferation was dependent on the space available within neonates. This conclusion was drawn from an experiment in which CFSE-labeled T cells were transferred into recipient of increasing ages. When T cells were transferred into 4 day old mice, the proportion of cells that had undergone >7 divisions was significantly decreased (Table 2). This decrease was more prominent as older recipients (7 or 15 day old) were used. Indeed, 15-day old recipients resembled adult recipients. It has been shown that thymic emigrants can control peripheral T cell homeostasis (Berzins et al., 1999). Thymic outputs were estimated as ~1-2 x 10^6 cells/day during early postnatal life

(Modigliani et al., 1994). Therefore, the space that early thymic emigrants see will be progressively filled by continued thymic output leading to a eventual cessation of early T cell proliferation. This concept is further supported by other experiments in which the transfer of additional CD4 T cells inhibited the proliferation of labeled CD4 T cells (Ernst et al., 1999; this paper). Recent work indicates that such bystander inhibition was mainly mediated by na ve rather than memory T cells (Dummer et al., 2001). It remains to be determined whether na ve T cells are responsible for cotransfer-mediated inhibition within neonates.

Table 2. CD4 T cell proliferation in recipients of different age[a]

Age of recipients	>7 divisions
1-day old	**18%**
4-day old	**8%**
7-day old	**5%**
15-day old	**<3%**

[a] CFSE labeled CD4 T cells were transferred into recipients of different ages. CFSE profiles were analyzed after 17-18 days of transfer.

The role of thymic output was directly assessed by neonatal thymectomy experiments. CFSE labeled T cells were transferred on day 1, and neonatal or sham thymectomy was performed on day 3. The CFSE profile of transferred CD4 T cells 17 days after transfer was striking in the neonatally thymectomized mice. Approximately 70% of the transferred cells have undergone >7 divisions, whereas ~20% of cells had undergone >7 divisions in sham-thymectomized mice (Table 3). This result provides evidence that thymic output controls the proliferation of T cells in the periphery.

Table 3. Role of thymic output in T cell proliferation in neonates[a]

Thymectomy	>7 divisions
Day 3	**68%**
Sham	**21%**

[a] Thymi were removed from 3-day old recipients that had received CFSE labeled CD4 T cells on the day of birth. CFSE profiles of transferred cells were analyzed after 16-18 days of transfer.

4. BEHAVIOR OF ENDOGENOUS CD4 T CELLS

The experiments described above analyze the division history and activation status of transferred adult and immature CD4 T cells. Our hypothesis is based on the assumption that endogenous cells behave in the same way as the transferred cells. In particular, the behavior of the cohort of endogenous cells emerging from the thymus on the day that Ly5.1 cells are transferred should be the same as that of the transferred cells.

This would predict that endogenous CD44bright cells should display a division history similar to that of the transferred cells. T cell receptor excision circles (TRECs) are stable DNA fragments deleted during TCR rearrangement in the thymus. These episomal circles are not replicated during cell division and thus are progressively diluted among a population of dividing cells. Thus, measuring TREC content in CD4 T cells can be used to determine the cell division history of a population (Chen et al., 2001; Hazenberg et al., 2001). Endogenous CD44dull and CD44high CD4 T cells were purified from 7-day old mice. TREC content was measured from those two populations. CD44high memory cells from 7-day old mice had no detectable TRECs in the cells, indicating that endogenous memory cells had undergone multiple cell divisions just as had the transferred cells. This finding provides evidence that multiple divisions by a subset of CD4 T cells occurs physiologically during early postnatal life.

5. CONCLUSIONS

Lymphopenia-driven proliferation of naïve CD4 T cells occurs in neonatal mice and provides these animals with a population of cells with memory phenotype. These "memory" cells are CD44bright, are found in lymphoid and non-lymphoid organs, and have an increased percent of IFNγ-producers. How do these cells emerge? Are they driven by exogenous antigen or by recognition of self-peptide/MHC complexes? Alternatively, are they a population of stochastically activated cells responding to cytokines, such as IL-7? The latter would predict a repertoire that is very broad while the former would lead to a narrow/skewed repertoire. We are currently examining the repertoire of the cells and also testing the role of TCR-mediated recognition in the expansion of the cells.

One may ask what is the role of this "memory population"? The forces driving their expansion will play an important part in formulating a hypothesis. They may be part of a life-long homeostatic mechanism that has no specific significance in the neonate. Alternatively, the relatively rapid generation of memory cells in the neonate may provide these mice with an "adaptive" T cell population that can respond promptly to attack by pathogenic microorganisms. Depending on the breadth of the repertoire of these cells, the existence of the memory population may be of particular importance in endowing the neonate with a population of cells capable of promptly responding to a broad array of pathogens. These memory cells may provide neonates with a more fully developed immune system than they could generate simply on the basis of their antigenic experience.

REFERENCES

Bender, J., Mitchell, T., Kappler, J., and Marrack, P., 1999, CD4 T cell division in irradiated mice requires peptides distinct from those responsible for thymic selection, *J. Exp. Med.* **190**:367.

Berzins, S. P., Godfrey, D. I., Miller, J. F. A. P., and Boyd, R. L., 1999, A central role for thymic emigrants in peripheral T cell homeostasis, *Proc. Natl. Acad. Sci. USA* **96**:9787.

Chen, B. J., Cui, X., Sempowski, G. D., Gooding, M. E., Liu, C., Haynes, B. F., and Chao, N. J., 2001, A comparison of murine T-cell-depleted adult bone marrow and full-term fetal blood cells in hematopoietic engraftment and immune reconstitution, *Blood* **99**:364.

Clarke, S. R. M. and Rudensky, A. Y., 2000, Survival and homeostatic proliferation of na ve peripheral CD4 T cells in the absence of self peptide:MHC complexes, *J. Immunol.* **165**:2458.

Dummer, W., Ernst, B., LeRoy, E., Lee, D-S., and Surh, C. D., 2001, Autologous regulation of na ve T cell homeostasis within the T cell compartment, *J. Immunol.* **166**:2460.

Ernst, B., Lee, D-S., Chang, J. M., Sprent, J., and Surh, C. D., 1999, The peptide ligands mediating positive selection in the thymus control T cell survival and homeostatic proliferation in the periphery, *Immunity* **11**:173.

Garcia, A. M., Fadel, S. A., Cao, S., and Sarzotti, M., 2000, T cell immunity in neonates, *Immunol. Res.* **22**:177.

Ge, Q., Rao, V. P., Cho, B. K., Eisen, H. N., and Chen, J., 2001, Dependence of lymphopenia-induced T cell proliferation on the abundance of peptide/MHC epitopes and strength of their interaction with T cell receptors, *Proc. Natl. Acad. Sci. USA* **9**:1728.

Goldrath, A. W. and Bevan, M. J., 1999, Selecting and maintaining a diverse T-cell repertoire, *Nature* **402**:255.

Hazenberg, M. D., Verschuren, M. C. M., Hamann, D., Miedema, F., and van Dongen, J. J. M., 2001, T cell receptor excision circles as markers for recent thymic emigrants: basis aspects, technical approach, and guidelines for interpretation, *J. Mol. Med.* **79**:631.

Marrack, P., Bender, J., Hildeman, D., Jordan, M., Mitchell, T., Murakami, M., Sakamoto, A., Schaefer, B. C., Swanson, B., and Kappler, J., 2000, Homeostasis of ab TCR+ T cells, *Nat. Immunol.* **1**:107.

Modigliani, Y., Continho, G., Burlen-Defranoux, O., Continho, A., and Bandeira, A., 1994, Differential distribution of thymic outputs and peripheral expansion in the development of peripheral T cell pools, *Eur. J. Immunol.* **24**:1223.

Schluns, K. S., Kieper, W. C., Jameson, S. C., and Lefrancois, L., 2000, Interleukin-7 mediates the homeostasis of na ve and memory CD8 T cells in vivo, *Nat. Immunol.* **1**:426.

Tan, J. T., Dudl, E., LeRoy, E., Murray, R., Sprent, J., Weinberg, K. I., and Surh, C. D., 2001, IL-7 is critical for homeostatic proliferation and survival of na ve T cells, *Proc. Natl. Acad. Sci. USA* **98**:8732.

SELF-RECOGNITION AND THE REGULATION OF CD4+ T CELL SURVIVAL

Ronald N. Germain, Irena Stefanova, and Jeffrey Dorfman*

1. SUMMARY

CD4+ T cells differentiate in the thymus from committed precursors to mature naive cells ready for peripheral circulation. Successful maturation depends on adequate but not excessive signaling upon T cell receptor (TCR) engagement of self-peptide/MHC class II molecule ligands present in the thymic environment. Persistent TCR signaling throughout development from the CD4+CD8+ to the CD4+ state is required for completion of the developmental process. Recent work has suggested that a continuation of this signaling is essential for sustained survival of CD4+ T cells once they leave the thymus but our studies suggest otherwise. Although we found clear evidence for active TCR signaling involving recognition of self-ligands in peripheral lymphoid tissues, we did not see a substantial effect of loss of such signaling on the life-time of naive CD4+ T cells. Based on a careful review of the literature, we conclude that essentially all previous claims that MHC class II recognition plays a significant role in the survival of CD4+ T cells can be re-interpreted as an effect of self-recognition on proliferation in lymphopenic environments, maintaining population numbers without a marked effect on individual cell viability. We propose a possible explanation for why, in contrast, the viability of naive CD8+ T cells appears to show such self-MHC dependence and suggest that a primary function of self-recognition by T cells may be to enhance responses to foreign antigen.

* Lymphocyte Biology Section, Laboratory of Immunology, National Institute of Allergy and Infectious Diseases, National Institutes of Health, Bethesda, MD 20892-1892

2. INTRODUCTION

CD4+ and CD8+ T lymphocytes bearing αβ receptors comprise the bulk of the peripheral lymphoid pool re-circulating in the blood and through secondary lymphoid organs such as lymph nodes and spleen. Naive CD4+ and CD8+ T cells arise in the thymus from precursors with the potential to enter either lineage. Shortly after generation of the clonotypic antigen receptor (TCR) by somatic gene rearrangement, these cells test the specificity of this receptor on the array of self-peptide/MHC ligands present on the thymic epithelium[1]. Inadequate intracellular signaling following such self-ligand scanning results in "death by neglect" over the next few days with the cells remaining in the precursor CD4+CD8+ (DP) state. Those lymphocytes receiving excessive signals die, either at this early stage, or more frequently, as they mature and contact medullary presenting cells. A small fraction of all newly generated DP receives just the right level of signaling to mature successfully (positive selection). Those with TCR recognizing self-MHC class II ligands most often develop into the CD4+ SP cells, while those with TCR binding self-MHC class I ligands typically become CD8+ SP.

The process of positive selection thus produces a cohort of peripheral lymphocytes with TCR selected to be capable of functionally significant self-recognition. The question is whether such self-recognition leads to active signaling in mature T cells within secondary lymphoid tissues and if so, what influence this has on the physiology of the cells. It is not necessarily predictable that mature T cell recognition of self-ligands will lead to biochemical changes in that lymphocyte. We have shown that as a thymocyte matures, its capacity for signal generation in response to low quality ligands of the type involved in positive selection decreases markedly because of developmentally regulated changes in how the TCR recognition subunits are coupled to the intracellular signal generation apparatus[2]. Such changes could extinguish the capacity of the TCR to initiate the response cascade in naïve peripheral T cells when confronted with self-ligands. Here we summarize our recent findings concerning the question of peripheral lymphocyte signaling upon self-recognition and the relationship between self-recognition and naive lymphocyte survival.

3. ACTIVE SELF-INDUCED SIGNALING

van Oers et al. first demonstrated that freshly isolated T cells showed partial tyrosine phosphorylation of the TCR-associated ζ chain[3]. Subsequent studies examined whether this was a consequence of autonomous kinase activity in the lymphocyte and unrelated to TCR engagement or if it represented active signaling induced by self-recognition. Initial studies in thymocytes suggested that most if not all ζ phosphorylation in these cells arose from spontaneous kinase function[4], although the data did not rule out a contribution from TCR ligand-driven signaling.

To address this issue more directly with peripheral T cells, we examined the effect of interfering with TCR binding to MHC class II molecules in vivo, using several experimental approaches[5]. Injection into naive mice of a monoclonal antibody known to block TCR/MHC class II interactions resulted in the loss of more than 70% of the normal level of TCR ζ chain phosphorylation in CD4+ T cells within 12-36 hrs. In contrast, the level of ζ chain phosphorylation in CD8+ T cells, whose selection is mediated by MHC class I molecules, was unchanged by this experimental manipulation, indicating that the effect on antibody injection was related to the MHC recognition specificity of the T cells.

In another series of experiments, naive T cells were recovered from normal mice and transferred to hosts that lacked wither MHC class II expression ($A\beta^{-/-}$), MHC class I expression ($\beta 2m^{-/-}$), MHC class II peptide diversity ($H\text{-}2M^{-/-}$), or MHC class I self peptide loading ($TAP^{-/-}$). The donor cells were dye labeled or marked by an Ly 5 congenic difference. After varying times ranging from 36 hrs. to two weeks, the transferred T cells were recovered, naive donor cells isolated, and the state of TCR ζ phosphorylation examined. Transfer of cells into mice lacking any MHC class II expression or with minimal peptide diversity resulted in loss of the majority of TCR ζ phosphorylation in CD4+ but not CD8+ T cells. Conversely, in mice deficient in MHC class I expression or peptide content, CD8+ but not CD4+ T cells showed a substantial decrease in the extent of TCR ζ phosphorylation. These findings were consistent with the antibody injection results and provided evidence for active signaling upon TCR recognition of self-ligands that was both peptide-specific and matched the MHC class involved in thymic positive selection of the T cell.

4. SURVIVAL STUDIES

Given this evidence for active TCR signaling by naive mature CD4+ and CD8+ T cells in response to self-ligands, we then explored the question of whether such signals contributed to the viability / lifetime of these cells. A number of groups had already reported that self-recognition was critical to maintaining high numbers of T cells in mice and had concluded this was because such self-recognition was essential to maintaining naive lymphocyte viability[6-15]. Based on the results that had already appeared in the literature, we imagined that in our transfer model, we would find that in MHC class II-deficient hosts, CD4+ T cells would disappear rapidly compared to T cells transferred to normal MHC class II-expressing recipients. However, this was not the case. The number of naive T cells decreased at indistinguishable rates in the two hosts, irrespective of the capacity of the transferred CD4+ T cells to engage and signal in response to MHC class II ligands[5], a result also obtained by others groups[16, 17].

In the course of these studies, we noted that even in putatively MHC class II-deficient mice, both CD4+ and CD8+ T cells underwent several rounds of cell division. Others had shown that proliferation of naive cells in a foreign antigen-free environment typically occurred when T cell numbers were subnormal (the mice were lymphopenic) and the animal expressed the peptide/MHC ligands of the environment in which the T cells differentiated[13, 14, 18-23]. We therefore checked whether there might be cryptic MHC class II expression in our recipient mice by injecting the blocking anti-MHC class II monoclonal antibody used previously to interfere with ζ chain phosphorylation. This reagent prevented division of CD4+ but not CD8+ transferred T cells in the recipients, confirming that the proliferation was based on unexpected MHC class II expression in the hosts, possibly from differentiation of MHC class II-negative precursors transferred with the T cells. But even under these conditions, the rate of loss of CD4+ T cells matched that of cells transferred into normal MHC class II expressing mice not treated with antibody. In all of these transfer experiments, TCR ζ phosphorylation was lost with 18-36 hrs., even though the CD4+ T cell half-life was greater than 3 weeks. Thus, no evidence for an effect of self-recognition on CD4+ T cell survival, as opposed to proliferation, could be seen.

5. RECONCILING THESE DATA WITH THE LITERATURE

How can these findings by reconciled with the substantial literature reaching the opposite conclusion[24]? A key clue came from the evidence that MHC class II expression was critical to CD4+ T cell division in the transfer model, even when it did not seem to matter for cell viability. The requirement for self-MHC recognition in what has been termed homeostatic proliferation of lymphocytes in lymphopenic hosts was in full agreement with these data[13, 14, 18-23]. We therefore carefully reviewed the published work to determine if the investigators had determined whether maintenance of cell number in MHC-replete hosts versus cell loss in MHC-deficient hosts was due to prolonged cell survival or to this type of self-MHC dependent proliferation. Our literature review found that in nearly all cases, one could better account for the results by postulating an MHC molecule-dependent difference in cell division rather than an effect of cell survival[25]. In some cases, there was direct evidence for extensive cell division in the MHC-expressing hosts as compared to the MHC-deficient recipients[10]. In others, there was no direct evidence for proliferation, but the experimental model involved lymphopenic recipients in which such proliferation is known to occur and the cell recoveries were best explained by such cell expansion in the MHC-expressing hosts[9, 11].

Two groups have used a distinct approach to examine the need for TCR signaling, presumably due to self-MHC recognition, in naive T cell survival. One group employed a Tet regulatable transgene to control TCR expression[26]. These investigators found that CD4+ T cell number declined with a $t_{1/2}$ of ~3 weeks once TCR expression was eliminated by drug administration. They concluded that these data agreed with their earlier results in MHC-deficient animals and supported the view that self-recognition maintains T cell viability. However, these experiments lack a control, namely a measure of the lifespan of T cells in an MHC-positive or TCR-positive environment. In fact, this 3-week half-life is what we and others[16, 17] observe for CD4+ naive cells whether MHC class II is present or not. Thus, these data do not argue for a faster death of T cells once TCR signaling is eliminated and the observations could be explained again by the maintenance of overall cells number by a balance of cell death and proliferation when the TCR is still present. No measurements of proliferation were made in this TCR-extinction model, so no direct evidence exists for or against this alternative view.

Another group used cre-lox recombination to eliminate TCR expression by already mature T cells[27]. Because the cre-mediated deletion only affected some and not other cells, this model had an internal control for lifespan, namely the cells maintaining TCR expression. These investigators also did careful studies of cell division using BrdU incorporation, to control for proliferation within in the TCR-expressing cohort in these increasingly lymphopenic mice. The data indicated that in the 3-4 weeks encompassed by our own studies, there was at best a very modest difference in the survival of TCR-expressing vs. TCR-low/negative CD4+ T cells, although rapid loss of TCR low/negative CD8+ T cells was observed ($t_{1/2}$ of 16 days). More prolonged observation indicated that the number of TCR-expressing CD4+ T cells declined with a $t_{1/2}$ of 78 days as compared with a $t_{1/2}$ of 46 days for TCR-low/negative cells. These findings led the investigators to conclude that ligand-dependent signaling influences CD4+ T cell lifespan, although cell death in the absence of self-recognition was not acute.

These are clearly the most compete set of data on this issue. However, as careful as the authors were in trying to distinguish effects on proliferation from those on viability, the issue remains cloudy. The half-life of CD4+ T cells unable to receive TCR signals resulting from self-recognition as calculated in this study is actually substantially greater than that seen in other work involving deprivation of MHC contact[5, 8, 15-17] or in the one other set of experiments in which TCR expression was extinguished[26]. The predicted half-life with such signals is only about 60% greater. Memory T cells divide on average every 30 days, with about 3% of cells dividing on any given day[28]. This results a potential doubling of the population in the absence of cell death every month. Because the number of memory cells stays constant, this means that the half-life of memory T cells is about 30 days, similar to naive cells in the absence of self-recognition, although the memory cell data were obtained MHC-expressing animals. This raises a question - is it logical that memory cells, typically considered to be long-lived, would actually have a shorter lifespan than naive cells when both can signal via the TCR? An alternative possibility is that in the cre-lox deletion experiment, the classification of cells as naive based on CD44 expression included a very small fraction of cells undergoing cell division. This marginal level of proliferation could fully explain the difference in apparent cell lifetime of TCR-expressing versus TCR low/negative cells in this study. It is thus still unclear if the loss of self-recognition does affect the lifespan of naive CD4+ T cells, reducing it by 50% at maximum or if this has little if any effect on viability but affects population maintenance through an influence on cell division. In addition, there is some evidence for crosstalk between cytokine receptors and the TCR[29]. It is therefore also possible that the modest differences seen in the cre-lox model reflect an impairment of cytokine receptor survival signaling that normally requires TCR surface expression, rather than representing evidence for an active role of TCR signaling in maintenance of CD4+ T cell viability.

What about CD8+ T cells? A very rapid loss (7 days $t_{1/2}$) of HY naive cells in the absence of the selecting MHC class I molecule has been reported[7]. This brief lifespan without suitable TCR signaling is about half of what is seen using the cre-lox TCR deletion model[27]. As a whole, the evidence for rapid death of CD8+ T cells deprived of MHC class I-dependent recognition and signaling is much stronger than the evidence for such an effect involving CD4+ T cells. Even though transfer experiments in the CD8+ case are made complicated by the greater potential for rejection as compared to the CD4+ T cell situation[30, 31] the data seem to clearly support the concept that self-signaling does markedly influence CD8+ T cell viability.

At present, there is no real explanation for this difference between the two T cell lineages. We have suggested one possibility[32], although it is yet to be supported by explicit evidence. Results from several laboratories show that selection of precursor DP thymocytes into the CD8+ lineage involves weaker / shorter signaling through the TCR than does selection into the CD4+ lineage[33-37]. We postulate that CD4+ T cells get a sufficient "kick" from the more vigorous signal that leads them into this lineage to eventually escape the need for additional TCR signals to sustain viability in the periphery. However, some cells just making it into the CD4 lineage might retain some necessity for such signaling, possibly contributing to the modest differences seen with TCR-expressing and TCR-low/negative cells in the cre-lox TCR deletion model. In contrast, cells entering the CD8 linage would get a less robust push, leaving them in need of TCR signals to sustain viability even after they exit the thymus. This would be most evident in the cells with the poorest match of TCR to selecting ligand, as is presumably the situation for the HY TCR[38, 39]. This model would account for all the data available at present and especially the CD4 / CD8 difference. It leaves unanswered why CD4+ and CD8+ T cells should differ in this fundamental manner, though one might speculate that CD8+ cells are inherently more dangerous to the host and need greater peripheral regulation.

6. CONCLUSION

Our data and those available from the literature suggest that self-recognition plays only a modest role in enhancing the viability of naive CD4+ T cells, though it may affect CD8+ T cells to a substantial degree. Most reports reaching a different conclusion seem to confound proliferation with control of individual cell survival. The argument that what matters is maintenance of population numbers, and hence, the combination of lifespan and proliferation, doesn't fit the available evidence. Cells that proliferate in lymphopenic hosts change their phenotype, becoming CD44 high and acquiring the capacity for enhanced effector cytokine secretion[19, 21, 40]. This can hardly be called homeostasis. By definition, this process involves maintenance of the pre-existing situation, which in this case is a population of naive phenotype cells without memory or effector properties.

This leaves the question of the significance of the active signaling by self-MHC class II recognition we and others have documented in CD4+ naive T cells. Rather than focusing on survival, we have instead considered the possibility that self-recognition contributes in a positive manner to foreign antigen sensitivity of naive T cells. A variety of studies now suggest that this presumption is correct and that TCR engagement of self augments the sensitivity of naive T cells when confronted with foreign ligands. This concept provides a satisfying alternative to the survival model and fits well with the fact that positive selection specifically ensures the creation of a TCR repertoire capable of effective self-recognition with an efficiency just below that suitable for overt T cell activation.

REFERENCES

1. E. Robey and B. J. Fowlkes, Selective events in T cell development, *Annu Rev Immunol* 12, 675-705 (1994).
2. B. Lucas, I. Stefanova, K. Yasutomo, N. Dautigny and R. N. Germain, Divergent changes in the sensitivity of maturing T cells to structurally related ligands underlies formation of a useful T cell repertoire, *Immunity* 10, 367-376 (1999).
3. N. S. van Oers, W. Tao, J. D. Watts, P. Johnson, R. Aebersold and H. S. Teh, Constitutive tyrosine phosphorylation of the T-cell receptor (TCR) ζ subunit: Regulation of TCR-associated protein tyrosine kinase activity by TCR ζ, *Mol Cell Biol* 13, 5771-5780 (1993).
4. N. S. C. van Oers, N. Killeen and A. Weiss, ZAP-70 is constitutively associated with tyrosine-phosphorylated TCR ζ in murine thymocytes and lymph node T cells., *Immunity* 1, 675-685 (1994).
5. J. R. Dorfman, I. Stefanova, K. Yasutomo and R. N. Germain, Cd4+ T cell survival is not directly linked to self-MHC-induced TCR signaling, 1, 329-335 (2000).
6. S. Takeda, H. R. Rodewald, H. Arakawa, H. Bluethmann and T. Shimizu, MHC class II molecules are not required for survival of newly generated CD4+ T cells, but affect their long-term life span, *Immunity* 5, 217-228 (1996).
7. C. Tanchot, F. A. Lemonnier, B. Perarnau, A. A. Freitas and B. Rocha, Differential requirements for survival and proliferation of cd8 naive or memory T cells, *Science* 276, 2057-2062 (1997).
8. R. Rooke, C. Waltzinger, C. Benoist and D. Mathis, Targeted complementation of MHC class II deficiency by intrathymic delivery of recombinant adenoviruses, *Immunity* 7, 123-134 (1997).
9. T. Brocker, Survival of mature CD4 T lymphocytes is dependent on major histocompatibility complex class II-expressing dendritic cells, *J Exp Med* 186, 1223-1232 (1997).
10. J. Kirberg, A. Berns and H. von Boehmer, Peripheral T cell survival requires continual ligation of the T cell receptor to major histocompatibility complex-encoded molecules, *J Exp Med* 186, 1269-1275 (1997).
11. D. Nesic and S. Vukmanovic, MHC class I is required for peripheral accumulation of CD8+ thymic emigrants, *J Immunol* 160, 3705-3712 (1998).
12. M. A. Markiewicz, C. Girao, J. T. Opferman, J. Sun, Q. Hu, A. A. Agulnik, C. E. Bishop, C. B. Thompson and P. G. Ashton-Rickardt, Long-term T cell memory requires the surface expression of self- peptide/major histocompatibility complex molecules, *Proc Natl Acad Sci U S A* 95, 3065-3070 (1998).
13. B. Ernst, D. S. Lee, J. M. Chang, J. Sprent and C. D. Surh, The peptide ligands mediating positive selection in the thymus control T cell survival and homeostatic proliferation in the periphery, *Immunity* 11, 173-181 (1999).
14. C. Viret, F. S. Wong and C. A. Janeway, Jr., Designing and maintaining the mature TCR repertoire: The continuum of self-peptide:self-MHC complex recognition, *Immunity* 10, 559-568 (1999).
15. D. Witherden, N. van Oers, C. Waltzinger, A. Weiss, C. Benoist and D. Mathis, Tetracycline-controllable selection of CD4(+) T cells: Half-life and survival signals in the absence of major histocompatibility complex class II molecules, *J Exp Med* 191, 355-364 (2000).
16. S. R. Clarke and A. Y. Rudensky, Survival and homeostatic proliferation of naive peripheral CD4+ T cells in the absence of self peptide:MHC complexes, *J Immunol* 165, 2458-2464. (2000).

17. B. Seddon, G. Legname, P. Tomlinson and R. Zamoyska, Long-term survival but impaired homeostatic proliferation of naive T cells in the absence of p56(lck), *Science* 290, 127-131. (2000).
18. U. Beutner and H. R. MacDonald, TCR-MHC class II interaction is required for peripheral expansion of CD4 cells in a T cell-deficient host, *Int Immunol* 10, 305-310 (1998).
19. S. Oehen and K. Brduscha-Riem, Naive cytotoxic T lymphocytes spontaneously acquire effector function in lymphocytopenic recipients: A pitfall for T cell memory studies?, *Eur J Immunol* 29, 608-614 (1999).
20. A. W. Goldrath and M. J. Bevan, Low-affinity ligands for the TCR drive proliferation of mature CD8+ T cells in lymphopenic hosts, *Immunity* 11, 183-190 (1999).
21. W. C. Kieper and S. C. Jameson, Homeostatic expansion and phenotypic conversion of naive T cells in response to self peptide/MHC ligands, *Proc Natl Acad Sci U S A* 96, 13306-13311 (1999).
22. P. Muranski, B. Chmielowski and L. Ignatowicz, Mature CD4+ T cells perceive a positively selecting class II MHC/peptide complex in the periphery, *J Immunol* 164, 3087-3094 (2000).
23. Q. Ge, V. P. Rao, B. K. Cho, H. N. Eisen and J. Chen, Dependence of lymphopenia-induced T cell proliferation on the abundance of peptide/ MHC epitopes and strength of their interaction with T cell receptors, *Proc Natl Acad Sci U S A* 98, 1728-1733. (2001).
24. J. Kirberg, H. von Boehmer, T. Brocker, H. R. Rodewald and S. Takeda, Class II essential for CD4 survival, *Nat Immunol* 2, 136-137 (2001).
25. J. R. Dorfman, I. I. Stefanova, K. Yasutomo and R. N. Germain, Response to 'Class II essential for CD4 survival', 2, 136-137 (2001).
26. N. Labrecque, L. S. Whitfield, R. Obst, C. Waltzinger, C. Benoist and D. Mathis, How much TCR does a T cell need?, *Immunity* 15, 71-82 (2001).
27. B. Polic, D. Kunkel, A. Scheffold and K. Rajewsky, How αβ t cells deal with induced TCR α ablation, *Proc Natl Acad Sci U S A* 98, 8744-8749 (2001).
28. K. Murali-Krishna, L. L. Lau, S. Sambhara, F. Lemonnier, J. Altman and R. Ahmed, Persistence of memory CD8 T cells in MHC class I-deficient mice, *Science* 286, 1377-1381 (1999).
29. E. F. Petricoin, 3rd, S. Ito, B. L. Williams, S. Audet, L. F. Stancato, A. Gamero, K. Clouse, P. Grimley, A. Weiss, J. Beeler, D. S. Finbloom, E. W. Shores, R. Abraham and A. C. Larner, Antiproliferative action of interferon-α requires components of T- cell-receptor signalling, *Nature* 390, 629-632 (1997).
30. H. G. Ljunggren, R. Glas, J. K. Sandberg and K. Karre, Reactivity and specificity of CD8+ T cells in mice with defects in the MHC class I antigen-presenting pathway, *Immunol Rev* 151, 123-148 (1996).
31. H. G. Ljunggren, L. Van Kaer, P. G. Ashton-Rickardt, S. Tonegawa and H. L. Ploegh, Differential reactivity of residual CD8+ T lymphocytes in TAP1 and β2-microglobulin mutant mice, *Eur J Immunol* 25, 174-178 (1995).
32. J. R. Dorfman and R. N. Germain, Mhc-dependent survival of naïve T cells? A complicated answer to a simple question, *Microbes and Infection* in press, (2002).
33. A. Itano, P. Salmon, D. Kioussis, M. Tolaini, P. Corbella and E. Robey, The cytoplasmic domain of CD4 promotes the development of CD4 lineage T cells, *J Exp Med* 183, 731-741 (1996).
34. E. O. Matechak, N. Killeen, S. M. Hedrick and B. J. Fowlkes, MHC class II-specific T cells can develop in the CD8 lineage when CD4 is absent, *Immunity* 4, 337-347 (1996).
35. K. Yasutomo, C. Doyle, L. Miele, C. Fuchs and R. N. Germain, The duration of antigen receptor signalling determines CD4+ versus CD8+ T-cell lineage fate, *Nature* 404, 506-510 (2000).

36. M. A. Basson and R. Zamoyska, The CD4/CD8 lineage decision: Integration of signalling pathways, *Immunol Today* 21, 509-514. (2000).
37. G. Hernandez-Hoyos, S. J. Sohn, E. V. Rothenberg and J. Alberola-Ila, Lck activity controls CD4/CD8 T cell lineage commitment, *Immunity* 12, 313-322 (2000).
38. E. W. Shores, T. Tran, A. Grinberg, C. L. Sommers, H. Shen and P. E. Love, Role of the multiple T cell receptor (TCR)-ζ chain signaling motifs in selection of the T cell repertoire, *J Exp Med* 185, 893-900 (1997).
39. P. E. Love, J. Lee and E. W. Shores, Critical relationship between TCR signaling potential and TCR affinity during thymocyte selection, *J Immunol* 165, 3080-3087 (2000).
40. B. K. Cho, V. P. Rao, Q. Ge, H. N. Eisen and J. Chen, Homeostasis-stimulated proliferation drives naive T cells to differentiate directly into memory T cells, *J Exp Med* 192, 549-556. (2000).

TCR-INDEPENDENT PROLIFERATION AND DIFFERENTIATION OF HUMAN $CD4^+$ T CELL SUBSETS INDUCED BY CYTOKINES

Jens Geginat, Stefania Campagnaro, Federica Sallusto and Antonio Lanzavecchia*

*Institute for Research in Biomedicine, Via Vela 6, 6500, Bellinzona, Switzerland

ABSTRACT

Naïve and memory T cells can divide in an antigen-independent manner *in vivo* maintaining independently a constant pool size. While naïve T cells require TCR tickling by self-MHC for homeostatic proliferation in lymphopenic mice, memory cells do not but respond to cytokines. Human naïve and memory $CD4^+$ T cell subsets can be selectively expanded *in vitro* with different cytokine combinations. Responsiveness of T cells to homeostatic cytokines is associated with the differentiation state. Thus, while memory cells respond directly to IL-7 and IL-15, naïve T cells require costimulation by dendritic cell-derived cytokines, and selectively respond to IL-4. This differential cytokine responsiveness is associated with the expression and modulation of the relevant cytokine receptors. Cytokine-driven proliferation is independent of TCR-stimulation and shows distinct signal transduction requirements. While cytokine-expanded naïve T cells maintain a naïve phenotype, memory cells differentiate acquiring new effector functions and switching expression of chemokine receptors. Thus human naive and memory T cell pools can be maintained with homeostatic cytokines in the absence of TCR stimulation.

INTRODUCTION

Maintenance of Naïve and Memory T Cells by Self MHC and Cytokines

Naïve and memory T cells have different tasks, since they protect the body from new and previously encountered pathogens, respectively. Homeostasis must therefore ensure that the two pools are independently kept constant in size (Freitas and Rocha 2000; Goldrath and Bevan 1999; Marrack et al. 2000) The naïve T cell pool is continuously replenished by the thymus. Naïve T cells therefore normally do not divide in the periphery unless thymic output is insufficient to maintain the pool size. In contrast, resident memory cells must replicate continuously in order to avoid that they are substituted by cells with different antigenic specificities (Tough and Sprent 1994). While antigenic stimulation leads to several divisions in a few days, homeostatic proliferation under steady state conditions is slow (Tough, Sprent 1994). Lymphopenia triggers a more rapid antigen-independent proliferation of naïve T cells, which acquire some characteristics of memory cells (Cho et al. 2000; Goldrath et al. 2000; Murali-Krishna and Ahmed 2000; Oehen and Brduscha-Riem 1999)

Naive $CD4^+$ and $CD8^+$ T cell homeostasis requires self-MHC, but if TCR tickling is required for long-term survival is under debate (Bender et al. 1999; Brocker 1997; Dorfman et al. 2000; Kirberg et al. 1997; Seddon et al. 2000; Takeda et al. 1996) Memory T cells survive and replicate in the absence of MHC expression in the periphery (Murali-Krishna et al. 1999; Swain et al. 1999; Tanchot et al. 1997) but require cytokines (Zhang et al. 1998). In particular, cytokines that bind to receptors containing the common γ-chain (γc-cytokines), such as IL-2, -4, -7 and –15, are important for T cell homeostasis. Indeed, γc-chain deficient mice show impaired T cell survival and homeostasis (Lantz et al. 2000; Nakajima et al. 1997). γc-cytokines have many redundant effects, but specificity can be achieved through different sites of their production and the expression of the relevant receptor components by the responding cells. IL-15 that is constitutively produced by a variety of cells plays an essential role in CD8 T cell homeostasis (Kanegane and Tosato 1996; Lodolce et al. 1998; Zhang et al. 1998). IL-2, the major growth factor of activated T cells, induces antigen-independent bystander proliferation of human $CD4^+$ T cells in the presence of inflammatory cytokines (Unutmaz et al. 1994). IL-7, which is produced by stromal cells, enhances homeostatic proliferation of memory $CD8^+$ T cells, is required for self-MHC driven proliferation of naïve T cells (Schluns et al. 2000) and its increased expression is associated with enhanced T cell turnover in HIV patients (Napolitano et al. 2001).

Memory T Cell Subsets Defined by CCR7 Expression

Naïve and memory cells have distinct capacities to traffic in lymphoid and non-lymphoid tissues (Mackay et al. 1990). Memory T cell subsets can be distinguished by the expression of lymph node homing receptors such as CD62L and CCR7. The lymph node homing $CCR7^+$ "central" memory T cells (T_{CM}) lack immediate effector functions such as effector cytokine production and homing capacity to inflamed tissues. In contrast, $CCR7^-$ "effector" memory cells (T_{EM}) produce IFN-γ and IL-4 and express receptors for inflammatory chemokines such as CCR5 (Sallusto et al. 1999). Furthermore while antigen-stimulated T_{CM} can differentiate into T_{EM}, antigen-stimulated T_{EM} do not revert. These findings suggest that T_{CM} are at an intermediate stage of differentiation as compared to naïve and T_{EM}. Since T_{CM} and T_{EM} have different migratory capacities (Iezzi et al. 2001; Manjunath et al. 1999; Masopust et al. 2001; Reinhardt et al. 2001) they have access to different environments and thus possibly occupy different homeostatic niches.

RESULTS AND DISCUSSION

Requirements of Human T Cell Subsets for Cytokine-Dependent Proliferation

The TCR and cytokine receptors are coupled to different intracellular signalling pathways. TCR and cytokine-dependent proliferation can therefore be selectively blocked with specific inhibitors. In particular the ERK and the p38 MAP kinase pathways are selectively required for cytokine and TCR-dependent proliferation, respectively. These findings indicate that cytokine-driven proliferation is independent of TCR stimulation and that specific inhibitors can be used to dissect cytokine- and TCR-dependent proliferation.

The capacity to proliferate in response to IL-7 and IL-15 is associated with the differentiation state of T cells, being low in naïve cells, intermediate in T_{CM} and high in T_{EM} (Table.1). DC can overcome the reduced cytokine responsiveness of naïve T cells. Three lines of evidence indicate that this enhancing effect is independent of self-peptide-MHC presentation but mediated by DC-derived cytokines. First, both inhibitors of TCR-dependent signalling such as the PKC inhibitor BIM and neutralizing anti-MHC class II antibodies did not inhibit the enhancing effect of DC on cytokine-driven proliferation. Second, live DC can be substituted by their cell-free supernatant after LPS-stimulation or by the recombinant cytokines TNF-α, IL-6 and IL-10. Third, TNF-α can be detected in the co-culture supernatants, and neutralizing anti-TNF-α antibodies inhibit the enhancing effect of DC (S. Campagnaro, unpublished observation).

DC-derived cytokines have redundant and additive effects, but while IL-10 and IL-12 preferentially stimulates T_{EM}, TNF-α and IL-6 have a stronger effect on naïve cells. IL-4 is a selective growth factor for naïve T cells but in order to induce proliferation it must act in combination with DC-derived cytokines.

Cytokine-dependent proliferation has a relative slow kinetic but is independent of TCR specificity. Thus, cytokine-dependent proliferation allows slow but polyclonal T cell expansion, while TCR tickling is influenced by the capacity of each TCR to interact with self-MHC molecules and probably leads to a skewing of the TCR repertoire. The selectivity of homeostatic cytokines for naïve and memory cells suggests an alternative possibility to TCR tickling for the independent regulation of naïve and memory homeostasis.

Table 1. Cytokine responsiveness of human $CD4^+$ T cell subsets (CFSE dilution)

	IL-4	IL-7	IL-15	IL-7, IL-15	TNF-α, IL-6, IL-10	TNF-α, IL-6, IL-10, IL-7, IL-15	TNFα, IL-6, IL-10, IL-4
Naive	–	–	–	+/-	–	+++	++
T_{CM}	–	+/-	+/-	+	–	+++	+/-
T_{EM}	–	+/-	+	++	–	+++	–

– no division, +/- <10% divided, + 10-30% divided, ++ 30-50% divided, +++ >50% divided

Naïve and Memory T Cell Subsets Express Different Cytokine Receptors

The differential requirement for DC-derived cytokines and cytokine responsiveness is associated with the expression of the relevant cytokine receptors. Thus, the expression of the IL-2/IL-15Rβ-chain reflects responsiveness to IL-15, being very low on naïve cells, intermediate on T_{CM} and at high levels on T_{EM} (Table 2). In contrast, the γc-chain is expressed at comparable levels on all three subsets. DC-derived cytokines up-regulate the expression of the IL-2/IL-15Rβ chain and the γc-chain, consistent with their enhancing effect on cytokine-driven proliferation.

The IL-4Rα chain is undetectable on un-stimulated cells but is induced by IL-4 selectively on naïve cells, consistent with its selective mitogenic effect. However, at the same time IL-4 down-regulates γc chain surface expression, explaining why IL-4 does not induce proliferation in the absence of DC-derived cytokines. Only IL-4Rα but not γc chain mRNA expression is modulated by IL-4, suggesting that IL-4 up-regulates IL-4Rα chain transcriptionally and induces IL-4R protein degradation at the same time. A possible role for receptor degradation in inhibiting cytokine-induced proliferation is further suggested by the fact that the differential IL-7 responsiveness of naïve and memory cells correlated inversely with IL-7Rα down-regulation.

The differential responsiveness to DC-derived cytokines was also reflected by receptor expression. Thus, while IL-6Rα is present only on naïve and T_{CM}, IL-10R is preferentially expressed on T_{EM}.

Table 2 Cytokine receptor modulation of CD4+ T cell subsets

	IL-4Rα +/- IL-4	IL-7Rα +/- IL-7	IL-2/15Rβ +/- TNF-α, IL-6, IL-10	γc +/- TNF-α, IL-6, IL-10
Naive	–/+	+++/+/–	+/–/++	+/++
T_{CM}	–/–	+++/+	+/++	+/++
T_{EM}	–/–	+++/+	++/+++	+/++

– undetectable, +/– MFI<3, + MFI 3–10, ++ MFI 10–20, +++ MFI >20

Antigen Versus Cytokine-Driven T Cell Differentiation

TCR and cytokine stimulation leads to distinct phenotypic changes of proliferating cells. TCR-stimulated naïve T cells lose CCR7, CD62L and CD45RA and acquire the effector/memory markers CD45RO, CD69, CD25 and CCR5. In contrast, cytokine-stimulated naïve T cells up-regulate CD45RA and CCR7, remain CD62L$^+$, express only low amounts of activation markers such as CD69 and CD25 and remain CCR5 negative. Interestingly, both naïve and memory cells up-regulate CD40L and OX40, which might induce cytokine secretion of DC. Cytokine-stimulated T_{CM} up-regulate CD69 and CD25 on a higher fraction of cells and to a higher level as compared to naive cells. Importantly, some cells down-regulate CCR7, CD62L and acquire CCR5, thus loosing the capacity to home to lymph nodes and acquiring the capacity to migrate to inflamed tissues.

TCR stimulation induces effector functions such as IL-4 and IFN-γ production in both naïve and memory T cells. While cytokine stimulation induces only a small number of naïve cells to acquire low IFN-γ producing capacity, T_{CM} acquire effector functions comparable to TCR-stimulated cells. Modulation of IL-4 and IFN-γ producing capacity differs further markedly in TCR- and cytokine-stimulated cells. Thus while IL-4 and IL-12 enhance IL-4 and IFN-γ production of TCR stimulated cells, they promote IFN-γ and both IFN-γ and IL-4 producing capacity in cytokine-stimulated cells, respectively.

Are T_{CM} Memory Stem Cell?

The above findings suggest that the T_{CM} pool is able to self-renew and to replenish the T_{EM} pool in response to homeostatic stimuli. Consistent with this hypothesis is the fact that T_{CM} incorporate BrdU ex vivo almost as efficiently as T_{EM}, which on the other hand have an enhanced rate of apoptosis (J. Geginat, unpublished observation). The high propensity of T_{CM} to undergo cytokine-driven differentiation suggests that at least some of the proliferating T_{EM} may be generated by homeostatic proliferation and differentiation of the T_{CM} pool. Under pathological conditions, a rapid and extensive differentiation of cytokine-driven memory cells may induce large numbers of aberrantly differentiated effector cells that may cause damage to sensitive organs.

ABBREVIATIONS

T_{CM} central memory T cell, T_{EM} effector memory T cell, BrdU 5-Bromo-2'-deoxyuridine, CCR chemokine receptor, ERK extracellular signal regulated kinase, MAP mitogen-activated protein, DC dendritic cell, BIM Bisindolylmaledeide, CD40L CD40 ligand, MFI mean fluorescence intensity.

REFERENCES

Bender, J., Mitchell, T., Kappler, J., and Marrack, P.: CD4+ T cell division in irradiated mice requires peptides distinct from those responsible for thymic selection. *J. Exp. Med. 190:* 367-374., 1999

Brocker, T.: Survival of mature CD4 T lymphocytes is dependent on major histocompatibility complex class II-expressing dendritic cells. *J. Exp. Med. 186:* 1223-1232, 1997.

Cho, B. K., Rao, V. P., Ge, Q., Eisen, H. N., and Chen, J.: Homeostasis-stimulated proliferation drives naive T cells to differentiate directly into memory T cells. *J. Exp. Med. 192:* 549-556., 2000.

Dorfman, J. R., Stefanova, I., Yasutomo, K., and Germain, R. N.: CD4+ T cell survival is not directly linked to self MHC-induced TCR signaling. *Nat. Immunol 1:* 329-335, 2000.

Freitas, A. A. and Rocha, B.: Population biology of lymphocytes: the flight for survival. *Annu. Rev. Immunol. 18:* 83-111, 2000.

Goldrath, A. W. and Bevan, M. J.: Selecting and maintaining a diverse T-cell repertoire. *Nature 402:* 255-262, 1999.

Goldrath, A. W., Bogatzki, L. Y., and Bevan, M. J.: Naive T cells transiently acquire a memory-like phenotype during homeostasis-driven proliferation. *J. Exp. Med. 192:* 557-564., 2000.

Iezzi, G., Scheidegger, D., and Lanzavecchia, A.: Migration and function of antigen primed nonpolarized T lymphocytes in vivo. *J. Exp. Med. in press*, 2001.

Kanegane, H. and Tosato, G.: Activation of naive and memory T cells by interleukin-15. *Blood 88:* 230-235., 1996.

Kirberg, J., Berns, A., and von Boehmer, H.: Peripheral T cell survival requires continual ligation of the T cell receptor to major histocompatibility complex-encoded molecules. *J. Exp. Med. 186:* 1269-1275., 1997.

Lantz, O., Grandjean, I., Matzinger, P., and Di Santo, J. P.: gamma chain required for naive CD4+ T cell survival but not for antigen proliferation. *Nat. Immunol. 1:* 54-58., 2000.

Lodolce, J. P., Boone, D. L., Chai, S., Swain, R. E., Dassopoulos, T., Trettin, S., and Ma, A.: IL-15 receptor maintains lymphoid homeostasis by supporting lymphocyte homing and proliferation. *Immunity 9:* 669-676., 1998.

Mackay, C. R., Marston, W. L., and Dudler, L.: Naive and memory T cells show distinct pathways of lymphocyte recirculation. *J. Exp. Med. 171:* 801-817, 1990.

Manjunath, N., Shankar, P., Stockton, B., Dubey, P. D., Lieberman, J., and von Andrian, U. H.: A transgenic mouse model to analyze CD8(+) effector T cell differentiation in vivo. *Proc. Natl. Acad. Sci. U. S. A. 96:* 13932-13937, 1999.
Marrack, P., Bender, J., Hildeman, D., Jordan, M., Mitchell, T., Murakami, M., Sakamoto, A., Schaefer, B. C., Swanson, B., and Kappler, J.: Homeostasis of alpha/beta TCR+ T cells. *Nat Immunol. 1:* 107-112, 2000.
Masopust, D., Vezys, V., Marzo, A. L., and Lefrancois, L.: Preferential localization of effector memory cells in nonlymphoid tissue. *Science 291:* 2413-2417., 2001.
Murali-Krishna, K. and Ahmed, R.: Cutting edge: naive T cells masquerading as memory cells. *J. Immunol. 165:* 1733-1737, 2000.
Murali-Krishna, K., Lau, L. L., Sambhara, S., Lemonnier, F., Altman, J., and Ahmed, R.: Persistence of memory CD8 T cells in MHC class I-deficient mice. *Science 286:* 1377-1381., 1999.
Nakajima, H., Shores, E. W., Noguchi, M., and Leonard, W. J.: The common cytokine receptor gamma chain plays an essential role in regulating lymphoid homeostasis. *J. Exp. Med. 185:* 189-195., 1997.
Napolitano, L. A., Grant, R. M., Deeks, S. G., Schmidt, D., De Rosa, S. C., Herzenberg, L. A., Herndier, B. G., Andersson, J., and McCune, J. M.: Increased production of IL-7 accompanies HIV-1-mediated T-cell depletion: implications for T-cell homeostasis. *Nat. Med. 7:* 73-79., 2001.
Oehen, S. and Brduscha-Riem, K.: Naive cytotoxic T lymphocytes spontaneously acquire effector function in lymphocytopenic recipients: A pitfall for T cell memory studies? *Eur. J. Immunol. 29:* 608-614, 1999.
Reinhardt, R. L., Khoruts, A., Merica, R., Zell, T., and Jenkins, M. K.: Visualizing the generation of memory CD4 T cells in the whole body. *Nature 410:* 101-105., 2001.
Sallusto, F., Lenig, D., Forster, R., Lipp, M., and Lanzavecchia, A.: Two subsets of memory T lymphocytes with distinct homing potentials and effector functions. *Nature 401:* 708-712, 1999.
Schluns, K. S., Kieper, W. C., Jameson, S. C., and Lefrancois, L.: Interleukin-7 mediates the homeostasis of naive and memory CD8 T cells in vivo. *Nat. Immunol. 1:* 426-432., 2000.
Seddon, B., Legname, G., Tomlinson, P., and Zamoyska, R.: Long-term survival but impaired homeostatic proliferation of Naive T cells in the absence of p56lck. *Science 290:* 127-131., 2000.
Swain, S. L., Hu, H., and Huston, G.: Class II-independent generation of CD4 memory T cells from effectors. *Science 286:* 1381-1383., 1999.
Takeda, S., Rodewald, H. R., Arakawa, H., Bluethmann, H., and Shimizu, T.: MHC class II molecules are not required for survival of newly generated CD4+ T cells, but affect their long-term life span. *Immunity 5:* 217-228, 1996.
Tanchot, C., Lemonnier, F. A., Perarnau, B., Freitas, A. A., and Rocha, B.: Differential requirements for survival and proliferation of CD8 naive or memory T cells. *Science 276:* 2057-2062, 1997.
Tough, D. F. and Sprent, J.: Turnover of naive- and memory-phenotype T cells. *J. Exp. Med. 179:* 1127-1135, 1994.
Unutmaz, D., Pileri, P., and Abrignani, S.: Antigen-independent activation of naive and memory resting T cells by a cytokine combination. *J. Exp. Med. 180:* 1159-1164, 1994.
Zhang, X., Sun, S., Hwang, I., Tough, D. F., and Sprent, J.: Potent and selective stimulation of memory-phenotype CD8+ T cells in vivo by IL-15. *Immunity 8:* 591-599, 1998

REGULATION OF MEMORY CD4 T CELLS: GENERATION, LOCALIZATION AND PERSISTENCE

Susan L. Swain, Javed N. Agrewala, Deborah M. Brown and Eulogia Román*

1. INTRODUCTION

Over the past few years several studies have influenced how we think about the generation and persistence of T cell memory. First, our studies have shown that CD4 memory T cells can be derived from effector cells, without further division.[1, 2] This implies that the differentiation events that are heritable and accompanied by epigenetic remodeling, all take place during the effector generation stage. Our results and those from Bradley's laboratory, further indicate that an important component of memory generation from effectors is the removal of antigen and cytokines which causes this transition.[2, 3]

Second, a number of studies have suggested there is heterogeneity of memory cells which can express different phenotypes and are to be found in different sites.[4-7] Lanzavecchia has suggested a dichotomy of central memory which consists of CCR7 positive cells in spleen and other secondary sites and effector memory consisting of CCR7 negative cells mainly found in peripheral tissues. It has been suggested that the peripheral or "effector" memory is most active in protection against re-infection[4, 6]. *In vitro* transfer models, in which effectors are generated *in vitro* under different conditions, such as different cytokines, and then transferred to adoptive hosts, can also recreate these two patterns of memory.[7, 8] Several recent studies with CD8 T cells, suggest there is an all-or-none aspect to effector generation and that if T cells are stimulated they undergo not only activation but also multiple rounds of division.[9-11]

Recent results have also established that recognition of self peptide/self MHC can support division of T cells. For naive T cells, this division seems to occur only when the pool of T cells is depleted, suggesting that homeostatic regulation occurs where T cells sense their own numbers and maintain them. The division that occurs when T cells are depleted has been termed "homeostasis-driven division or proliferation" (HDD).[12, 13] Regulation of the pool of memory cells may be more complex. Memory T cells divide at a faster rate than naive T cells, even when there is no T cell depletion[14]. Memory CD8 T

* Susan L. Swain, Javed N. Agrewala, Deborah M. Brown and Eulogia Román, Trudeau Institute, Saranac Lake, New York 12983.

cell division can also be driven by cytokines, especially IL-15[15] and may be downregulated by IL-2[16] and IFNα/β dependent mechanisms.[17] Unlike their CD8 counterparts, resting CD4 memory T cells do not respond directly to cytokines in the absence of antigen[14] (and our unpublished results). Maintenance of memory T cell population size may also depend on cytokine-independent HDD. The rules governing the regulation of memory T cell populations are poorly understood and the underlying mechanisms are unknown.

Here we will briefly review the studies supporting the direct default of effector to memory CD4 T cells. We will then present data that there is a continuum of effector heterogeneity and that only the most differentiated effectors can migrate into an inflammatory site. Migration to other non-lymphoid sites is also restricted to effector CD4 T cells. We suggest that memory heterogeneity and location reflects that of the *in vivo* generated effector population, that is present when antigen is cleared. Finally, we present data indicating that while memory CD4 T cell survival is independent of Class II recognition, the size of the population is regulated by Class II dependent HDD. Moreover, the T cells which homeostatically regulate HDD (and thus pool size) are mainly memory CD4 T cells themselves.

2. RESULTS AND DISCUSSION

In the experiments we describe we have used TcR transgenic mice as sources of naive CD4 T cells with a defined antigen specificity. These naive cells have either been used to generate effectors in vitro, using conditions we have optimized over the past decade, or they have been transferred to nontransgenic hosts which are stimulated with antigen or influenza virus. We routinely transfer cells from Thy1.1$^+$ donors into Thy1.2 hosts, so we can detect the donor cells readily by FACS analysis.

2.1. Requirements for Memory Generation

To evaluate whether there was a direct transition from effector cell to memory cell and whether further division or differentiation is necessary, we used the effector transfer model. In earlier studies with this model we found that Th1 and Th2 effectors gave rise to memory cells with the cytokine polarization of the population from which they were derived.[18] We also found that neither the generation of memory nor it's persistence required Class II expression by the host.[1] These results were compatible with a direct pathway from effector to memory CD4 T cells, but because the effector population could be heterogeneous, it was possible to consider that only a small cohort of pre-committed memory precursors distinct from effectors within the total population, were the actual precursors of the memory cells. We found that transferred effectors underwent a short period of rapid division (Class II independent) immediately after transfer,[2] which increased the uncertainty about the lineage relationships.

We were, however, able to re-culture washed effectors in media without additives and found that Th2 effectors survive very well and become resting cells. Th1 effectors can also be rested but many undergo programmed cell death,[19] which would limit our ability to detect lineage relationships. Therefore further studies utilized exclusively Th2 "rested effectors". Fortunately only a fraction of the rested effectors (about 35-40%) divided after transfer to Class II KO mice[2]. The rested effectors gave rise to persistent "memory" CD4 T cells with very high efficiency. Taking into account cells lost by transfer (not detectable anywhere in the host mouse at 18 hr post-transfer), at least one third of the transferred effectors gave rise to a memory cell. Since the proportion of undivided memory cells (not labeled by a 2-day pulse of BrdU, immediately after transfer, and cells remaining undivided by CFSE-staining) remained constant for several months, we can be confident that a large portion of the effector cell population has the potential to become long-lived memory cells.[2] The memory cells generated retained the potential to make the same pattern of cytokines as the transferred effector population, upon restimulation. Similar studies with Th1 rested effectors by ourselves and by Bradley's group[3] support the concept that the direct transition is applicable regardless of CD4 T cell subset.

Thus we conclude that when effector cells can escape programmed or activation induced cell death, and when they no longer are exposed to antigen or ambient cytokines, they undergo a transition to a resting stage and can become memory cells without further signals from antigen recognition and without division. This implies that their differentiation has been completed by the effector stage, with heritable epigenetic changes that will persist in the memory cells which are their descendants, already accomplished. This further suggests that the signals and factors which regulate CD4 T cell effector generation will, in so doing, determine the kind of memory CD4 T cells that develop.

2.2. Effector Generation *In Vivo*: Heterogeneity

We and others have learned a great deal over the years by using *in vitro* models to study the regulation of naive T cell response from activation to effector generation (reviewed).[20] *In vivo* models have only recently been developed in which such responses can be clearly visualized.[21] Certainly the optimized responses we have favored *in vitro*, depend on high doses of peptide antigen, presented by antigen-presenting cells (APC) which express high levels of costimulatory molecules[22]. Furthermore, maximum effector expansion and optimum polarization to the Th1 and Th2 phenotypes, depends on exogenous IL-2 which supports growth and drives full differentiation.[22, 23] Polarization also depends on IL-4 or IFNγ/IL-12 for Th2 and Th1 differentiation, respectively.[18] Since the *in vivo* conditions are unlikely to reproduce the *in vitro* cultures in so many ways, we must turn to in situ models to confirm the details of effector and memory generation.

To this end, we have used a model similar to Jenkin's,[21] employing the transfer of a cohort of naive, TcR transgenic T cells to intact hosts. We have used the HNT TcR transgenic (BALB/c background), in which the CD4 T cells express a receptor specific for influenza hemagglutinin peptide[24] with I-A^d. We have crossed the transgenic with BALB/c.Thy1.1 mice, so we can visualize the donor cells with anti-Thy1.1 staining.

Following inoculation of live influenza (PR8), the donor naive CD4 T cells commence activation (expression of CD69, CD44) and then division (by both BrdU labeling and loss of CFSE) between days 2-4. The peak of effector response, in terms of number of donor cells, is detected between 6-8 days and donor cell numbers then decline rapidly. Viral clearance begins between 6 and 8 days and live virus is undetectable by 10 days.

Division and phenotypic changes are first noted in the draining, mediastinal lymph nodes between days 2 and 4 and subsequently in spleen at day 4. Donor cells only migrate to the lung and bronchial-alveolar lavage (BAL) after day 5.

The effector response is very heterogeneous. At 5-8 days a spectrum of response is seen in spleen, draining lymph nodes and peripheral blood in which we detect cells that are at different stages of phenotypic and functional progression. Donor cells detected in the secondary lymphoid organs have undergone from 0-8 divisions, and they range from CD62L high to low, CCR5 low to high, CCR7 high to low, and CD49d low to high (almost all are CD44 high), these changes running in concert with division (See Figure 1). This profile represents a continuum rather than easily discerned subpopulations.

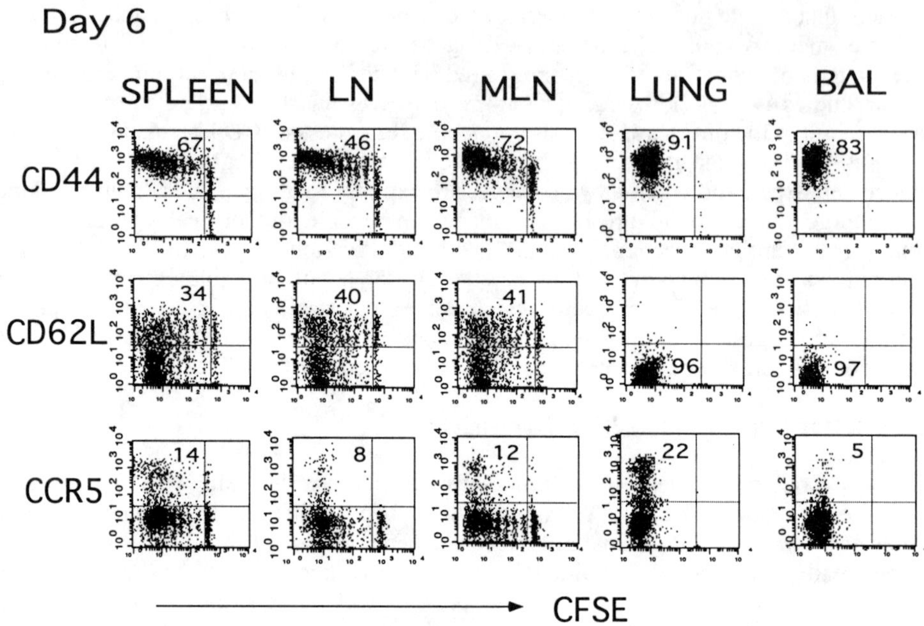

Figure 1. Heterogeneity of Effector Response. CFSE-labeled Thy1.1 donor HNT naive CD4 T cells (5 X 10^6) were adoptively transferred to host mice (Thy1.2). with were infected with influenza the following day. 6 days later organs were harvested including spleen, pooled, non draining, lymph nodes (LN), draining mediastinal lymph nodes (MLN), whole lung and the bronchoalveolar lavage (BAL) taken before lung harvest. The CD4+/Thy1.1$^+$ donor population was analyzed by staining and FACS analysis for a panel of markers, three of which are shown here.

Clearly, the cells which are recruited to lung and BAL only represent the most divided, most phenotypically changed and presumably most highly differentiated cohort. Functionally a similar heterogeneity is observed, with increasing IFNγ production seen as more divisions occur and an enriched higher cytokine-producing cohort in the lung and BAL. The cohort of effectors in lung/BAL at 6 days postinfection is also unique in that it is the only population with high expression of CD25. Donor cells in lung and BAL also produce IFNγ constitutively, such that intracellular cytokine staining reveals IFNγ production in the absence of restimulation with PMA and ionomycin, whereas in other organs no IFNγ staining is detectable except after restimulation *ex vivo*. Both of these features suggest that the cohort of cells in the lung and BAL have been recently exposed to stimulation with antigen. Influenza replicates almost exclusively in lung epithelium, therefore the lung should be the site of inflammation caused by infection, the site of lysis of targets and the site of greatest antigen density.

Preliminary experiments indicate that memory donor CD4 T cells recovered at 5 weeks after primary infection, also reflect this heterogeneous effector pattern, with the spleen displaying heterogeneity and containing cells with a spectrum of phenotype and the lung/BAL population expressing only the "highly differentiated" phenotype.

These studies lead us to make several conclusions. 1) In contrast to the all-or-none model suggested previously for CD8 T cells, we see an extremely heterogeneous response of CD4 naive T cells to influenza infection in situ. 2) In contrast to a dichotomy of effector distribution with two homogeneous populations that represent central and peripheral effectors, in secondary lymphoid organs we see a continuous spectrum of effector phenotypes and find only a defined, presumably highly differentiated cohort, in the inflammatory site of the lung. 3) Our studies are compatible with the prevailing model, that responses are initiated in draining nodes and spleen, and that after effectors are generated they can traffic to a site of inflammation.

2.3. Populations Which Migrate to Inflammatory and Non-Lymphoid Sites

Recently several laboratories have stressed the "diaspora" of effector responses.[5, 25, 26] We wanted to further investigate what stages of CD4 T cells were recruited to both inflammatory and not obviously involved, non lymphoid sites. Earlier studies had indicated that almost all naive cells were restricted to lymphoid locations or found in the recirculating pool[2, 20, 27]. To get enough cells to adequately visualize potentially infrequent donor cells, we have begun by generating effectors *in vitro*[1, 2] and then transferring them to adoptive hosts. In this case the hosts have been T-depleted (ATXBM) to aid us in visualizing the very small number of cells in some of the non lymphoid sites. Four days after iv transfer, we detect donor effectors not only in the spleen but also in significant numbers among the peritoneal exudate cells (PEC) and in the fat pads. Th1 and Th2 effectors are distributed in the same pattern. When we examine hosts at 4 weeks or more, we find that donor cells of memory phenotype persist in the spleen, but the recovery of donor cells in the tertiary sites (PEC and fat pads), has declined to nearly undetectable levels. When we instead transfer rested effectors that have been washed and cultured in media alone for 2-3 days before transfer, they can only rarely can be detected in tertiary sites 4 days after transfer. The distribution of rested effectors at 4 days resembles that of the memory cells developed from effectors and analyzed at 4 weeks.

When we challenge hosts that have received such rested effectors with specific antigen we observe a re-call response. Four days after antigen boost, we again detect donor cells in the PEC and fat pads, suggesting that secondary memory effectors regain the ability to migrate to non lymphoid tissues. Table 1 summarizes these results.

Table 1. Migration of CD4 T Cell Subsets to Lymphoid & Non Lymphoid Sites

	Spleen	Lymph Nodes	Peritoneal Exudate	Fat Pads
Naive	+++	++++	---	---
Active Effectors	+++	++	+++	++
Rested Effectors	++++	+	---	---
Restimulated Effectors	+++	+	+++	++

These results complement those obtained in the influenza model and re-enforce the concept that the activated effector state is unique but transient, allowing recruitment to inflammatory sites only during an acute phase of high activation. We have observed that the decline of the transferred effector sites in tertiary non lymphoid sites is fairly slow, whereas 2-3 day rested effectors are no longer recruited to such sites. It thus seems likely that once effectors have entered sites they can transition to more resting cells which are retained at the site and may not be part of the recirculating lymphoid pool. More experiments need to be done to establish the validity of these hypotheses.

2.4. Regulation of Memory Cell Pool Size

In the transfer studies on the development of memory described above, we noted that once formed, the memory cell pool was remarkably stable in Class II KO hosts, but that it declined in intact hosts and increased slightly in ATXBM hosts. This suggested that homeostatic regulation by other T cells might be controlling the size of the memory pool, as it is known to do for naive T cells.

To further investigate the rules governing the memory pool size we transferred rested effectors to hosts that have different T cell subsets and then examined both donor cell recovery and the HDD of the donor cells. Results are summarized in Table II. In Class II KO mice, there was a stable population of memory cells, with very little division which was proportional to cell input over a wide range.[2] This suggested that when the donor memory cells could not recognize self MHC, they persisted for long times but did not expand to levels dictated by homeostasis, because they could not divide. The presence of host CD8 T cells seemed to have little effect.

When the same cells were titrated into intact mice, they did divide but very slowly (presumably because they encountered Class II, but also were downregulated by host CD4 T cells). In ATXBM cells, depleted of host T cells, the donor cells divided quite rapidly and expanded (presumably because of the host CD4 deficit). As greater numbers of rested effectors were transferred, they reached a similar plateau number and division decreased, indicating they could regulate each other.

The most interesting result is seen in transferring rested effectors to AND or OT-II TcR hosts. The transgenic mice have Class II and have abundant CD4 T cells, but most of these are of naive phenotype.

Table 2. Endogenous T cells and MHC Effect the Persistence of Transferred T cells

HOST	NAIVE CD4 T CELLS		RESTED EFFECTOR CELLS	
	Pool size	HDD	Pool size	HDD
Intact Control	+	---	++	+/-
Class II KO	+	---	+++	+
AND or OT-II Tg	++	++	++++	++
ATXBM	+++	+++	++++	+++
CD4-depleted	+++	+++		
CD8-depleted	+	---		
Adult TX (20 Weeks)	+	---		

In these naive-enriched hosts, the donor cells expanded and formed a stable pool larger than that in intact mice. This is compatible with the concept that only antigen-experienced CD4 T cells negatively regulate the memory CD4 pool size by suppressing

HDD. Experiments with transfer of naive CD4 T cells, which appear to be regulated by a similar mechanism, indicate that in situ CD4 but not CD8 depletion of intact mice, release the population and allows them to undergo HDD. These results are not compatible with the conclusions of Surh and colleagues who interpreted their results to mean that either CD4 or CD8 T cells could regulate CD4 HDD and that naive CD4 T cells were able to carry out this homeostatic regulation. One difference in their model is that the used polyclonal populations of donor T cells, while we used a population with a single TcR. From our studies, we conclude that two factors contribute to CD4 memory pool size. The first is the size of the initial memory population formed at the end of the effector response when antigen is cleared. The second is the size of the CD4 memory pool in the animal, which as it increases will suppress homeostasis driven expansion. It may be that other factors such as TcR avidity for self MHC/self peptide also play a role. We do not think cytokines play a major role in regulating CD4 memory pool size, although they do seem to play an important role in regulating CD8 memory pool size. We would expect that in older intact mice, where the memory pool is already established, little HDD of memory cells would occur and the pool size of newly formed memory cells would remain fairly constant.

3. CONCLUSIONS

The formation and maintenance of memory CD4 T cells have recently received much attention and several new paradigms have emerged. First we argue that from our work, it is clear CD4 effector cells develop into memory cells without needing further signals which drive division or differentiation. Much of the process seems to be simply the transition to a resting state. Second, it is clear that CD4 effector response can produce a heterogeneous spectrum of effectors of different phenotypes and functional maturity. We suspect this leads to a comparably heterogeneous memory population.

Finally it is apparent that recruitment of effectors to inflammatory sites (the lung during influenza infection) or other non lymphoid sites, requires a set of defined characteristics which are associated with the most highly differentiated effectors. The expression of particular array of chemokine receptors is almost certainly a factor here, and we will investigate which are necessary in the future. As soon as effectors become resting they lose the potential to migrate to these sites, but it is regained when resting memory cells are restimulated to become memory effectors.

These observations argue against an all-or-none model of CD4 T cell responses and they argue for a continuous spectrum of effectors/memory, with only a cohort in non-lymphoid organs rather than a dichotomy of central Vs peripheral memory.

REFERENCES

1. S. L. Swain, H. Hu, and G. Huston, Class II independent generation of CD4 memory T cells from effectors, *Science*. 286, 1377-1383 (1999).
2. H. Hu, G. Huston, D. Duso, N. Lepak, E. Roman, and S. L. Swain, CD4 T cell effectors can become memory cells with high efficiency and without further division, *Nat. Immunol.* 2, 705-710 (2001).
3. J. Harbertson, E. Biederman, K. E. Bennett, R. M. Kondrack, and L. M. Bradley, Withdrawal of stimulation may initiate the transition of effector to memory CD4 cells, *J Immunol.* 168, 1095-1102 (2002).

4. F. Sallusto, D. Lenig, R. Foster, M. Lipp, and A. Lanzavecchia, Two subsets of memory T lymphocytes with distinct homing potentials and effector functions, *Nature.* 401, 708-712 (1999).
5. D. Masopust, V. Vezys, A. L. Marzo, and L. Lefrancois, Preferential localization of effector memory cells in nonlymphoid tissue, *Science.* 291, 2413-2417 (2001).
6. R. J. Hogan, W. Zhong, E. J. Usherwood, T. Cookenham, A. D. Roberts, and D. L. Woodland, Protection from respiratory virus infections can be mediated by antigen-specific $CD4^+$ T cells that persist in the lungs, *J Exp. Med.* 193(8), 981-6 (2001).
7. N. Manjunath, P. Shankar, J. Wan, W. Weninger, M. A. Crowley, K. Hieshima, T. A. Springer, X. Fan, H. Shen, J. Lieberman, and U. H. von Andrian, Effector differentiation is not prerequisite for generation of memory cytotoxic T lymphocytes, *J Clin. Invest.* 108, 871-878 (2001).
8. W. Weninger, M. A. Crowley, N. Manjunath, and U. H. von Andrian,. Migratory properties of naive, effector, and memory $CD8^+$ T cells, *J Exp. Med.* 194, 953-966 (2001).
9. S. M. Kaech, and R. Ahmed, Memory $CD8^+$ T cell differentiation: initial antigen encounter triggers a developmental program in naive cells, *Nat. Immunol.* May, 2(5), 415-422 (2001).
10. P. Wong, and E. G. Pamer, Cutting edge: antigen-independent CD8 T cell proliferation, *J Immunol.* 166, 5864-5868 (2001).
11. M. J. van Stipdonk, E. E. Lemmens, and S. P. Schoenberger, Naive CTLs require a single brief period of antigenic stimulation for clonal expansion and differentiation, *Nat Immunol.* 2, 423-429 (2001).
12. C. F. Viret, S. Wong and C. A. Janeway, Jr., Designing and maintaining the mature TCR repertoire: the continuum of self-peptide: self-MHC complex recognition, *Immunity.* 10, 559-568 (1999).
13. J. Bender, T. Mitchell, J. Kappler and P. Marrack, $CD4^+$ T cell division in irradiated mice requires peptides distinct from those responsible for thymic selection, *J. Exp. Med.* 190, 367-374 (1999).
14. J. Sprent, and D. F. Tough, Lymphocyte life-span and memory. *Science.* 265, 1395-1400 (1994).
15. X. Zhang, S. Sun, I. Hwang, D. F. Tough, and J. Sprent, Potent and selective stimulation of memory-phenotype $CD8^+$ T cells *in vivo* by IL-15, *Immunity.* 8, 591-599 (1998).
16. C. C. Ku, M. Murakami, A. Sakamoto, J. Kappler, and P. Marrack, Control of homeostasis of CD8+ memory T cells by opposing cytokines, *Science.* 288, 675-678 (2002).
17. D. F. Tough, X. Zhang, and J. Sprent, An IFN-gamma-dependent pathway controls stimulation of memory phenotype $CD8^+$ T cell turnover *in vivo* by IL-12, IL-18, and IFN-gamma, *J Immunol.* 166, 6007-6011 (2001).
18. S. L. Swain, Generation and *in vivo* persistence of polarized Th1 and Th2 memory cells, *Immunity.* 1, 543-552 (1994).
19. X. Zhang, T. Brunner, L. Carter, R. W. Dutton, P. Rogers, L. Bradley, J. Reed, T. Sato, D. Green, and S. L. Swain, Unequal Death in Th1 and Th2 Effectors: Th1 but not Th2, Effectors undergo rapid Fas/FasL-mediated apoptosis, *J Exp Med* 185, 3738-3749 (1997).
20. S. L. Swain, M. Croft, C. Dubey, L. Haynes, M. P. Rogers, X. Zhang, and L. M. Bradley, From naive to memory T cells, *Immunol. Rev.* 150, 143-167 (1996).
21. M. K. Jenkins, A. Khoruts, E. Ingulli, D. L. Mueller, S. J. McSorley, R. L. Reinhardt, A. Itano, and K. A. Pape, *In vivo* activation of antigen-specific CD4 T cells. *Annu. Rev. Immunol.* 19, 23-45 (2001).
22. P. R. Rogers, G. Huston, and S. L. Swain, High antigen density and IL-2 are required for generation of CD4 effectors secreting Th1 rather than Th0 cytokines, *J. Immunol.* 161, 3844-3852 (1998).
23. L. Haynes, P. J. Linton, S. M. Eaton, S. Tonkongy, and S. L. Swain, IL-2, but not other common γ chain (γc)-binding cytokines, can reverse the defect in generation of CD4 effector T cells from naive T cells of aged mice, *J. Exp. Med.* 190, 1013-1024 (1999).
24. B. Scott, R. Liblau, S. Degermann, L. A. Marconi L. Ogata A. J. Caton H. O. McDevitt, and D. Lo, A role for non-MHC genetic polymorphism in susceptibility to spontaneous autoimmunity, *Immunity.* 1, 73-83 (1994).
25. D. R. Marshall, S. J. Turner, G. T. Belz, S. Wingo, S. Andreansky, M. Y. Sangster, J. M. Riberdy, T. Liu, M. Tan, and P. C. Doherty, Measuring the diaspora for virus-specific $CD8^+$ T cells, *Proc. Natl. Acad. Sci. U. S. A.* 98(11), 6313-6318 (2001).
26. R. L. Reinhardt, A. Khoruts, R. Merica, T. Zell, and M. K. Jenkins, Visualizing the generation of memory CD4 T cells in the whole body, *Nature.* 410(6824), 101-105 (2001).
27. E. C. Butcher, H. Williams, K. Youngman, L. Rott, and M. Briskin 1999. Lymphocyte trafficking and regional immunity, *Adv. in Immunol.* 72, 209-253 (1999).
28. W. Dummer, B. Ernst, E. LeRoy, D. Lee, and C. Surh, Autologous regulation of naive T cell homeostasis within the T cell compartment. *J Immunol.* 166(4), 2460-2468 (2001).

SOME PROPERTIES OF T CELLS IN ANIMALS

Angela Rico de Souza, Bradley Swanson, Jennifer Robertson, Jeremy Bender, John Kappler and Philippa Marrack*[1]

1. INTRODUCTION

T cells in animals have several unexpected properties. The work described here concerns two of these properties. The first property to be discussed concerns the fact that T cells divide in the apparent absence of antigen when transferred to T cell deficient mice. The second investigates the finding that memory T cells contain very large amounts of mRNA for certain chemokines. These two phenomena will be discussed in separate sections below.

2. ANTIGEN SPECIFICITY OF THE T CELLS WHICH DIVIDE RAPIDLY AFTER TRANSFER TO T CELL DEFICIENT HOSTS

Many studies have shown that, in apparent contradiction to the rules of the clonal selection theory, T cells divide after transfer to T cell deficient hosts[1-8]. Recent experiments suggest that, at least in the case of naïve T cells, proliferation is driven by IL-7, a cytokine which is made constitutively by nonlymphoid cells of the host[6]. Interleukin 7 is known to enable the survival of naïve T cells in vitro[9,10]. It also allows these cells to survive better in normal, T cell sufficient mice[11]. At high concentrations, IL-7 drives division of T cells in culture. Its ability to drive naïve T cell proliferation in animals may be due to a similar phenomenon. Due to the lack of T cells to consume it, IL-7 may be present at higher than normal concentrations in T cell deficient mice and this increased availability may drive naïve T cells into cycle.

Other experiments indicate that the proliferation of T cells in T cell deficient animals requires, in addition, interaction between the TCRs on the dividing cells and MHC proteins of the host [3,12-16]. The characteristics of the peptide bound to the stimulating

---

[1] * All authors, Department of Immunology, National Jewish Medical and Research Center and University of Colorado Health Sciences Center, Denver, CO 80206. Brad Swanson, Jennifer Robertson, John Kappler and Philippa Marrack, Howard Hughes Medical Institute, National Jewish Medical and Research Center, Denver, CO 80206.

MHC proteins remain obscure, however. Some investigators suggest that the peptide should be that involved in positive selection of the cells in the thymus [3]. Others have found that the selecting peptide is not sufficient and therefore suggest that the MHC/peptide combination with which the dividing cells are interacting may have to have an affinity for the TCRs on the dividing cells which is greater than that usually involved in positive selection, but not so high as to drive division of the cells in normal animals [15].

Several sets of investigators including ourselves have noticed that the dividing T cells in T cell deficient mice appear to fall into two classes. One group divides very rapidly, readily acquires surface markers characteristic of activated T cells, is increased in amount by transfer of dendritic cells to the T cells deficient hosts, and not reduced in amount or division rate by the presence of bystander T cells. The other group divides and acquires activation markers slowly, is less affected by increased numbers of dendritic cells and is reduced by competing bystander T cells [17,18] personal observations). Thus the rapidly dividing cells resemble T cells undergoing a true antigen response, whereas the slowly dividing cells are less like cells responding to true antigen.

2.1. Under Some Circumstances The Rapidly Dividing Cells May Be Responding To Mixed Isotype Class II Proteins

In one series of experiments which studied the role of MHC/peptide in T cell division in T cell deficient mice, CFSE labeled normal T cells from B6.PL (Thy1.1+) mice were transferred into sublethally irradiated C57BL/6 (Thy1.2+) mice or into mice which were genetically deficient in IAβ^b. [19] These latter mice are thought not to express class II MHC proteins, since normal C57BL/6 mice are able to express only IAα^b, IAβ^b and IEβ^b and do not contain a functional IEα^b gene.

The data in Figure 1 show that some of the transferred CD4$^+$T cells divided rapidly when transferred into either of these irradiated hosts. Moreover, the numbers of rapidly dividing cells were much greater in the IAβ^b deficient hosts than in the C57BL/6 animals.

To find out whether or not this was due to some unexpected class II expression in the recipients, the CFSE labeled cells were transferred also into animals which lacked both IAβ^b and invariant chain (Ii) [20]. The numbers of rapidly dividing CD4$^+$ T cells were reduced in these animals by comparison with those in mice lacking IAβ^b only (Figure 1).

These results suggest that IAβ^b deficient animals express a mixed isotype class II protein, made up of IAα^b and IEβ^b and that this can be detected by a few CD4$^+$ T cells from normal B6.PL mice. This occurs even though such mice should also, theoretically, be able to express the same protein, and therefore their CD4$^+$ T cells should be tolerant to this mixed isotype protein. Perhaps, however, lack of IAβ^b allows expression of a higher concentration of the mixed isotype protein, because of increased availability of the otherwise unpaired IAα^b chain. The CD4$^+$ T cells from normal B6.PL mice may not be tolerant to this higher level of expression of IAα^b/IEβ^b. Lack of Ii may reduce the amount of IAα^b/IEβ^b that appears on the surface of the antigen presenting cells, in the same way that lack of Ii reduces the level of expression of other, more conventional class II proteins.

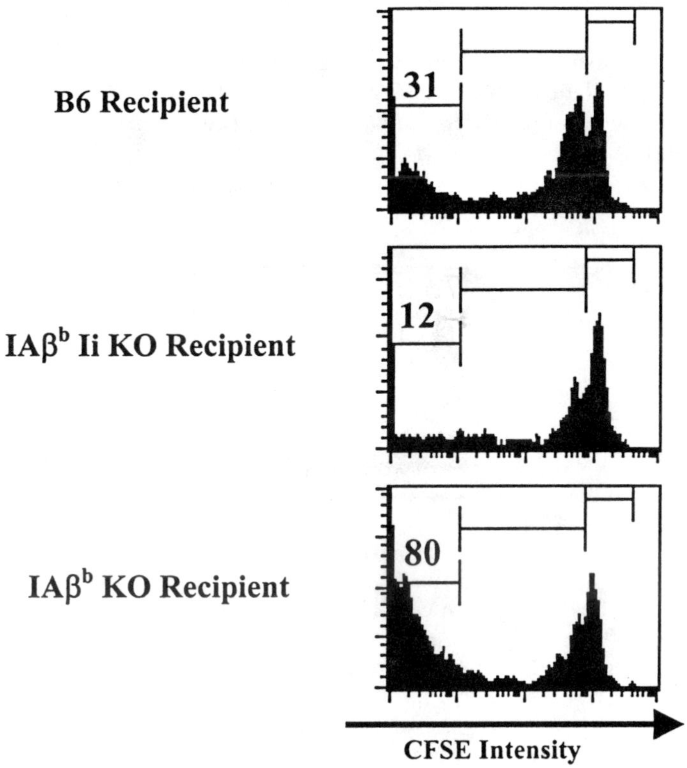

Figure 1. Some CD4+ T cells may respond to the IAα^b/IEβ^b mixed isotype protein. CFSE labeled T cells from B6.PL mice were transferred to sublethally irradiated B6 congenic recipients of the indicated genotypes. Seven days later spleen and lymph node cells were isolated and donor (Thy1.1+) CD4+ T cells analyzed for their content of CFSE. Numbers of the graphs indicate the percentages of the recovered donor CD4+ T cells which have divided rapidly (>6 times since transfer).

2.2. T cells can detect antigenic differences between donor and host

Because of these and other experiments, we predicted that the rapidly dividing cells in syngeneic sublethally irradiated hosts might be responding to autoantigens, having been released from elements which normally limit their expansion as a consequence of their transfer to T cell deficient animals. To test this idea we decided to isolate the rapidly dividing cells and examine whether or not the TCRs they bore were autoreactive. One simple method to do this is to convert them to T cell hybridomas, cells which can easily be produced, maintained and tested for antigen specificity.

In one early set of experiments T cells were isolated from 129S/SvEv animals from Taconic Corporation, labeled with CFSE and transferred to RAG 2 deficient 1296/SvEv animals from the same source [21]. Seven days later, cells were purified from the spleens and lymph nodes of the recipients. High speed sorting was used to isolate the CD4+ T cells which had divided more than 6 times, and the CD4+ T cells which had not divided at all (Figure 2).

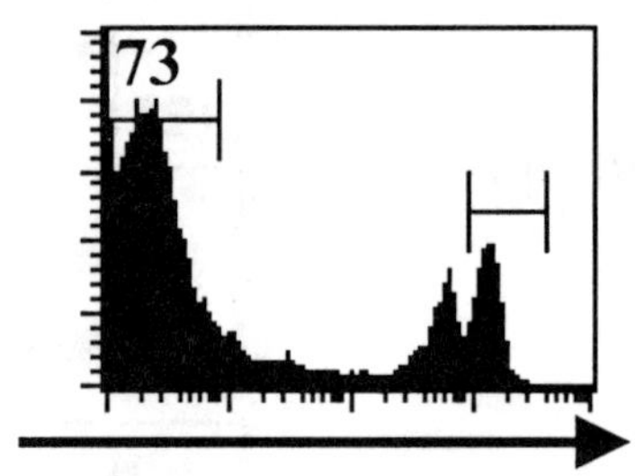

Figure 2. Sorting gates for 1296/SvEv CD4+ T cells which divided rapidly in 1296/SvEv RAG-2^{tml} hosts. The number on the figure shows the percentage of dodnor derived CD4+ T cells which had divided more than 6 times during the 7 days since the cells were transferred.

These cells were than expanded for a short time in culture and fused to produce hybridomas. The hybrids were then tested for their ability to react with spleen cells from donor and host and also from C57BL/6, a strain which bears the same MHC type ($H2^b$) as 129 mice. The data in Table 1 show that none of the hybridomas from T cells which had not divided at all reacted with any of the test spleen cells. A high percentage of the hybrids from the rapidly dividing cells, on the other hand, reacted with spleen cells from the recipient strain, or C57BL/6 mice. These cells were not truly autoreactive, however, since they did not react with spleen cells from the donor. Sixty nine percent of the hybridomas which reacted with 1296/SvEv RAG-2^{tml} bore the same Vβ, Vβ6 suggesting that they had related TCRs and therefore related antigen specificities.

Table 1. 129S/SvEvTac and 129S/SvEvTac RAG-2^{tm1} animals probably differ genetically at sites other than the RAG-2 locus.

	Hybridomas from 129S/SvEv CD4+ T cells which divided after transfer:					
	0 times			>6 times		
Presenting cells from:	# Hybrids	# Reactive	% Reactive	# Hybrids	# Reactive	% Reactive
129S/SvEvTac	60	0	0	62	0	0
129S/SvEvTac RAG-2 KO	60	0	0	62	36	58
BL/6	60	0	0	62	36	58

Enquiry revealed that, at the time that the embryonic stem cell (ES) line from which the RAG deficient animals were produced was made, the 129S/SvEv mouse strain was not completely inbred. The ES line and 129S/SvEv mice at Taconic differ at at least one locus, that for glucose 6 phosphate isomerase (Gpi). The ES line expresses Gpi^{a}, and 129S/SvEv mice express Gpi^{c}. C57Bl/6 mice express a different allele of this same protein. We believe therefore, that many of the rapidly dividing cells in this particular combination were actually responding to a real allelic difference between the 1296/SvEv and the 1296/SvEv RAG-2^{tm1} mice, perhaps at the Gpi locus. Thus the T cells were, in fact, responding to genuine antigen. Therefore if donor and host do differ at a peptide which can be presented by class II, this can easily be discerned by the transferred T cells in such an assay.

2.3 Many of the rapidly dividing cells may not be responding to autoantigens

Examination of the specificities of transferred T cells requires that they be distinguishable from host. This is not difficult to manage if the host is deficient in RAG. However, the 129S/SvEv RAG-2^{tm1} mice could not be used in such experiments, for the reason discussed above. Since the complete relatedness of the C57BL/6J and C57BL/6J RAG-1^{tm1} available from the Jackson Laboratory has also not been unequivocally established we were reluctant to commit effort to examination of T cells specificities in this combination lest some unsuspected antigen difference interfere with interpretation of the results. We therefore decided that further experiments would be performed by transferring (C57BL/6 x B6.PL)F1 T cells into sublethally irradiated C57Bl/6 hosts. In this combination the donor T cells should be fully tolerant to the host, and yet distinguishable from their hosts by their expression of Thy1.1.

T cells were therefore isolated from (B6 x B6.PL)F1 donors, labeled with CFSE, and transferred to sublethally irradiated B6 animals. Seven days later T cells were isolated and sorted into the rapidly dividing and nondividing CD4+ populations as described above. The sorted cells were expanded, fused, and the hybridomas thus produced examined for their specificity for donor and host. Again, none of the hybridomas produced from nondividing cells reacted with their donor or host. A few of the hybridomas produced from the rapidly dividing cells reacted with both donor and host (Table 2). The frequency of these autoreactive cells was low and was not significantly different from the frequency of autoreactive T cell hybrids (4%) obtained from a fusion of random T cells from C57BL/6 mice.

Table 2. The T cells which divide rapidly after transfer to T cell deficient hosts are not detectably autoreactive.

	Hybridomas from (B6 x B6.PL)F1 CD4+ T cells which divided after transfer:					
	0 times			>6 times		
Presenting cells from:	# Hybrids	# Reactive	% Reactive	# Hybrids	# Reactive	% Reactive
BL/6	71	0	0	67	3	4.5
B6.PL	71	0	0	67	3	4.5

This experiment leads to one of three hypotheses. The rapidly dividing cells may indeed be autoreactive, however, the autoantigen they are responding to may not be expressed on spleen cells or dendritic cells. Alternatively, these T cells may be autoreactive, but their avidity for self antigen may be too low to be detected by hybridoma assay. The fact that hybridomas are often less sensitive to antigen than the T cells from which they are derived supports this notion. Finally, most of the rapidly dividing T cells may not be responding to autoantigen at all, but rather driven into rapid proliferation by some other factor.

3. CD8+ MEMORY T CELLS CONTAIN ABUNDANT RANTES mRNA BUT DO NOT SECRETE THE PROTEIN

In a recent Affymetrix gene array comparison of gene expression in CD4+ or CD8+ naïve or memory T cells we noticed that memory T cells contain very large amounts of mRNA for several chemokines, most notably RANTES, whereas naïve T cells do not (Table 3). To study the significance of this, ELISAs were used to assay RANTES secretion by naïve and memory CD8+ T cells. Neither type of cell secreted RANTES spontaneously. Even when they were stimulated to divide by culture in IL-15, CD8+ memory T cells failed to secrete RANTES. CD8+ memory T cells did secrete the chemokine rapidly, however, after engagement or their TCRs. Costimulation was not required for this to occur.

Table 3. Memory T cells contain high levels of mRNA for certain chemokines

	Affymetrix Score for:			
	CD4+ T cells		CD8+ T cells	
Chemokines	Memory	Naïve	Memory	Naive
MIP-1β	1453	281	3427	628
Lymphotactin	1909	-51	2361	101
Rantes	11311	628	18408	319
IP-10	1228	249	141	101

To study the ways in which this secretion is controlled, we examined the ability of CD8+ memory T cells to secrete RANTES in the presence of inhibitors of RNA or protein synthesis. Protein synthesis inhibitors completely abolished RANTES secretion. RNA synthesis inhibitors did not. Preliminary experiments with CD8+ memory T cells produced in vitro indicated that these cells contained only a small amount of RANTES protein, hence the inhibitors of protein synthesis could not have been acting entirely via RANTES mRNA itself.

These experiments lead us to suggest that memory CD8+ T cells control secretion of stored proteins by translational inhibition of the mRNA for RANTES as well as for some presently unknown protein which is required for secretion. Such a system allows memory cells to respond rapidly to antigen stimulation, in the absence of costimulation, at nonlymphoid sites.

ACKNOWLEDGEMENTS

This work was supported by USPHS grants AI-17134, AI-18785 and AI-22295. ARS was supported by a fellowship from the Fundacao de Amparo a Pesquisa de Sao Paulo (FAPESP).

REFERENCES

1. B. Rocha, N. Dautigny, and P. Pereira, Peripheral T lymphocytes: expansion potential and homeostatic regulation of pool sizes and CD4/CD8 ratios in vivo. *Eur J Immunol* **19** (5), 905-11 (1989).
2. W. C. Kieper and S. C. Jameson, Homeostatic expansion and phenotypic conversion of naive T cells in response to self peptide/MHC ligands. *Proc Natl Acad Sci U S A* **96** (23), 13306-11 (1999).
3. B. Ernst, D. S. Lee, J. M. Chang, J. Sprent and C. D. Surh, The peptide ligands mediating positive selection in the thymus control T cell survival and homeostatic proliferation in the periphery. *Immunity* **11** (2), 173-81 (1999).
4. S. R. Clarke and A. Y. Rudensky, Survival and homeostatic proliferation of naive peripheral CD4+ T cells in the absence of self peptide:MHC complexes. *J Immunol* **165** (5), 2458-64 (2000).
5. A. W. Goldrath, L. Y. Bogatzki and M. J. Bevan, Naive T cells transiently acquire a memory-like phenotype during homeostasis-driven proliferation. *J Exp Med* **192** (4),557-64 (2000).
6. K. S. Schluns, W. C. Kieper, S. C. Jameson and L. Lefrancois, L. Interleukin-7 mediates the homeostasis of naive and memory CD8 T cells in vivo. *Nat Immunol* **1** (5), 426-32 (2000).
7. W. C. Kieper, M. Prlic, M., C. S. Schmidt, M. F. Mescher and S. C. Jameson, IL-12 enhances CD8 T cell homeostatic expansion. *J Immunol* **166** (9), 5515-21 (2001).
8. O. Annacker, et al. CD25+ CD4+ T cells regulate the expansion of peripheral CD4 T cells through the production of IL-10. *J Immunol* **166** (5), 3008-18 (2001).
9. A. Vella, T. K. Teague, J. Ihle, J. Kappler and P. Marrack, Interleukin 4 (IL-4) or IL-7 prevents the death of resting T cells: stat6 is probably not required for the effect of IL-4. *J Exp Med* **186** (2), 325-30 (1997).
10. M. V. Soares, et al. IL-7-dependent extrathymic expansion of CD45RA+ T cells enables preservation of a naive repertoire. *J Immunol* **161** (11), 5909-17 (1998).
11. T. E. Boursalian, T. E. and K. Bottomly, Survival of naive CD4 T cells: roles of restricting versus selecting MHC class II and cytokine milieu. *J Immunol* **162** (7), 3795-801 (1999).

12. T. Brocker, Survival of mature CD4 T lymphocytes is dependent on major histocompatibility complex class II-expressing dendritic cells. *J Exp Med* **186** (8), 1223-32 (1997).
13. J. Kirberg, A. Berns and H. von Boehmer, Peripheral T cell survival requires continual ligation of the T cell receptor to major histocompatibility complex-encoded molecules. *J Exp Med* **186** (8), 1269-75 (1997).
14. D. Nesic, D. and S. Vukmanovic, MHC class I is required for peripheral accumulation of CD8+ thymic emigrants. *J Immunol* **160** (8), 3705-12 (1998).
15. J. Bender, T. Mitchell, J. Kappler, J. and P. Marrack, CD4+ T cell division in irradiated mice requires peptides distinct from those responsible for thymic selection. *J Exp Med* **190** (3), 367-74 (1999).
16. C. Viret, F. S. Wong and C. A. Janeway Jr. Designing and maintaining the mature TCR repertoire: the continuum of self-peptide:self-MHC complex recognition. *Immunity* **10** (5), 559-68 (1999).
17. B. K. Cho, V. P. Rao, Q. Ge, H. N. Eisen and J. Chen, Homeostasis-stimulated proliferation drives naive T cells to differentiate directly into memory T cells. *J Exp Med* **192** (4), 549-56 (2000).
18. K. Murali-Krishna, K. and R. Ahmed, Cutting edge: naive T cells masquerading as memory cells. *J Immunol* **165** (4), 1733-7 (2000).
19. D. Cosgrove, et al. Mice lacking MHC class II molecules. *Cell* **66** (5), 1051-66 (1991).
20. E. K. Bikoff, et al. Defective major histocompatibility complex class II assembly, transport, peptide acquisition, and CD4+ T cell selection in mice lacking invariant chain expression. *J Exp Med* **177** (6), 1699-712 (1993).
21. Y. Shinkai, et al. RAG-2-deficient mice lack mature lymphocytes owing to inability to initiate V(D)J rearrangement. *Cell* **68** (5), 855-67 (1992).

GENERATION AND CHARACTERIZATION OF MEMORY CD4 T CELLS

S. Z. Ben-Sasson*#, Irena Zukovsky*, Aliza Biton*, Ron Vogel#, Gilles Foucras#, Nobuki Hayashi# and William E. Paul#

1. INTRODUCTION

Upon exposure to antigen, specific CD4 T cells are activated and differentiate into cytokine-producing effector cells that ultimately determine the outcome of the immune challenge. However, the activated T cells are relatively short lived and little is known about the factors that determine the proportion that develop into long-lived memory cells. Furthermore, the behavior of memory cells upon re-encounter with antigen is poorly understood.

Detailed analysis of the generation and responsiveness of specific memory CD4 cells is often hampered by their low numbers. Moreover, in conventional systems, it is difficult to follow specific T cells *in vivo* before and after exposure to antigen. The availability of T cell receptor transgenic mice makes possible a more critical examination of the fate of an antigen–specific lymphocyte populations *in vivo*. This experimental system allows the evaluation of memory cell dynamics and the definition of the characteristics of the memory T cells in immunized mice. The experimental model we have used involves activation of TCR transgenic T cells *in vivo*, followed by adoptive transfer of these cells into syngeneic, non-transgenic recipients.

In order to obtain large number of memory cells, we initially examined priming conditions that activate most naive antigen specific cells to become effectors and to maximize survival of memory cells after the removal of the antigen challenge. We choose mini-osmotic pumps for antigen delivery. They provide a time-limited source of antigen, which is constant over the duration of antigen delivery. To minimize activation-induced cell death and increase the number of memory cells, we choose to examine the effect of IL-6, a pro-inflammatory cytokine, on the life span

* The Lautenberg Center for General and Tumor Immunology, The Hebrew University, Hadassah Medical School, Jerusalem, Israel and the #Laboratory of Immunology, National Institute of Allergy and Infectious Diseases, National Institutes of Health, Bethesda, MD 20892

of activated CD4 T cells. IL-6 was chosen since there is *in vitro* evidence that it promotes survival of activated naïve CD4 T cells.[1]

2. RESULTS

2.1 Antigen Administered in Mini-osmotic Pumps Elicits an Efficient Response by CD4 T Cells

We found that cytochrome C induced a powerful antigenic response when admin-istered in mini-osmotic pumps to transgenic mice expressing TCR's specific for this antigen (5C.C7 mice) or to syngeneic (B10.A) recipients of CD4 T cells from these mice. More then 99% of the antigen specific cells replicated in response to challenge with a pump containing 1 mg of antigen. Within 2-3 weeks, more then 50% of the surviving cells became capable of producing IL-4 and/or IFNγ.

2.2 IL-6 Strikingly Augments Survival and IL-4-Producing Capacity of ActivatedT Cells

Addition of IL-6 to the antigen-containing pump increased the numbers of antigen-specific cells beginning at 10 days. It also increased the percent of IL-4-producing cells in the population and the number of long-lived memory cells. Thus, ten days after immunization, the numbers of TCR transgenic cells in the lymph nodes was increased 10-fold and in the spleen ~6.5-fold in mice that received IL-6 in addition to cytochrome C compared to those that received cytochrome C only. At 16 days after priming, 67% of the TCR transgenic cells produced IL-4 when challenged *in vitro* with immobilized anti-CD3 and anti-CD28 compared to 35% in mice immunized with cytochrome C only. At 90 days after immunization, 77% of the TCR transgenic cells were $CD44^{high}$, implying that they were memory cells, whereas only 40% of the TCR transgenic cells in the cyctochrome C-primed mice had this phenotype.

2.3 Analysis of IL-6 Effects in a Transfer Model

To study the dynamics of TCR transgenic cells immunized with mini-osmotic pumps and IL-6 in a setting in which thier frequency is substantially less than in transgenic mice, we used a transfer system. Twenty million TCR transgenic cells from 5C.C7 donors were transferred to normal B10.A mice. Mini-osmotic pumps containing cytochrome C (1 mg) with or without IL-6 (100 μg) were implanted in these mice and the percentage of antigen specific cells (Vα11,Vß3) CD4 T cells was monitored. Five days after pump implantation, the percentage of Vα11,Vß3 cells in the blood of mice immunized with cytochrome C alone was similar to that in mice immunized with cytochrome C and IL-6 (16.5% Vs. 18.9%, respectively). By day 19, the percentage of Vα11,Vß3 cells in the blood of the mice immunized with cytochrome C had fallen precipitously, to 0.8%, while the proportion of Vα11Vß3 cells in the blood of mice immunized with cytochrome C plus IL-6 had risen to

46%. Two months after priming, mice immunized with cytochrome C plus IL-6 still had high percentage of Vα11,Vβ3 cells in their blood (14%). Similar differences in the number of antigen specific cells were found in the lymph nodes and the spleens of these mice. Thus, B10.A mice that received T cells from naive TCR transgenic mice and were then immunized 13 days earlier with cytochrome C plus IL-6 had 17-fold more Vα11,Vβ3 cells in their lymph nodes and 6.5 fold more in the spleens than mice immunized with cytochrome C alone.

The initial rate of replication of Vα11,Vβ3 cells in mice primed in the presence of IL-6 is only slightly faster than cells primed in the absence of IL-6. CFSE analysis revealed that >99% of the transferred naïve TCR transgenic cells replicated by day 3 in response to either immunization protocol. The calculated cell yield at that time, assuming complete survival, in mice immunized with IL-6 was only 15% greater than in mice immunized in the absence of IL-6.

2.4 Characteristics of *In Vivo* Generated Memory Cells

Three months after priming TCR transgenic mice, the surviving antigen specific cells failed to secrete IL-4 or IFNγ acute stimulation *in vitro*. However after short period of stimulation *in vitro* (2 days), a considerable fraction of the antigen specific cells (40%) from mice that had been immunized with cytochrome C plus IL-6 produced IL-4 compared with <5% of antigen-specific cells that had been obtained from un-immunized mice. Thus, surviving antigen specific cells retain their cytokine producing potential.

Memory cells and naïve 5C.C7 were transferred into naïve B10.A mice. The animals were then challenged with a mini-osmotic pump containing cytochrome C. Mice that had received memory cells produced IgG anti-cytochrome C antibodies 7 days after priming. Mice that have received a similar number of naïve transgenic cells and then primed under the same conditions display a very low IgG anti-cytochrome C response at that time.

Secondary stimulation of *in vivo* generated memory cells in B10.A recipients with a cytochrome C-containing pump induced a modest and transient increase in their cell number; by contrast, mice that had received naïve 5C.C7 cells displayed a striking elevation in the numbers of TCR transgenic cells. Thus, 13 days after pump implantation, there was a 1.5-fold increase in the pool of stimulated memory cells in the spleen whereas there was a 5-fold increase in the pool of stimulated naive cells. The difference was greater at later times. Fifty-three days after pump implantation, the number of memory cells in the spleen was 25% lower than in animals that were not challenged. By contrast, the number of antigen specific cells in immunized mice that had received naïve cells was 22.6-fold greater than in their un-immunized controls.

The re-challenged memory cells replicate vigorously; indeed, based on CFSE analysis, all of the cells responded and they replicate slightly more rapidly than do the naïve cells. The low cell yield of stimulated memory cells in the blood and the lymphoid organs is not due to their preferential migration to the tissues. Thirteen days after immunization, the percentage of Vα11,Vß3 cells in the livers of B10.A mice that had received memory cells was less than 50% that observed in their un-immunized controls. By contrast, the percentage of Vα11,Vß3 cells in the livers of immunized B10.A mice that received naïve transgenic cells was 2 fold greater than their un-immunized controls. Thus, long-lived memory cells appear to be more susceptible to activation-induced cell death than naïve cells and appear to contribute quite poorly to the memory cell pool after secondary challenge. By contrast, naive cells show massive expansion and a substantial portion of these cells enter the memory pool. However, the memory cells do show striking immediate responses consisting of the production of effector cytokines and the mediation of helper function.

3. DISCUSSION

The increase in cell yield in mice that were primed with mini-osmotic pumps containing cytochrome C and IL-6 are in agreement with our observation that a pure population of naïve CD4 cells can be rescued from activation-induced apoptotic cell death *in vitro* by adding either IL-4 or IL-6 to the priming culture.[1] In the presence of APC, these effects are minimized, possibly through APC production of cytokines. Nonetheless, striking increases in cell yields are observed in the presence of IL-6 in cultures of naive 5C.C7 CD4 cells with cytochrome C peptide and macrophages, B cells or dendritic cells as APC.

IL-6 has been reported to enhance survival of resting T cells[2] and to rescue them from apoptosis upon activation.[3] This effect appears to be important *in vivo* as well as *in vitro*. IL-6-deficient mice have a deficit in the induction of experimental allergic encephalomyelitis,[4] pristane-induced lupus[5] and antigen-induced arthritis.[6] In MOG-induced EAE, this appears to be due to decreased priming rather than to deviation of the response to a TH1 or TH2 polarity.[7] IL-6 has also been reported to induce IL-4 production in certain circumstances.[8]

The dependence of TCR stimulated naïve CD4 T cells on cytokines for survival may be part of the "fail-safe" mechanism that determines whether it is appropriate for T cells to respond to their cognate antigen. It has been increasingly recognized that the context of the antigen encounter is critical in determining the nature of the response. Various pathogens may induce conditions that favor a response on the part of T cells, a mechanism that insures that potentially tissue–damaging immune responses are only elicited in the face of authentic threats to the organism, as would be posed by infectious agents. Much of this may be expressed at the interface between the innate and the acquired limbs of the immune response.[9, 10] Based on the extreme cytokine requirements of naïve cells, a key element of the context that determines the appropriateness of responses is the cytokine milieu present at the time of antigen-recognition. Indeed, IL-6 is not a product of quiescent APC; it is made when these cells are appropriately activated by microbial stimulants;[11-16] IL-4 is not generally produced by APC, but it can be made by other hematopoietic cells such as basophils;[17] such production may also be determined indirectly by the cytokine milieu.

What is the physiological significance of the rapid activation and expression of effector function followed by elimination when resting memory cells are challenged with antigen? In particular, how is memory sustained after secondary challenge if activated memory cells show such striking elimination? One possibility is that memory is a very dynamic process in which there are cohorts of cells that are initially primed, enter the memory pool, show effector function upon re-challenge and die. They may be replaced by newly emerging naïve cells of the same specificity that respond to the challenge with a primary response. These cells show striking expansion, more modest effector function, but populate the memory pool for extended periods of time.

Do naïve cells participate in secondary responses? There is relatively little information on this point using methods in which naïve and memory cells can be adequately enumerated. However, it should be noted that the response of the memory cells, with their attendant production of cytokines, could produce an environment that was quite favorable for a vigorous primary response so that the naïve cells could undergo a striking expansion, perhaps even greater than they do under typical primary conditions. Direct studies of this issue are needed.

These ideas need to be more extensively tested both in a highly controlled models in which the numbers of memory and naïve cells can be directly and unambiguously measured and in more physiologic settings, particularly in responses to challenges with infectious agents.

REFERENCES

1. S. Z. Ben-Sasson, K. Makedonski, J. Hu-Li, and W. E. Paul, Survival and cytokine polarization of naive CD4(+) T cells in vitro is largely dependent on exogenous cytokines, *Eur. J. Immunol.* **30,** 1308-1317 (2000).
2. T. K. Teague, P. Marrack, J. W. Kappler, and A. T. Vella, IL-6 rescues resting mouse T cells from apoptosis, *J. Immunol.* **158,** 5791-5796 (197).
3. B. Adkins, K. Chun, K. Hamilton, and M. Nassiri, Naive murine neonatal T cells undergo apoptosis in response to primary stimulation. *J. Immunol.* **157,** 1343-1349 (1996).
4. Y. Okuda, S. Sakoda, C. C. Bernard, H. Fujumura, Y. Saeki, T. Koshimoto, and T. Yanagihara, IL-6-deficient mice are resistant to the induction of experimental autoimmune encephalomyelitis provoked by myelin oligodendrocyte glycoprotein. *Int. Immunol.* **10,** 703-708 (1998).
5. H. B. Richards, M. Satoh, M. Shaw, C. Libert, V. Poli, and W. H. Reeves, Interleukin 6 dependence of anti-DNA antibody production: evidence for two pathways of autoantibody formation in pristane-induced lupus. *J. Exp. Med.* **188,** 985-990 (1998).
6. S. Ohshima, Y. Saeki, T. Mima, M. Sasal, K. Nishioka, S. Nomura, M. Kopf, Y. Katada, T. Tanaka, M. Suemura, and T. Kishimoto, Interleukin 6 plays a key role in the development of antigen-induced arthritis. *Proc. Natl. Acad. Sci. USA* **95,** 8222-8226 (1998).
7. E. B. Samoilova, J. L. Horton, B. Hilliard, T. S. Liu, and Y. Chen, IL-6-deficient mice are resistant to experimental autoimmune encephalomyelitis: roles of IL-6 in the activation and differentiation of autoreactive T cells. **161,** 6480-6486 (1998).
8. M. Rincon, J. Anguita, T. Nakamura, E. Fikrig, and R. A. Flavell, Interleukin (IL)-6 directs the

differentiation of IL-4-producing CD4+ T cells. *J. Exp. Med.* **185,** 461-469 (1997).

9. D. T. Fearon, and R. M. Locksley, The instructive role of innate immunity in the acquired immune response. *Science* **272,** 50-53 (1996).
10. R. Medzhitov, and C. A. Janeway, Jr., Innate immunity: the virtues of a nonclonal system of recognition. *Cell* **91,** 295-298 (1997).
11. C. S. Hsieh, S. E. Macatonia, C. S. Tripp, S. F. Wolf, A. O'Garra, and K. M. Murphy, Development of TH1 CD4+ T cells through IL-12 produced by Listeria induced macrophages. *Science* **260,** 547-549 (1993).
12. C. R. Sousa, S. Hieny, T. Scharton-Kersten, D. Jankovic, H. Charest, R. N. Germain, and A. Sher, In vivo microbial stimulation induces rapid CD40 ligand independent production of interleukin 12 by dendritic cells and their redistribution to T cell areas. *J. Exp. Med.* **186,** 1819-1829 (1997).
13. S. Gessani, U. Testa, B. Varano, P. Di Marzio, P. Borghi, L. Conti, T. Barberi, E. Tritarelli, R. Martucci, and D. Seripa, Enhanced production of LPS induced cytokines during differentiation of human monocytes to macrophages. Role of LPS receptors. *J. Immunol.* **151,** 3758-3766 (1993).
14. J. Krutmann, A. Kirnbauer, A. Kock, T. Schwarz, E. Schopf, L. T. May, P. B. Sehgal, and T. A. Luger, Cross-linking Fc receptors on monocytes triggers IL-6 production. Role of anti-CD3-induced T cell activation. *J. Immunol.* **145,** 1337-1342 (1990).
15. J. Gosselin, L. Flamand, M. D'Addario, J. Hiscott, and J. Menezes, Infection of peripheral blood mononuclear cells by herpes simplex and Epstein-Barr viruses. Differential induction of interleukin 6 and tumor necrosis factor-alpha. *J. Clin. Invest.* **89,** 1849-1856 (1992).
16. K. Nakajima, O. Martinez-Maza, T. Hirano, E. C. Breen, P. G. Nishanian, J. F. Salazar-Gonzalez, J. L. Fahey, and T. Kishimoto, Induction of IL-6 (B cell stimulatory factor-2/IFN-beta2) production by HIV. *J. Immunol.* **142,** 531-536 (1989).
17. J. J. Bird, D. R. Brown, A. C. Mullen, N. Z. Moskowitz, M. A. Mahowald, J. R. Sider, T. F. Gajewski, C. R. Wang, and S. L. Reiner, Helper T cell differentiation is controlled by the cell cycle. *Immunity* **9,** 229-237 (1998).

T CELL PROLIFERATION, DIFFERENTIATION, AND RESTORATION IN LYMPHOPENIC INDIVIDUALS

For Contributed Volumes

Jianzhu Chen, Qing Ge, Bryan Cho, and Herman N. Eisen*

1. INTRODUCTION

When a small number of naïve T cells are adoptively transferred into syngeneic lymphopenic hosts, such as mice deficient in recombination activating gene-1 ($RAG^{-/-}$) or sublethally irradiated normal mice, the transferred T cells proliferate, a phenomenon referred to as homeostasis-driven or lymphopenia-induced proliferation. Although deliberate antigen stimulation is not required, homeostatic T cell proliferation requires i) the engagement of T cell receptor (TCR) with self peptide-major histocompatibility complexes (pepMHC) and ii) the presence of lymphopenia (or "space") (Tanchot et al., 1997, Ernst et al., 1999, Viret et al., 1999, Goldrath and Bevan, 1999, Bender et al., 1999). We have shown that following homeostatic proliferation naïve CD8 T cells acquire the phenotype and functional properties characteristic of antigen-induced memory T cells. We have also developed an in vitro system for studying homeostatic T cell proliferation. These findings and their implications for memory T cell development and the stable maintenance of naïve and memory T cell compartments are discussed.

2. HOMEOSTATIC MEMORY T CELL DEVELOPMENT

In the course of our studies of memory CD8 T cell development, we developed an adoptive transfer procedure for generating large numbers of homogenous memory T cells. In this procedure, naive CD8 T cells from transgenic mice on the $RAG1^{-/-}$ background, in which all T cells express the 2C TCR, were transferred into syngeneic $RAG1^{-/-}$ recipients. 2-3 days later, the recipients were challenged with antigenic peptide in complete Freund's adjuvant. Persisting 2C cells in the recipients expressed the characteristic cell surface markers and functional properties of memory CD8 T cells (antigen-induced memory T cells) (Cho et al., 1999). During the course of this study, we found that transfer of naïve 2C T cells into the $RAG1^{-/-}$ mice but without immunization

*Jianzhu Chen, Qing Ge, Herman N. Eisen, Center for Cancer Research and Department of Biology, Massachusetts Institute of Technology, Cambridge, MA, 02139. Bryan Cho, Medical Scholars Program and Department of Biochemistry, University of Illinois, Urbana, IL 61801.

also leads to their acquisition of memory T cell phenotype (homeostatic memory T cells or memory-like T cells) (Cho et al., 2000).

Naïve 2C T cells from transgenic mice were negative or low for CD44, IL-2Rβ, and Ly-6C. In contrast, persisting 2C cells from both immunized and non-immunized $RAG1^{-/-}$ recipients expressed high levels of CD44, IL-2Rβ, and Ly-6C. Both types of memory T cells responded much faster than naïve T cells to re-stimulation: They were induced to express cytolytic activity and IFN-γ in a few hours in vitro and were much more effective in rejecting antigen-bearing tumor cells in mice (Cho et al., 2000). The spontaneous memory T cell differentiation was not a result of expansion of few contaminating memory cells in the transferred population because transfer of purified $CD8^{+}CD44^{-}$ 2C cells also resulted in their differentiation into memory cells. Homeostatic memory T cell differentiation has also been observed with transgenic T cells expressing different TCR, including the 2C, P14, and OT-1 TCR, and polyclonal CD8 T cells from B6 mice (Cho et al., 2000, Kieper and Jameson, 1999, Goldrath et al., 2000, Murali-Krishna and Ahmed, 2000). Therefore, acquisition of memory phenotype and function by transferred T cells in lymphopenic hosts is a general phenomenon.

The critical requirement for homeostatic memory T cell differentiation is apparently the TCR-dependent proliferation. When different numbers of naïve CD8 T cells were transferred into lymphopenic hosts, they proliferated at different rates and extents. Correlating with the extent of proliferation, different proportions of persisting T cells were induced to express IFN-γ after a brief stimulation. Furthermore, as the transferred T cells divided they progressively acquired the capacity to express IFN-γ, exhibiting a linear relationship between the numbers of cell division and percentages of cells that were able to express IFN-γ (Cho et al., 2000). Thus, the proliferation and differentiation appear to be coupled. The fundamental difference between naïve and memory T cells lies in their different pattern of gene expression, which is achieved through modulating chromatin structures by epigenetic mechanisms, such as DNA methylation, histone acetylation, and binding of transcription factors. Proliferation probably facilitates chromatin remodeling of genes whose expression is critical for memory T cell development and maintenance.

Homeostatic memory T cell differentiation represents a novel pathway of memory T cell development. The requirement for TCR-mediated proliferation for homeostatic memory T cell differentiation is the same as antigen-induced memory T cell differentiation, except in the latter case the peptides are derived from exogenous antigens. There are, however, significant differences between the antigen- and homeostasis-driven memory T cell differentiation. Antigen-induced T cell proliferation is much more rapid than homeostatic T cell proliferation. Antigen-induced memory T cell differentiation goes through an effector cell stage whereas proliferating T cells under lymphopenic conditions do not become overtly activated to express CD69 and CD25 (Ernst et al., 1999, Cho et al., 2000, Murali-Krishna and Ahmed, 2000, Kieper and Jameson, 1999, Goldrath and Bevan, 1999). Although no significant differences have been reported, further studies are necessary to unequivocally demonstrate that homeostatic memory T cells are the same as antigen-induced memory T cells.

3. STABLE MAINTENANCE OF HOMEOSTATIC MEMORY T CELLS

Many studies have shown that antigen-induced memory T cells persist for prolonged periods, with memory phenotype and function stably maintained, even in the absence of MHC molecules (Murali-Krishna et al., 1999, Swain et al., 1999). Similarly, following homeostasis-driven differentiation, memory phenotype of 2C and OT-I T cells in RAG1$^{-/-}$ recipients or P14 T cells in sublethally irradiated C57BL/6 (B6) recipients is stable (Cho et al., 2000, Goldrath et al., 2000, Murali-Krishna and Ahmed, 2000). However, Goldrath et al. reported that following transfer of spleen and lymph node cells from OT-I transgenic mice into sublethally irradiated B6 recipients, naïve OT-I T cells appear to acquire memory phenotype only transiently and persisting T cells with naïve phenotype progressively accumulated in the recipients (Goldrath et al., 2000). Because the appearance of naïve OT-1 T cells correlated with the cessation of their proliferation in the recipients, homeostatic memory T cells were thought to revert to naïve phenotype when they cease to proliferate. Whether homeostatic memory T cells can actually revert to naïve T cells is a critical issue with two profound implications: One is that the memory T cell phenotype may not be as stable as once thought. The other is that the naïve T cell compartment can be restored by homeostasis-driven proliferation independent of thymopoiesis.

We reexamined the apparent phenotype reversion of homeostatic memory T cells in sublethally irradiated B6 recipients (Ge et al., 2002a). First, we showed that the number of homeostatic memory T cells of donor origin was stable in both RAG1$^{-/-}$ and irradiated B6 hosts and that their phenotype was stably maintained even after re-transfer into non-irradiated B6 mice for 30 days. Second, when naïve and memory T cells were considered separately, memory T cells proliferated at the same rate in RAG1$^{-/-}$ and irradiated B6 hosts. Third, accumulation of naïve CD8 T cells of donor origin only occurred in recipients that had been irradiated and had an intact thymus. Finally, small numbers of hematopoietic stem cells in the total spleen and lymph node cell populations used to reconstitute the recipients can repopulate the bone marrow of the irradiated mice and then differentiate in the irradiated recipients' thymus to give rise to naïve T cells. These findings demonstrate that the accumulation of donor-derived naïve T cells in irradiated recipients results from de novo T cell development in the thymus. Homeostatic memory T cells do not revert to naïve phenotype. Thus, after severe T cell depletion homeostatic proliferation restores only the memory T cell compartment whereas thymopoiesis is required for the reconstitution of the naïve compartment.

4. MOLECULAR BASIS OF "SPACE"

Homeostatic T cell proliferation depends upon the presence of lymphopenia (or "space"), since it occurs only in T cell-deficient hosts but not in hosts with normal numbers of T cells. As survival of naïve T cells requires contact with self pepMHC, why do these cells not proliferate in normal individuals? In the broad term, there are two mechanisms that may prevent T cell proliferation in the absence of exogenous antigen in normal hosts: First, homeostatic T cell proliferation requires stimulating signals that are limiting. In normal individuals, T cells compete for the same limiting signals and no cells receive enough of them to proliferate. In lymphopenic hosts, however, small numbers of naïve cells are able to receive sufficient stimulation to proliferate. Possibly, self pepMHC complexes may be limiting and other positive signals may also be limiting. Recently, cytokine IL-7 and IL-12 were shown to promote homeostatic CD8 T cell proliferation in lymphopenic mice (Schluns et al., 2000, Kieper et al., 2001), indicating that soluble stimulating factors may provide a concrete basis for the notion of "space". Second, T cells may be mutually inhibitory when T cell density is high as in normal individuals. When small numbers of naïve T cells are transferred into lymphopenic hosts, there is less opportunity for inhibition.

To investigate these various possibilities, we have established an in vitro T cell-dendritic cell coculture system in which T cell undergo homeostatic proliferation (Ge et al., 2002b). Just as in lymphopenic mice, T cells proliferated in the cocultures only in the presence of syngeneic dendritic cells (DCs) that express self pepMHC but not DCs from mice deficient in β2m. Addition of the Fab fragment of antibody specific for the 2C TCR blocked 2C T cell proliferation. In addition, $CD8^-$ 2C cells proliferated very little and anti-CD8 antibody blocked the proliferation of $CD8^+$ 2C cells.

Using the in vitro model system, we showed that DCs, but not other antigen presenting cells, provide the required self-pepMHC complexes for homeostatic T cell proliferation. Besides providing self pepMHC complexes, DCs secrete IL-15 that promotes T cell proliferation in the coculture system. T cells actively competed for limiting amounts of self pepMHC as well as for soluble stimulating factors, suggesting that competition for stimulating positive signals limits T cell proliferation in normal individuals. In addition, we showed that $CD4^+CD25^+$ regulatory T cells actively inhibited homeostatic T cell proliferation. This inhibition was abolished when the regulator T cells were separated from naïve CD8 T cells, indicating a requirement for cell-cell contact. Thus, it appears that inhibitory T cells in normal individuals can also prevent homeostatic T cell proliferation.

In summary, homeostatic T cell proliferation requires stimulation through positive signals (self pepMHC and soluble factors) and relief from inhibitory signals. Following homeostatic proliferation, naïve T cells acquire stable memory T cell phenotype and functions. Thus, after severely T cell depletion, homeostatic proliferation restores the memory T cell compartment but not the naïve T cell compartment, which requires for its restoration de novo T cell development in the thymus.

REFERENCES

Bender, J., Mitchell, T., Kappler, J., and Marrack, P. (1999). $CD4^+$ T cell division in irradiated mice requires peptides distinct from those responsible for thymic selection. J. Exp. Med. *190*, 367-373.

Cho, B., Varada, R., Ge, Q., Eisen, H. N., and Chen, J. (2000). Homeostasis-stimulated proliferation drives naïve T cells to differentiate directly into memory cells. J. Exp. Med. *192*, 549-556.

Cho, B. K., Wang, C., Sugawa, S., Eisen, H. N., and Chen, J. (1999). Functional differences between memory and naive CD8 T cells. Proc. Natl. Acad. Sci. USA *96*, 2976-2981.

Ernst, B., Lee, D.-S., Chang, J. M., Sprent, J., and Surh, C. D. (1999). The peptide ligands mediating positive selection in the thymus control T cell survival and homeostatic proliferation in the periphery. Immunity *11*, 173-181.

Ge, Q., Hu, H., Eisen, H. N., and Chen, J. (2002a). Different contributions of thymopoiesis and homeostasis-driven proliferation to the reconstitution of naive and memory T cell compartments. Proc. Natl. Acad. Sci. USA, In press.

Ge, Q., Palliser, D., Eisen, H. N., and Chen, J. (2002b). Homeostatic T cell proliferation in a T-dendritic cell coculture system. Proc. Natl. Acad. Sci. USA, In press.

Goldrath, A. W., and Bevan, M. J. (1999). Low-affinity ligands for the TCR drive proliferation of mature $CD8^+$ T cells in lymphopenic hosts. Immunity *11*, 183-190.

Goldrath, A. W., and Bevan, M. J. (1999). Selecting and maintaining a diverse T-cell repertoire. Nature *402*, 255-261.

Goldrath, A. W., Bogatzki, L. Y., and Bevan, M. J. (2000). Naive T cells transiently acquire a memory-like phenotype during homeostasis-driven proliferation. J. Exp. Med. *192*, 557-564.

Kieper, W. C., and Jameson, S. C. (1999). Homeostatic expansion and phenotypic conversion of naive T cells in response to self peptide/MHC ligands. Proc. Natl. Acad. Sci. USA *96*, 13306-13311.

Kieper, W. C., Prlic, M., Schmidt, C. S., Mescher, M. F., and Jameson, S. C. (2001). IL-12 enhances CD8 T cell homeostatic proliferation. J. Immunol. *166*, 5515-5521.

Murali-Krishna, K., and Ahmed, R. (2000). Cutting edge: naive T cells masquerading as memory cells. J. Immunol. *165*, 1733-1737.

Murali-Krishna, K., Lau, L. L., Sambhara, S., Lemonnier, F., Altman, J., and Ahmed, R. (1999). Persistence of memory CD8 T cells in MHC class I-deficient mice. Science *286*, 1377-1381.

Schluns, K. S., W.C., K., Jameson, S. C., and Lefrancois, L. (2000). Interleukin-7 mediates the homeostasis of naive and memory CD8 T cells in vivo. Nat. Immunol. *1*, 426-432.

Swain, S., Hu, H., and Huston, G. (1999). Class II-independent generation of CD4 memory T cells from effectors. Science *286*, 1381-1383.

Tanchot, C., Lemonnier, F. A., Pérarnau, B., Freitas, A. A., and Rocha, B. (1997). Differential requirements for survival and proliferation of CD8 naive or memory T cells. Science *276*, 2057-2062.

Viret, C., Wong, F. S., and Janeway, J. C. A. (1999). Designing and maintaining the mature TCR repertoire: the continuum of self-peptide:self-MHC complex recognition. Immunity *10*, 559-568.

MIGRATION OF PRIMARY AND MEMORY CD8 T CELLS

Leo Lefrançois, Amanda L. Marzo, David Masopust, Kimberly S. Schluns, and Vaiva Vezys*

1. INTRODUCTION

Infection with viruses and bacteria generally results in robust induction of the innate immune response followed by the generation of an antigen-specific adaptive immune response comprised of CD4 and CD8 T cells and B cells. Thus, there are multiple cellular players in the induction of the response and although our knowledge has increased dramatically in recent years, there is much to learn regarding the complex interplay of the various cell types necessary in mounting a productive, and potentially protective, immune response. Long-term protection following infection or vaccination is mediated by memory T and B cells, as well as by circulating neutralizing antibody. The molecular and cellular requirements for generation and maintenance of immunological memory are largely unknown, although recent studies have begun to yield some intriguing clues in this area.

Activation of T cells in vivo results in a complex program of events including clonal expansion and induction of effector function as well as alteration of the migration of activated T cells.[1-3] In the case of CD8 T cells activated during virus or bacteria infections, effector cells acquire the ability to migrate to a wide variety of non-lymphoid tissues, including the liver, lung and intestinal mucosa.[4,5] This finding also applies to CD4 T cells activated with soluble antigen in the presence of adjuvant.[6] It can be hypothesized that the raison d'etre for this diaspora of activated T cells is the patrolling of the body for the purpose of clearance of the infectious agent. It remains unclear whether distinct subsets of tissue homing cells are generated during the primary response or whether multiple homing receptors are available on all responding cells. That is, it is possible that cells activated in a particular lymph node (LN), e.g., a LN draining the lung, would be induced in that site to preferentially express homing receptors geared toward migration to lung tissue. Recent data showing that programming of CD4 T cells for a particular chemokine responsiveness occurs in distinct lymph node environments supports this concept.[7] It is also possible that different functional programs could be induced upon migration of T cells into a particular tissue. To wit, T cells, initially

*University of Connecticut Health Center, Division of Immunology, Department of Medicine, Farmington, CT 06030-1319.

activated in a LN, upon entry into the intestinal mucosa might be impinged upon by the cytokine environment of the gut or by additional APC interactions in the mucosa to preferentially produce a given array of cytokines, while cells derived from the same initial activation but migrating to another tissue would be "programmed" in that site to produce a different array of cytokines. Distinguishing between the different possibilities of where and when programming for effector function occurs is oftentimes difficult to discern given the complex patterns of trafficking and the vast array of cytokines, chemokines and adhesion molecules involved in immune response induction and regulation. Nevertheless, the systems we and others are now employing have allowed us to begin to address these issues. For example, following activation in vivo, adoptively transferred TCR transgenic CD8 T cells utilize CD40L to undergo secondary activation and further proliferation upon entry into the intestinal mucosa.[8]

2. DETECTION OF MEMORY CELLS FOLLOWING INFECTION

Following the peak of the primary response to infection a dramatic contraction phase occurs in which most antigen-specific T cells die by apoptosis.[9-11] The mechanism by which this occurs is not fully understood but may be due in part to growth factor withdrawal.[12] Subsequent to the contraction of the response a numerically small but stable population of memory T cells can be detected.[11,13-15] Available data suggests that memory T cells arise from a subset of surviving effector cells,[16-19] although it remains possible that some memory cells could represent a distinct lineage that has not passed through an effector stage.[20,21] In any case, with the use of the adoptive transfer system as well as MHC tetramers much of the dynamics of the CD8 response to viral and bacterial infections has been elucidated. In some infections of mice such as with lymphocytic choriomeningitis virus (LCMV),[2,3] vesicular stomatitis virus (VSV)[22] or Listeria monocytogenes (LM),[23] 20–50% of the CD8 T cell population in the spleen can be virus specific at the peak of the response. Following these and other infections, memory CD8 T cells can be detected directly ex vivo using MHC class I tetramers containing the appropriate peptides.[24]

3. MIGRATION OF ACTIVATED CD8 T CELLS TO MULTIPLE TISSUES

While most studies have focused on analysis of memory cells present in secondary lymphoid tissues, we analyzed the primary and memory CD8 response to VSV and LM in non-lymphoid tissues.[5] In both the effector and memory phases of the response, antigen-specific T cells were at least as plentiful in the non-lymphoid tissues (lung, liver, LP, kidney, etc) as they were in the spleen and LN. In a later study, similar results were obtained after induction of a recall CD8 response to influenza infection.[33] In addition, analysis of the response of adoptively transferred TCR transgenic CD4 T cells to soluble antigen and adjuvant immunization also demonstrated that a substantial proportion of the primary and memory cells were present outside of secondary lymphoid tissues.[6]

It is unknown whether the memory cells present in non-lymphoid tissues are part of a larger circulating pool or represent in whole or part separate populations of memory cells. Our data suggests that the latter may be the case, at least with respect to memory CD8 T cells present in the intestinal LP. A subset of LP memory CD8 T cells express the αEβ7 integrin at high levels while memory cells outside the intestinal mucosa lack this integrin.[4,5] Also, nearly all LP memory cells express the CD43-associated carbohydrate antigen, 1B11, while only a subset of extra-intestinal memory cells express this marker.[5] While this data is circumstantial, the implication is that intestinal LP memory cells may represent a separate pool of memory cells.

4. FUNCTIONAL DIFFERENCES BETWEEN LYMPHOID AND NON-LYMPHOID MEMORY T CELLS

Perhaps more importantly than the quantitative aspects of the non-lymphoid memory response, is the finding that these memory cells display heightened effector function as compared to memory cells in LN and spleen. We demonstrated that CD8 memory cells from lung, liver or intestinal LP exhibited high levels of constitutive lytic activity, while splenic memory cells were poorly cytolytic. Similarly, Rheinhardt et al showed that memory ova-specific D011.10 CD4 T cells in the lung rapidly responded to peptide challenge by producing higher levels of IFN-γ than did LN memory cells. These results again suggest the possibility that at least two distinct memory pools exist as proposed by Sallusto et al based on analysis of human peripheral blood lymphocytes: effector memory cells residing in tertiary tissues and central memory cells present in lymphoid tissue.[34] The latter require reactivation to exert effector function while the former are constitutive effectors.

5. REGULATION OF CD8 MEMORY T CELL DEVELOPMENT BY IL-7

The factors involved in generation and maintenance of CD8 memory T cells are largely unknown. Interleukin-7 (IL-7) is a cytokine important in T and B cell development in part as a survival factor.[35-38] Memory CD8 T cells express the IL-7R but whether the receptor is important for memory T cell development or survival is unclear. In order to test the role of IL-7 in generation of CD8 memory T cells we utilized an adoptive transfer system in which ovalbumin specific TCR transgenic CD8 T cells (OT-I) either capable of IL-7R expression or not, are activated in vivo by virus infection.[39] A mixture containing equal numbers of OT-I and OT-I IL7R$^{-/-}$ cells were transferred to normal mice which were then infected with VSV-ova. We examined the ratio of OT-I and OT-I IL7R$^{-/-}$ cells at the peak of the primary response four days after infection. Substantial expansion of normal and OT-I IL7R$^{-/-}$ T cells had occurred on day four after infection. At this time, the ratio of normal OT-I T cells to OT-I IL7R$^{-/-}$ cells was ~60:40 in the spleen and LN indicating that the OT-I IL7R$^{-/-}$ cells had greatly expanded in response to the virus infection. In contrast, 40 days after infection the OT-I memory population was made up primarily of normal cells due to the loss of OT-I IL7R$^{-/-}$ cells.

This data demonstrated that IL-7R-deficient OT-I T cells were less likely to survive and differentiate into memory T cells.

The relative decrease in the number of OT-I IL7R$^{-/-}$ T cells at day four after infection could have been due to decreased ability to respond to the infection or to a decreased survival of OT-I IL7R$^{-/-}$ T cells. Therefore, the experiment was repeated in a similar manner except that the OT-I T cells were labeled with CFSE prior to transfer and an earlier time point was analyzed. Three days after infection with VSV-ova, most of the OT-I IL7R$^{-/-}$ T cells and the normal OT-I T cells had undergone more than seven divisions in the spleen. In the peripheral LN, the proliferation of the OT-I T cells and the OT-I IL7R$^{-/-}$ T cells had lagged behind that detected in the spleen with the cells having undergone four-eight divisions. In both tissues the proliferation pattern of the OT-I IL7R$^{-/-}$ T cells was the same as normal OT-I T cells. These results suggested that the decrease in the ratio of mutant to control T cells early during the immune response was due to lack of IL-7 mediated survival signals.

6. SUMMARY

Our results indicate that a substantial proportion of the antimicrobial CD8 and CD4 T cell response is focused in non-lymphoid tissues. This finding makes teleological sense since maximum protection against infection is better served by the widespread presence of effector and memory cells. In the case of CD8 T cells, it appears that irrespective of the site at which initial activation of naive cells occurs, the end result is production of effector cells with broad migratory capabilities. Memory T cells perhaps have more restricted migratory abilities as compared to effector cells, although this needs to be tested definitively. Our results suggest that memory CD8 T cells in the intestinal LP may not be part of the recirculating pool of memory cells, though our data does not preclude the possibility that migrants from outside the mucosa contribute to the LP memory pool. Our data also demonstrates that CD8 memory T cells in non-lymphoid tissue exhibit heightened effector function as compared to their splenic counterparts. Whether these findings indicate the existence of distinct lineages of memory cells remains to be seen. The functional abilities of migrating memory cells could be modulated by the migration process and/or by the environmental milieu of a particular tissue. Although the development of CD8 memory T cells is a complex process requiring multiple signalling pathways, we identified IL-7 as an important player in memory generation. Much further work is needed to decipher the cellular and molecular mechanisms of memory induction as well as to learn the in vivo functional significance of memory cell subsets.

7. ACKNOWLEDGMENTS

This work was supported by NIH grants AI41576, DK45260, AI35917 (LL), a collaborative grant from the Edward Jenner Institute for Vaccine Research and a C. J. Martin Fellowship (007151) (A.L.M.) and a USPHS training grant T32-AI07080 (VV and DM).

REFERENCES

1. E. C. Butcher, M. Williams, K. Youngman, L. Rott, and M. Briskin, Lymphocyte trafficking and regional immunity, *Adv. Immunol.* **72**, 209-253 (1999).
2. K. Murali-Krishna, J. D. Altman, M. Suresh, D. J. Sourdive, A. J. Zajac, J. D. Miller, J. Slansky, and R. Ahmed, Counting antigen-specific CD8 T cells: a reevaluation of bystander activation during viral infection, *Immunity* **8**, 177-187 (1998).
3. E. A. Butz and M. J. Bevan, Massive expansion of antigen-specific CD8+ T cells during an acute virus infection, *Immunity* **8**, 167-175 (1998).
4. S. K. Kim, K. S. Schluns, and L. Lefrançois, Induction and visualization of mucosal memory CD8 T cells following systemic virus infection, *J. Immunol.* **163**, 4125-4132 (1999).
5. D. Masopust, V. Vezys, A. L. Marzo, and L. Lefrançois, Preferential localization of effector memory cells in nonlymphoid tissue, *Science* **291**, 2413-2417 (2001).
6. R. L. Reinhardt, A. Khoruts, R. Merica, T. Zell, and M. K. Jenkins, Visualizing the generation of memory CD4 T cells in the whole body, *Nature* **410**, 101-105 (2001).
7. R. Ishimitsu, H. Nishimura, T. Yajima, T. Watase, H. Kawauchi, and Y. Yoshikai, Overexpression of IL-15 in vivo enhances Tc1 response, which inhibits allergic inflammation in a murine model of asthma, *J. Immunol.* **166**, 1991-2001 (2001).
8. L. Lefrançois, S. Olson, and D. Masopust, A critical role for CD40-CD40 ligand interactions in amplification of the mucosal CD8 T cell response, *J. Exp. Med.* **190**, 1275-1284 (1999).
9. C. Zimmermann, K. Brduscha-Riem, C. Blaser, R. M. Zinkernagel, and H. Pircher, Visualization, characterization, and turnover of CD8+ memory T cells in virus-infected hosts., *J. Exp. Med.* **183**, 1367-1375 (1996).
10. P. C. Doherty, D. J. Topham, and R. A. Tripp, Establishment and persistence of virus-specific CD4+ and CD8+ T cell memory., *Immunol. Rev.* **150:23-44**, 23-44 (1996).
11. R. Ahmed and D. Gray, Immunological memory and protective immunity: understanding their relation., *Science* **272**, 54-60 (1996).
12. M. Lenardo, K. M. Chan, F. Hornung, H. McFarland, R. Siegel, J. Wang, and L. Zheng, Mature T lymphocyte apoptosis--immune regulation in a dynamic and unpredictable antigenic environment, *Annu. Rev. Immunol.* **17**, 221-253 (1999).
13. J. Sprent and C. D. Surh, Generation and maintenance of memory T cells, *Curr. Opin. Immunol.* **13**, 248-254 (2001).
14. K. J. Flynn, J. M. Riberdy, J. P. Christensen, J. D. Altman, and P. C. Doherty, In vivo proliferation of naive and memory influenza-specific CD8(+) T cells., *Proc. Natl. Acad. Sci. USA* **96**, 8597-8602 (1999).
15. R. W. Dutton, L. M. Bradley, and S. L. Swain, T cell memory., *Annu. Rev. Immunol.* **16**, 201-223 (1998).
16. J. T. Opferman, B. T. Ober, and P. G. Ashton-Rickardt, Linear differentiation of cytotoxic effectors into memory T lymphocytes, *Science* **283**, 1745-1748 (1999).
17. S. M. Kaech and R. Ahmed, Memory CD8+ T cell differentiation: initial antigen encounter triggers a developmental program in naive cells, *Nat. Immunol* **2**, 415-422 (2001).
18. S. Ehl, P. Klenerman, P. Aichele, H. Hengartner, and R. M. Zinkernagel, A functional and kinetic comparison of antiviral effector and memory cytotoxic T lymphocyte populations in vivo and in vitro, *Eur. J. Immunol* **27**, 3404-3413 (1997).
19. M. J. Van Stipdonk, E. E. Lemmens, and S. P. Schoenberger, Naive CTLs require a single brief period of antigenic stimulation for clonal expansion and differentiation, *Nat. Immunol* **2**, 423-429 (2001).
20. N. Manjunath, P. Shankar, J. Wan, W. Weninger, M. A. Crowley, K. Hieshima, T. A. Springer, X. Fan, H. Shen, J. Lieberman, and U. H. von Andrian, Effector differentiation is not prerequisite for generation of memory cytotoxic T lymphocytes, *J. Clin. Invest* **108**, 871-878 (2001).
21. J. Sprent and D. F. Tough, T cell death and memory, *Science* **293**, 245-248 (2001).
22. D. Masopust, J. Jiang, H. Shen, and L. Lefrançois, Direct analysis of the dynamics of the intestinal mucosa CD8 T cell response to systemic virus infection, *J. Immunol.* **166**, 2348-2356 (2001).
23. D. H. Busch, I. M. Pilip, S. Vijh, and E. G. Pamer, Coordinate regulation of complex T cell populations responding to bacterial infection, *Immunity* **8**, 353-362 (1998).
24. J. D. Altman, P. A. H. Moss, P. J. R. Goulder, D. H. Barouch, M. G. McHeyzer-Williams, J. I. Bell, A. J. McMichael, and M. M. Davis, Phenotypic analysis of antigen-specific T lymphocytes, *Science* **274**, 94-96 (1996).

25. J. J. Peschon, P. J. Morrissey, K. H. Grabstein, F. J. Ramsdell, E. Maraskovsky, B. C. Gliniak, L. S. Park, S. F. Ziegler, D. E. Williams, C. B. Ware, J. D. Meyer, and B. L. Davison, Early lymphocyte expansion is severely impaired in interleukin 7 receptor-deficient mice, *J. Exp. Med.* **180**, 1955-1960 (1994).
26. K. A. Hogquist, S. C. Jameson, W. R. Heath, J. L. Howard, M. J. Bevan, and F. R. Carbone, T cell receptor antagonistic peptides induce positive selection., *Cell* **76**, 17-27 (1994).
27. K. Laky, L. Lefrançois, and L. Puddington, Age-dependent intestinal lymphoproliferative disorder due to stem cell factor receptor deficiency: parameters in small and large intestine., *J. Immunol.* **158**, 1417-1427 (1997).
28. J. W. Yewdell, J. R. Bennink, M. Mackett, L. Lefrançois, D. S. Lyles, and B. Moss, Recognition of cloned vesicular stomatitis virus internal and external gene products by cytotoxic T lymphocytes, *J. Exp. Med.* **163**, 1529-1538 (1986).
29. S. K. Kim, D. S. Reed, S. Olson, M. J. Schnell, J. K. Rose, P. A. Morton, and L. Lefrançois, Generation of mucosal cytotoxic T cells against soluble protein by tissue-specific environmental and costimulatory signals., *Proc. Natl. Acad. Sci. USA* **95**, 10814-10819 (1998).
30. G. M. Van Bleek and S. G. Nathenson, Isolation of an endogenously processed immunodominant viral peptide from the class I H-2K^b molecule, *Nature* **348**, 213-216 (1990).
31. H. Shen, M. K. Slifka, M. Matloubian, E. R. Jensen, R. Ahmed, and J. F. Miller, Recombinant Listeria monocytogenes as a live vaccine vehicle for the induction of protective anti-viral cell-mediated immunity, *Proc Natl Acad Sci U S A* **92**, 3987-3991 (1995).
32. H. Shen, J. F. Miller, X. Fan, D. Kolwyck, R. Ahmed, and J. T. Harty, Compartmentalization of bacterial antigens - differential effects on priming of CD8 T cells and protective immunity, *Cell* **92**, 535-545 (1998).
33. D. R. Marshall, S. J. Turner, G. T. Belz, S. Wingo, S. Andreansky, M. Y. Sangster, J. M. Riberdy, T. Liu, M. Tan, and P. C. Doherty, Measuring the diaspora for virus-specific CD8+ T cells, *Proc. Natl. Acad. Sci. U. S. A* (2001).
34. F. Sallusto, D. Lenig, R. Forster, M. Lipp, and A. Lanzavecchia, Two subsets of memory T lymphocytes with distinct homing potentials and effector functions [see comments], *Nature* **401**, 708-712 (1999).
35. U. von Freeden-Jeffry, P. Vieira, L. A. Lucian, T. McNeil, S. E. G. Burdach, and R. Murray, Lymphopenia in interleukin (IL)-7 gene-deleted mice identifies IL-7 as a nonredundant cytokine, *J. Exp. Med.* **181**, 1519-1526 (1995).
36. N. B. Perumal, T. W. Kenniston Jr., D. J. Tweardy, K. F. Dyer, R. Hoffman, J. Peschon, and P. M. Appasamy, TCR-g Genes Are Rearranged but Not Transcribed in IL-7Ra-Deficient Mice., *Journal of Immunology* **158**, 5744-5750 (1997).
37. E. Maraskovsky, M. Teepe, P. J. Morrissey, S. Braddy, R. E. Miller, D. H. Lynch, and J. J. Peschon, Impaired survival and proliferation in IL-7 receptor-deficient peripheral T cells, *J. Immunol.* **157**, 5315-5323 (1996).
38. J. Hassan and D. J. Reen, IL-7 promotes the survival and maturation but not differentiation of human post-thymic CD4+ T cells, *Eur J Immunol* **28**, 3057-3655 (1998).
39. K. S. Schluns, W. C. Kieper, S. C. Jameson, and L. Lefrançois, Interleukin-7 mediates the homeostasis of naive and memory CD8 T cells in vivo, *Nat. Immunol.* **1**, 426-432 (2000).

CYTOKINES AND MEMORY-PHENOTYPE $CD8^+$ CELLS

Jonathan Sprent, Adam D. Judge, and Xiaohong Zhang*

1. INTRODUCTION

In searching for pathogens, T cells migrate continuously between blood and lymph through defined regions of the secondary lymphoid tissues.[1] Prior to contact with antigen, T cells are maintained in a quiescent state and rarely divide.[2, 3] These "virgin" (naive) cells comprise the vast majority of T cells in young life and display a resting phenotype typified by low expression of CD44. With exposure to foreign antigens, specifically-reactive naive $CD44^{lo}$ T cells respond to these antigens and differentiate into T cells with an activated/memory $CD44^{hi}$ phenotype.[3-8] At a population level, $CD44^{hi}$ T cells are presumed to carry immunological memory for a wide variety of commonly-encountered environmental antigens. In addition, some memory-phenotype $CD44^{hi}$ T cells may represent the progeny of naive T cells responding to self antigens.[9] This process, termed homeostatic expansion of T cells, is prominent when T cell numbers are reduced, e.g. after irradiation, but is generally inconspicuous in normal animals.

A remarkable feature of the immune system is that total numbers of T cells remain relatively constant throughout life.[2, 3, 10, 11] T cell numbers increase in parallel with body weight during young life but then reach a stable equilibrium after puberty. Maintenance of T cells at a constant level applies despite a marked shift in the proportions of naive- and memory-phenotype T cells with age. Thus, whereas naive-phenotype T cells predominate in young life, memory-phenotype T cells gradually increase thereafter and become the majority population in old age.

The homeostatic mechanisms that control total T cell numbers are still poorly understood. Regulation at the level of initial T cell formation is unlikely because release of newly-formed T cells from the thymus reaches a peak at puberty and then falls to a low level thereafter.[12] Currently, there is increasing evidence that contact with cytokines plays a key role in T cell homeostasis.[11] In this article we review evidence that the turnover and survival of memory-phenotype $CD8^+$ T cells in mice is controlled by IL-15.

* Jonathan Sprent, The Scripps Research Institute, La Jolla, CA 92037

1.1. Features of Memory-phenotype $CD8^+$ Cells

Like $CD4^+$ cells, memory-phenotype $CD8^+$ cells display a characteristic pattern of surface markers that distinguishes these cells from naive T cells.[2-11] Thus, whereas naive T cells have a $CD44^{lo}$ $CD45RB^{hi}$ $CD62L^{hi}$ phenotype, most memory $CD4^+$ and $CD8^+$ cells are $CD44^{hi}$ $CD45RB^{lo}$ $CD62L^{lo}$. Nevertheless, some memory cells, especially $CD8^+$ cells, can retain or reacquire a $CD45RB^{hi}$ $CD62L^{hi}$ phenotype and thus resemble naive T cells. Currently, high expression of CD44 seems to be the most reliable marker for memory cells. In addition, most memory $CD8^+$ cells are $Ly6C^{hi}$ and $CD122^{hi}$. As discussed below, the $CD122^{hi}$ phenotype of memory $CD8^+$ cells is important because CD122 (IL-2Rβ) is an essential component of the receptor for IL-15 and IL-2.[13]

1.2. Rate of Turnover

As for $CD4^+$ cells, the background rate of proliferation (turnover) of memory-phenotype $CD8^+$ cells is appreciably higher than for naive-phenotype cells.[3] Thus, whereas naive-phenotype T cells can remain in interphase for weeks or months, memory-phenotype cells divide intermittently, in the order of once every 2-3 weeks.

1.3. Stimuli for Turnover

As discussed in the chapter of Charles Surh, the long-term survival of naive T cells requires continuous TCR contact with ubiquitous self peptides bound to MHC molecules.[11] In conjunction with exposure to a cytokine, IL-7, TCR interaction with self peptide/MHC ligands is thought to provide a low-level signal that stimulates T cell survival, probably by maintaining expression of anti-apoptotic molecules. This signal is normally too weak to induce entry into cell cycle. However, the intensity of TCR signalling appears to increase when total T cell numbers drop below a certain threshold, e.g. when T cells are transferred to irradiated hosts. Under these conditions, contact with self ligands plus IL-7 drives naive T cells to undergo "homeostatic" proliferation and expansion.

In light of these findings on naive T cells, one could envisage that the normal turnover and survival of memory-phenotype T cells reflects contact with self ligands. However, this is not the case. Thus, based on the results of transferring T cells to $MHC^{-/-}$ hosts (MHC class $II^{-/-}$ hosts for $CD4^+$ cells and MHC class $I^{-/-}$ hosts for $CD8^+$ cells), the survival and turnover of memory-phenotype T cells is MHC independent.[14, 15]

In considering other stimuli for memory-phenotype T cells, useful information has come from studies on the effects of injecting mice with compounds that activate the innate immune system.

1.4. Bystander Stimulation

Past studies showed that injecting mice with Poly I:C, lipopolysaccharide or CpG DNA induced a brief burst of T cell proliferation.[16-18] Proliferation was restricted to memory-phenotype T cells and was most prominent for $CD8^+$ cells. Based on studies with $MHC^{-/-}$ mice and mice lacking expression of type I interferon (IFN-I) receptors and IFN-γ, the proliferation induced by the above compounds was MHC class I independent but IFN dependent. In agreement with this notion, comparable proliferation occurred following injection of purified IFN-I[16] or IFN-γ.[19] Proliferation thus seemed to reflect a "bystander" (TCR-independent) response mediated by IFNs. Significantly, however, despite strong proliferation induced by IFNs and IFN-inducing agents *in vivo*, proliferation failed to occur when purified T cells were exposed to IFNs *in vitro.*[20]

In light of these findings, IFN-induced proliferation of memory-phenotype ($CD44^{hi}$) $CD8^+$ cells *in vivo* was presumably mediated by a different cytokine. IL-15 seemed a likely candidate since expression of CD122 was found to be much higher on $CD44^{hi}$ $CD8^+$ cells than on $CD44^{hi}$ $CD4^+$ cells or on naive T cells.[20] In favor of this idea, stimulation of macrophages by IFNs and IFN-inducing agents led to strong synthesis of IL-15 mRNA. More importantly, unlike IFNs, IL-15 elicited potent proliferation of purified $CD44^{hi}$ $CD8^+$ cells *in vitro*. IL-15 also stimulated $CD44^{hi}$ $CD4^+$ cells, but much less effectively than $CD44^{hi}$ $CD8^+$ cells. By contrast, IL-2 induced equivalent stimulation of $CD44^{hi}$ $CD4^+$ and $CD44^{hi}$ $CD8^+$ cells. More recently, proof that bystander proliferation is elicited by IL-15 came from the finding that only very limited proliferation occurred when T cells were exposed to IFN-inducing agents after transfer to $IL\text{-}15^{-/-}$ hosts.[21]

1.5. Normal Turnover

The above evidence indicates that stimulation of the innate immune system by products of pathogens leads to production of IFNs which, in turn, cause IL-15 synthesis and selective stimulation of $CD44^{hi}$ $CD8^+$ cells. This scenario raises the question whether the normal background turnover of $CD44^{hi}$ $CD8^+$ cells *in vivo* is IL-15 dependent. In favor of this idea, injecting mice with anti-CD122 mAb reduces the turnover of $CD44^{hi}$ $CD8^+$ cells but does not affect $CD44^{hi}$ $CD4^+$ cells.[22] Interestingly, injection of anti-IL-2 mAb has the opposite effect and augments $CD44^{hi}$ $CD8^+$ cell turnover. The implication therefore is that the normal turnover of $CD44^{hi}$ $CD8^+$ cells is indeed IL-15 dependent but is negatively regulated by IL-2.

Although IL-2 is made almost exclusively by T cells, the tissue source of IL-15 is unknown. With the notable exception of T cells, many cell types can synthesize IL-15, including dendritic cells (DC).[13, 23] Since T cells make frequent contact with DC during their migration through the lymphoid tissues, DC could be a prime source of IL-15 for mediating normal turnover of $CD44^{hi}$ $CD8^+$ cells. Direct evidence on this question, however, is lacking.

1.6. Survival

In addition to controlling $CD44^{hi}$ $CD8^+$ cell turnover, IL-15 may also promote the generation and/or survival of these cells. Strong support for this possibility is provided by the finding that $IL\text{-}15^{-/-}$ mice[24] and also $IL\text{-}15R\alpha^{-/-}$ mice[25] are selectively depleted of $CD44^{hi}$ $CD8^+$ cells. A priori, this finding might indicate that contact with IL-15 is required for the initial formation of $CD44^{hi}$ $CD8^+$ cells. This possibility is unlikely because antigen-driven differentiation of naïve $CD44^{lo}$ T cells into $CD44^{hi}$ $CD8^+$ cells can occur in $IL\text{-}15^{-/-}$ hosts (A. Judge and J. Sprent, unpublished observations). It would therefore seem more probable that IL-15 is required for the maintenance rather than the generation of $CD44^{hi}$ $CD8^+$ cells. Direct support for this notion has come from the finding that CD44^{hi} CD8^{+} cells disappear within 1-2 weeks after transfer to IL-15$^{-/-}$ hosts.[21]

It should be noted that the selective depletion of CD44^{hi} CD8^{+} cells in IL-15$^{-/-}$ and IL-15R$\alpha^{-/-}$ mice is partial (60-70% of normal) rather than complete.[24, 25] This finding could indicate that the survival of CD44^{hi} CD8^{+} cells is only partially dependent on IL-15. Alternatively, the residual CD44^{hi} CD8^{+} cells in IL-15$^{-/-}$ and IL-15R$\alpha^{-/-}$ mice could represent a separate subset of IL-15-independent cells. This latter idea is supported by the finding that the CD44^{hi} CD8^{+} cells found in IL-15$^{-/-}$ mice are nearly all CD122^{lo} cells.[21] By contrast, in normal mice, the majority of CD44^{hi} CD8^{+} cells (about 70%) are CD122^{hi} cells. Since responsiveness to IL-15 requires high expression of CD122, it would therefore follow that the role of IL-15 in controlling the survival of CD44^{hi} CD8^{+} cells is limited to the CD122^{hi} subset of these cells. In favor of this idea, transferring CD44^{hi} CD8^{+} cells to IL-15$^{-/-}$ hosts leads to disappearance of CD122^{hi} cells but survival of CD122^{lo} cells.[21]

The finding that the CD122^{lo} subset of CD44^{hi} CD8^{+} cells is IL-15 independent raises the question whether these cells are maintained by other cytokines. Although this question has yet to be resolved, it is of interest that the background turnover of CD44^{hi} CD8^{+} cells is much higher for CD122^{lo} cells than for CD122^{hi} cells.[21] It is conceivable that this high rate of turnover is not controlled by cytokines but reflects chronic TCR stimulation by environmental antigens. To test this possibility, we are in the process of examining the consequences of transferring CD122^{lo} CD44^{hi} CD8^{+} cells to MHC I$^{-/-}$ hosts.

The above data indicate that CD44^{hi} CD8^{+} cells comprise two discrete subsets of cells, namely IL-15-dependent CD122^{hi} cells and IL-15-independent CD122^{lo} cells. For CD122^{hi} cells, continuous contact with low levels of IL-15 serves to maintain T cell viability and also to cause the cells to undergo intermittent cell division. Curiously, homeostasis of CD122^{hi} cells is distinctly different in old mice.

1.7. CD8^{+} Cells in Aged Mice

As discussed earlier, the proportion of memory-phenotype CD44^{hi} T cells increases considerably in old age. For CD44^{hi} CD8^{+} cells, the physiology of these cells in young vs old mice is different in three respects. First, old mice often contain expanded clones of CD8^{+} cells, presumably reflecting monoclonal responses to unknown environmental antigens.[26] Second, the proportion of CD44^{hi} CD8^{+} cells with a CD122^{hi} phenotype is appreciably higher in old mice (90-95%) than in young mice (60-70%).[27]

With regard to why aged CD122^{hi} CD8^{+} cells have a slow turnover, levels of Bcl-2 expression may be important. In young mice, it is well documented that Bcl-2 expression is higher in CD44^{hi} CD8^{+} cells than in other T cell subsets.[28] Curiously, the elevation of Bcl-2 expression in CD44^{hi} (and CD122^{hi}) CD8^{+} cells is even more pronounced (3-fold higher) in old mice than in young mice.[27] Since Bcl-2 overexpression can retard cell division,[29, 30] the reduced turnover of old CD44^{hi} CD8^{+} cells could in part reflect the elevated expression of Bcl-2 in these cells.

Another possibility is that the slow turnover of aged CD44^{hi} CD8^{+} cells is a reflection of reduced levels of IL-15 in old mice. This idea seems unlikely because measuring IL-15 mRNA levels in spleen revealed no obvious difference between young and old mice.[27] Interestingly, however, examination of other cytokines showed that levels of IFN-I are about 3-fold higher in old mice than in young mice.[27] This finding is of interest because, in addition to eliciting IL-15 synthesis, IFN-I can have a direct anti-proliferative effect on T cells.[31] Hence, the raised levels of IFN-I in old mice could inhibit responsiveness to IL-15 and thus reduce the tempo of CD44^{hi} CD8$^+$ cell turnover. Three lines of evidence support this possibility.[27] First, as for background turnover, bystander proliferation of CD44^{hi} CD8$^+$ cells elicited by injection of Poly I:C or IL-15 is reduced in old mice but can be restored by coinjecting anti-IFN-I antibody. Second, the reduced turnover of aged CD44^{hi} CD8$^+$ cells is not intrinsic and can be overcome by transferring these cells to young mice or by culturing the cells *in vitro*. Third, both for young and old CD44^{hi} CD8$^+$ cells, proliferative responses of these cells to IL-15 *in vitro* can be inhibited by even very low concentrations of IFN-I.

In light of the above findings, the slow turnover of aged CD44^{hi} CD8$^+$ cells *in vivo* may reflect the combined effects of two inhibitory mechanisms, namely Bcl-2 overexpression and exposure to elevated levels of IFN-I.

2. CONCLUSION

As outlined above, transition of naive CD44^{lo} CD8$^+$ cells into memory CD44^{hi} CD8$^+$ cells causes the latter cells to become heavily dependent on IL-15. Dependency on IL-15 is unique to the CD122^{hi} subset of CD44^{hi} CD8$^+$ cells, which correlates with the crucial importance of CD122 as a receptor for IL-15.

In normal animals, the dependency of CD122^{hi} CD8$^+$ cells on IL-15 applies not only to proliferation but also to survival. How exposure to IL-15 maintains T cell survival is unclear, although it is notable that, even in low concentrations, IL-15 causes strong upregulation of Bcl-2 in CD44^{hi} CD8$^+$ cells *in vitro*.[27] Hence, IL-15 may act largely by maintaining levels of Bcl-2 and perhaps other anti-apoptotic molecules.

Since certain other γ_c cytokines, namely IL-4 and IL-7, can mimic the capacity of IL-15 to keep CD44^{hi} CD8$^+$ cells alive *in vitro*, it may seem surprising that only one cytokine, IL-15, seems to be important for maintaining T cell survival *in vivo*. Here, the simplest possibility is that, with the exception of IL-15, the concentration of γ_c cytokines *in vivo* is normally too low to affect CD44^{hi} CD8$^+$ cell homeostasis. In favor of this idea, Charles Surh (this volume) has recently found that the strong dependency of CD122^{hi} CD8$^+$ cells on IL-15 can be overcome by raising the *in vivo* concentration of IL-7. Thus, in marked contrast to the near-total absence of CD122^{hi} CD8$^+$ cells in IL-15$^{-/-}$ mice, very large numbers of these cells are found in IL-7 transgenic (Tg) mice, including IL-15$^{-/-}$ IL-7 Tg mice.

In light of this finding, homeostatic control of total CD122^{hi} CD8$^+$ cell numbers may be determined by the relative concentration of a number of different γ_c cytokines. Levels of these cytokines may also limit the tempo of T cell turnover. In this respect, it is notable that the turnover of CD44^{hi} CD8$^+$ cells *in vivo* is inhibited by injection of anti-CD122 antibody but enhanced by anti-IL-12 antibody.[26] Since only two cytokines, IL-2 and IL-15, bind to CD122, the implication therefore is that the normal tempo of CD44^{hi} CD8$^+$ cell turnover reflects a balance in the positive effects of IL-15 and the inhibitory influence of IL-2. In addition to γ_c cytokines, several other cytokines may influence CD44^{hi} CD8$^+$ cell homeostasis. IFN-I is of particular interest because this cytokine can have a direct anti-proliferative effect on T cells but can also stimulate proliferation through synthesis of IL-15. For old mice where IFN-I levels are increased, the inhibitory influence of IFN-I seems to dominate, thereby causing a reduction in $CD44^{hi}$ $CD8^{+}$ cell turnover.

It is important to emphasize that some $CD44^{hi}$ $CD8^{+}$ cells, namely $CD122^{lo}$ cells, are IL-15 independent. This is apparent from the finding that $IL\text{-}15^{-/-}$ mice contain near-normal numbers of $CD122^{lo}$ $CD44^{hi}$ $CD8^{+}$ cells and that, unlike $CD122^{hi}$ cells, $CD122^{lo}$ cells from normal mice survive when transferred to $IL\text{-}15^{-/-}$ hosts. As mentioned earlier, whether $CD122^{lo}$ $CD44^{hi}$ $CD8^{+}$ cells receive stimulation from other cytokines or are maintained through TCR recognition of environmental antigens is still unclear. On this point it is of interest that, for memory $CD8^{+}$ cells generated through contact with antigen *in vitro*, $CD122^{hi}$ cells survive almost indefinitely after adoptive transfer (presumably through contact with IL-15) whereas $CD122^{lo}$ cells die after several weeks.[32] Why $CD122^{lo}$ cells have a much shorter lifespan than $CD122^{hi}$ cells is unclear.

In contrast to $CD8^{+}$ cells, almost nothing is known about the factors maintaining the survival of memory $CD4^{+}$ cells. Since these cells persist after transfer to MHC class $II^{-/-}$ mice,[14] the survival of memory $CD4^{+}$ cells appears to be MHC independent. By analogy with $CD8^{+}$ cells, it would seem highly likely that memory $CD4^{+}$ cells are kept alive, and in intermittent cell cycle, through contact with cytokines. However, contact with IL-15 and other γ_c cytokines seems to be irrelevant.[33] The influence of other cytokines, and also chemokines, has yet to be studied.

REFERENCES

1. Butcher, E. C. and L. J. Picker, Lymphocyte homing and homeostasis, *Science* **272**, 60-66 (1996).
2. Sprent, J., Lifespans of naive, memory and effector lymphocytes, *Cur. Opin. Immunol.* **5**, 433-438 (1993).
3. Tough, D. F. and J. Sprent, Turnover of naive- and memory-phenotype T cells, *J. Exp. Med.* **179**, 1127-1135 (1994).
4. Ahmed, R. and D. Gray, Immunological memory and protective immunity: Understanding their relation, *Science* **272**, 54-60 (1996).
5. Zinkernagel, R. M., M. F. Bachmann, T. E. Kundig, S. Oehen, H. Pircher and H. Hengartner, On immunological memory, *Annu. Rev. Immunol.* **14**, 333-368 (1996).
6. Dutton, R. W., L. M. Bradley and S. L. Swain, T cell memory, *Annu. Rev. Immunol* **16**, 201-223 (1998).
7. Lin, M. Y., L. K. Selin and R. M. Welsh, Evolution of the CD8 T-cell repertoire during infections, *Microbes Infect.* **2**, 1025-1039 (2000).
8. Sprent, J. and C. D. Surh, Generation and maintenance of memory T cells, *Curr. Opin. Immunol.* **13**, 248-254 (2001).
9. Surh, C. D. and J. Sprent, Homeostatic T Cell Proliferation. How far can T cells be activated to self-ligands?, *J. Exp. Med.* **192**, F9-F14 (2000).
10. Tanchot, C., H. V. Fernandes and B. Rocha, The organization of mature T-cell pools, *Philos. Trans. R. Soc. Lond. B. Biol. Sci.* **355**, 323-328 (2000).
11. Sprent, J. and C. D. Surh, T cell memory, *Annu. Rev. Immunol.*, (2002) In press.
12. Scollay, R. and K. Shortman, 1985, Cell traffic in the adult thymus: Cell entry and exit, cell birth and death, in: *Recognition and Regulation in Cell Mediated Immunity*, Watson, J. D. and J. Marbrook, ed., Dekker, New York, pp. 3-30.
13. Waldmann, T. A. and Y. Tagaya, The multifaceted regulation of interleukin-15 expression and the role of this cytokine in NK cell differentiation and host response to intracellular pathogens, *Annu. Rev. Immunol.* **17**, 19-49 (1999).
14. Swain, S. L., H. Hu and G. Huston, Class II-independent generation of CD4 memory T cells from effectors, *Science* **286**, 1381-1383 (1999).

15. Murali-Krishna, K., L. L. Lau, S. Sambhara, F. Lemonnier, J. Altman and R. Ahmed, Persistence of memory CD8 T cells in MHC class I-deficient mice., *Science* **286**, 1377-1381 (1999).
16. Tough, D. F., P. Borrow and J. Sprent, Induction of bystander T cell proliferation by viruses and type I interferon in vivo, *Science* **272**, 1947-1950 (1996).
17. Tough, D. F., S. Sun and J. Sprent, T cell stimulation in vivo by lipopolysaccharide (LPS), *J. Exp. Med.* **185**, 2089-2094 (1997).
18. Sprent, J., X. Zhang, S. Sun and D. Tough, T cell proliferation in vivo and the role of cytokines, *Phil. Trans. R. Roc. Lond. Biol. Sci.* **355**, 317-322 (2000).
19. Tough, D. F., X. Zhang and J. Sprent, An IFN-γ-dependent pathway controls stimulation of memory-phenotype CD8+ T cell turnover in vivo by IL-12, IL-18 and IFN-γ, *J. Immunol.* **166**, 6007-6011 (2001).
20. Zhang, X., S. Sun, I. Hwang, D. F. Tough and J. Sprent, Potent and selective stimulation of memory-phenotype CD8+ T cells in vivo by IL-15, *Immunity* **8**, 591-599 (1998).
21. Judge, A., X. Zhang, C. D. Surh and J. Sprent, IL-15 controls both proliferation and survival of a subset of memory-phenotype CD8+ T cells, *Manuscript submitted for publication*, (2002).
22. Ku, C. C., M. Murakami, A. Sakamoto, J. Kappler and P. Marrack, Control of homeostasis of CD8+ memory T cells by opposing cytokines, *Science* **288**, 675-678 (2000).
23. Cella, M., D. Jarrossay, F. Facchetti, O. Alebardi, H. Nakajima, A. Lanzavecchia and M. Colonna, Plasmacytoid monocytes migrate to inflamed lymph nodes and produce large amounts of type I interferon, *Nat. Med.* **5**, 919-923 (1999).
24. Kennedy, M. K., M. Glaccum, S. N. Brown, E. A. Butz, J. L. Viney, M. Embers, N. Matsuki, K. Charrier, L. Sedger, C. R. Willis, K. Brasel, P. J. Morrissey, K. Stocking, J. C. Schuh, S. Joyce and J. J. Peschon, Reversible defects in natural killer and memory CD8 T cell lineages in interleukin 15-deficient mice, *J. Exp. Med.* **191**, 771-780 (2000).
25. Lodolce, J. P., D. L. Boone, S. Chai, R. E. Swain, T. Dassopoulos, S. Trettin and A. Ma, IL-15 receptor maintains lymphoid homeostasis by supporting lymphocyte homing and proliferation, *Immunity* **9**, 669-676 (1998).
26. Ku, C. C., J. Kappler and P. Marrack, The growth of the very large CD8+ T cell clones in older mice is controlled by cytokines, *J. Immunol.* **166**, 2186-2193. (2001).
27. Zhang, X., H. Fujii, H. Kishimoto, E. LeRoy, C. D. Surh and J. Sprent, Aging leads to disturbed homeostasis of memory-phenotype CD8+ cells, *J. Exp. Med.* **195**, 283-293 (2002).
28. Grayson, J. M., A. J. Zagac, J. D. Altman and R. Ahmed, Cutting Edge: Increased expression of Bcl-2 in antigen-specific memory CD8+ T cells., *J. Immunol.* **164**, 3950-3954 (2000).
29. Cory, S., Regulation of lymphocyte survival by the bcl-2 gene family, *Annu. Rev. Immunol.* **13**, 513-544 (1995).
30. O'Reilly, L. A., D. C. Huang and A. Strasser, The cell death inhibitor Bcl-2 and its homologues influence control of cell cycle entry, *Embo. J.* **15**, 6979-6990 (1996).
31. Petricoin, E. F., III, S. Ito, B. L. Williams, S. Audet, L. F. Stancato, A. Gamero, K. Clouse, P. Grimley, A. Weiss, J. Beeler, D. S. Finbloom, E. W. Shores, R. Abraham and A. C. Larner, Antiproliferative action of interferon-α requires components of T-cell-receptor signalling, *Nature* **390**, 629-632 (1997).
32. Huang, L. R., F. L. Chen, Y. T. Chen, Y. M. Lin and J. T. Kung, Potent induction of long-term CD8+ T cell memory by short-term IL-4 exposure during T cell receptor stimulation, *Proc. Natl. Acad. Sci. USA* **97**, 3406-3411 (2000).
33. Lantz, O., I. Grandjean, P. Matzinger and J. P. Di Santo, Gamma chain required for naive CD4+ T cell survival but not for antigen proliferation, *Nat. Immunol.* **1**, 54-58 (2000).

ANTIVIRAL MEMORY T CELL RESPONSES

Correlation with Protective Immunity and Implication for Vaccine Development

Alexandre Harari, Kim Ellefsen, Patrick Champagne, Massimo Nobile, and Giuseppe Pantaleo*

INTRODUCTION

In the past 4-6 years, the advances in the characterization of the HIV-specific immune response have drowned the attention of the scientific community on the importance of the cell-mediated immune response in the protection against HIV infection. There are at least three major evidences for the central role played by cytotoxic T lymphocytes (CTLs) in controlling HIV replication. Firstly, primary HIV infection is associated with a very potent CTL response that generally coincides with the peak in viremia and precedes the neutralizing antibody response (1-4). The advent of CTL response is temporally associated with the downregulation of viremia. This observation indicated a role for CTLs, which are mostly contained in the T lymphocyte population characterized by the expression of the CD8 surface molecule, in the initial control of virus replication. However, direct demonstration of the important role played by CD8 CTLs comes from the SIV monkey model of HIV infection. In fact, it has been demonstrated that the depletion of CD8 T cells following infusion with a specific ant-CD8 mAb resulted in a failure to control the early peak of viremia in the infected animals (5). Along the same line, the depletion of CD8 CTLs in SIV chronically infected monkeys was associated with transient rises in the viremia levels (6), and loss of immune control occurs following the emergence of virus mutants *in vivo* that are not recognized by CTLs (7, 8). Secondly, it has been shown that the human leukocyte antigen (HLA) type may significantly influence the rate of HIV disease progression (9, 10). In particular, HLA types such as HLA-B27 and HLA-B57 are associated with slow disease progression while HLA-B35 with faster disease progression. Since CTLs recognized virus peptides presented by HLA class I molecules, different HLA types may have different ability to present peptides and thus substantially influence the quality of the immune response elicited. Thirdly, HIV-specific CD8 T-cell responses have been found in virus exposed

*Laboratory of AIDS Immunopathogenesis, Divisions of Immunology and Allergy, Department of Medicine, Centre Hospitalier Universitaire Vaudois, University of Lausanne, 1011 Lausanne, Switzerland.

uninfected subjects (11) and/or animals (12, 13). Therefore, the above observations together with a large number of recently accumulated experimental evidences indicate that the development of a preventative HIV vaccine based on the induction of CD8 CTLs may potentially confer partial and/or complete control from HIV infection.

MAINTENANCE OF CD8 T CELL IMMUNE RESPONSE

The maintenance of the CD8 T cell immune response over time, i.e. maintenance of memory CD8 T cells, is dependent upon the presence of antigen-specific helper T cells which are phenotypically characterized by the expression of the surface CD4 molecule. The quantitative as well as the qualitative defects of CD4 T cells are the major abnormalities associated with HIV infection (14). HIV-specific lymphoproliferative responses are generally not detected in patients with chronic HIV infection. The lack of these responses is probably due to the rapid elimination of the precursors of virus-specific helper CD4 T cells during primary infection (15, 16). In fact, even though HIV-specific CD4 T cells with effector function, i.e. IFN-γ secreting cells can be detected in patients with chronic infection (17), HIV-specific lymphoproliferative CD4 T cell responses are generally not recovered after prolonged antiretroviral treatment and effective virus suppression (18, 19). Vigorous HIV-specific CD4 helper T cell responses are however found in a small percentage (1%) of HIV-infected subjects with chronic infection who did not experience disease progression after 10-15 years of documented infection (15). Furthermore, it has been recently shown that HIV-specific helper CD4 T cells may be preserved in patients who received treatment during the early phase of primary infection (15, 20). More importantly, the preservation of these responses seems to be associated with partial but effective control of HIV replication following interruption of antiviral therapy (16). These HIV-specific helper CD4 T cell responses are associated also with a vigorous virus-specific CD8 T cell response. These observations indicate that immune control of HIV infection can be achieved and suggest that an HIV vaccine based on the cell-mediated immune response should induce both HIV-specific CD8 and CD4 T cells.

HETEROGENEITY IN THE COMPOSITION OF THE MEMORY CD8 T CELL POOL AND RELATIONSHIP BETWEEN FUNCTION AND MATURATION

Although the detection of HIV-specific helper CD4 T cell response seems to be a reliable marker of the ability of the host immune system to control HIV infection, virus-specific CD8 CTLs are detected already during primary infection (1-4), persist for several years and may be even present in the advanced stages of HIV disease (21-23). These latter observations indicate that the detection of CD8 CTLs does not necessarily reflect the effectiveness of the immune response in the control of HIV infection. CD8 T cells may mediate antiviral activity either by the production of soluble factors such as the cytokine IFN-γ (24-26), the chemokines MIP-1α, MIP-1β and RANTES (27-29), and the partially characterized CD8 T-cell antiviral factor or CAF (30-32) and/or by lytic mechanisms. HIV-specific CD8 CTLs are able to produce the above antiviral factors. However, recent studies have strongly suggested that the HIV-specific CD8 T cells present in HIV-infected subjects may have major abnormalities at both functional and maturational levels. Appay *et al.* have shown that HIV-specific CTLs have a selective defect in the levels of intracellular perforin that may significantly affect their lytic capacity (33). Therefore, although HIV-specific CTLs have been shown to lyse HIV-infected CD4 T cells following activation *in vitro* (34), freshly isolated virus-specific T cells, show poor lytic activity (35, 36). Furthermore, recent studies have also demonstrated that the pool of memory T lymphocytes is composed of several populations of memory CD4 and CD8 T lymphocytes with different functional capacities and at different stages of maturation (37-39). Studies aimed at the functional characterization of the different populations of memory HIV- and CMV-specific CD8 T cells have allowed to develop a lineage differentiation pattern for memory CD8 T lymphocytes and have shown major differences within the composition of the HIV- and CMV-specific memory CD8 T cell pools (39). The memory HIV-specific CD8 T cell pool is predominantly composed of pre-terminally differentiated CD8 T cells as compared to the memory CMV-specific CD8 T cell pool which is mostly composed by terminally differentiated CD8 T cells (39). The differences in the maturation between HIV-specific and CMV-specific CD8 T cells are likely associated to a different lytic capacity since CMV- but not HIV-specific CD8 T cells have normal intracellular levels of perforin (33). More importantly, the different composition in the pool of memory CD8 T cells seems to translate into a different efficacy in the control of the two virus infections: effective control of CMV infection versus poor control of HIV infection and progressive disease. Taken together these observations indicate that the detection of large number of antigen-specific CD8 T cells by tetramer staining as well as a high frequency of IFN-γ secreting cells are not necessarily an indicator of an effective immune response.

PHENOTYPIC CHARACTERIZATION OF MEMORY CD8 T CELLS

During the last two years we have performed extensive characterization of memory CD8 and CD4 T cell responses and of the correlates of the efficacy of these immune responses. Studies have been performed in patients with chronic infection at early stages of HIV disease and in patients with primary infection. All the phenotypic and functional analyses have been performed on blood samples collected prior to the initiation of antiviral therapy. The above research projects have taken advantage of a study by Sallusto *et al.* (37) which showed that the chemokine receptor CCR7 defines distinct subsets of naive and memory T lymphocytes with different homing and effector capacities. In particular, CD4 and CD8 T lymphocyte populations lacking CCR7 contain the majority of cells with potential effector function as defined by the secretion of IFN-γ and by the expression of perforin. On the basis of this observation, the phenotypic characterization of blood mononuclear cells obtained from HIV-infected subjects with chronic infection and with no previous antiretroviral therapy, was performed with mAbs to CD8 and CCR7, and with HLA tetrameric complexes loaded with the relevant peptides that allow direct evaluation of antigen-specific memory CD8 T cells. This analysis demonstrated that most of HIV-specific tetramer $(Tet)^+$ cells (~70-80%) were contained in the $CCR7^-$ population thus indicating that the majority of HIV-specific CD8 T cells are potential effector cells. The analysis with anti-CD45RA mAb allowed to identify four cell populations of HIV-specific CD8 T cells: a) $CD45RA^+CCR7^+$; b) $CD45RA^-CCR7^+$; c) $CD45RA^-CCR7^-$; and d) $CD45RA^+CCR7^-$. The $CD45RA^+CCR7^+$ cells represented 15% of the total HIV-specific Tet^+ cells, the $CD45RA^-CCR7^+$ cells 5%, the $CD45RA^-CCR7^-$ cells 70%, and the $CD45RA^+CCR7^-$ cells 10% (Figure 1).

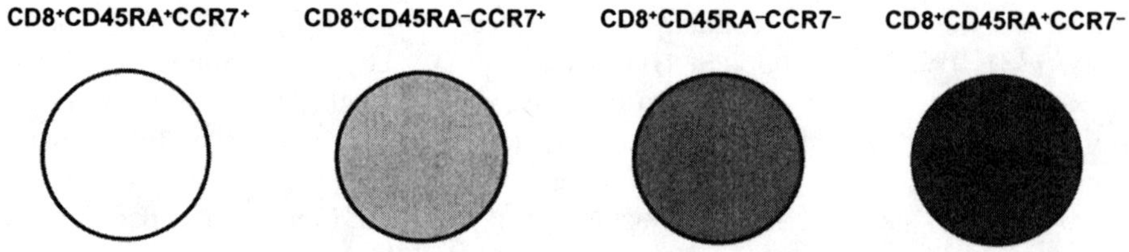

Figure 1: populations of HIV-specific memory CD8 T cells with different homing and functional capacities

These results became particularly interesting after the comparison with similar analysis of CMV-specific CD8 T cells. The $CD45RA^{+}CCR7^{-}$ cell population was the mostly represented, i.e. 50%, while only 40% of cells were $CD45RA^{-}CCR7^{-}$. There are a series of consideration that can be made on the basis of these results: a) a significant proportion of memory cells have a phenotype typical of naïve T lymphocytes (40); b) the pool of virus-specific memory CD8 T cells is heterogeneous, i.e. composed by at least four cell populations whose distribution within the pool is different; and c) the distribution of the memory cell populations is different between HIV and CMV.

Another surface marker, the CD27 molecule, has been used in other studies to distinguish among different subsets of memory CD8 T cells (38). Preliminary investigation on the potential overlapping between the memory cell populations identified by CD27 and CCR7, have clearly shown that these markers do not identify the same cell populations. $CD27^{+}$ and $CD27^{-}$ cells are present in each of the four populations identified by the expression of CCR7. The rationale for using CCR7 instead of CD27 resides on the fact that CCR7 has very well characterized immune functions.

FUNCTIONAL CHARACTERIZATION OF MEMORY CD8 T CELLS

In order to determine the lineage differentiation pattern of these four subsets of $CD8^{+}$ T lymphocytes, we have followed two strategies: a) *ex vivo* and *in vitro* assessment of the cell division capacity of the four cell subsets; and b) the differentiation patterns of purified $CD8^{+}CCR7^{-}CD45RA^{-}$ and $CD8^{+}CCR7^{-}CD45RA^{+}$ cell populations following peptide-specific stimulation. Both *ex vivo* and *in vitro* analyses have demonstrated that cell division was mostly confined to the populations of CD8 T lymphocytes expressing CCR7. Therefore, the $CCR7^{+}CD8^{+}$ T cells and, in particular the $CD8^{+}CD45RA^{+}CCR7^{+}$ cells, function as precursors of the $CCR7^{-}CD8^{+}$ T cells. In contrast, there was no evidence for proliferative capacity of HIV-specific CD8 T cells within the $CCR7^{-}$ cell populations. These findings demonstrate that there is a dichotomy with regard to the cell division and proliferative capacities amongst the $CCR7^{+}$ and $CCR7^{-}$ memory $CD8^{+}$ T cell compartments: high cell division capacity/expansion of the $CCR7^{+}$ versus limited of the $CCR7^{-}$ $CD8^{+}$ T cell populations. Therefore, we have shown that there are two types of memory CD8 T cells. The minority of memory cells that function as precursors retain proliferative capacity while the majority of memory cells proliferate poorly. A series of experiments involving antigen-specific stimulation of purified cell subsets of memory CD8 T cells has allowed us to demonstrate the following lineage differentiation pattern for antigen-specific CD8 T lymphocytes: CD45RA^{+}CCR7^{+} → CD45RA^{-}CCR7^{+} →· CD45RA^{-}CCR7^{-} → CD45RA^{+}CCR7^{-} (Figure 2).

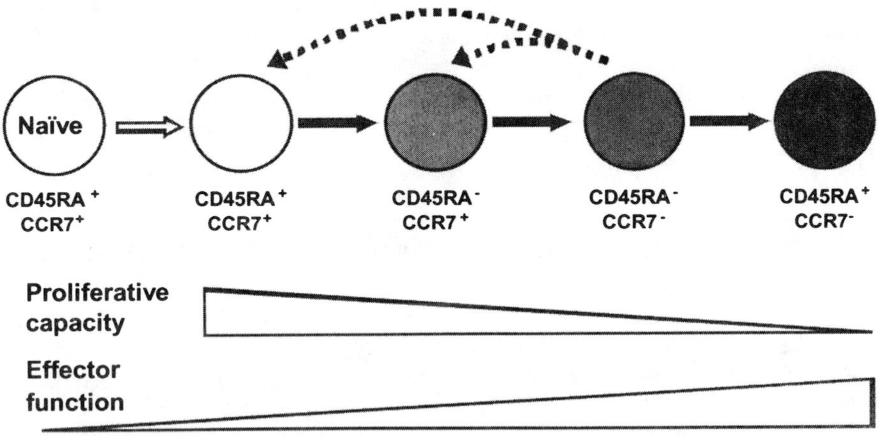

Figure 2: lineage differentation pattern of HIV- specific memory CD8 T cells; proliferative capacity is progressively lost while effector function is acquired during differentiation

Based on our model the CD45RA^{+}CCR7^{-} cell population is composed of terminally differentiated cells. From a functional standpoint, there are some differences in the ability of these four cell populations to mediate effector function. Both IFN-γ secretion and perforin expression were predominantly confined to the CCR7^{-} cell populations although also a significant percentage of CCR7^{+} cells secreted IFN-γ.

DISTRIBUTION OF VIRUS-SPECIFIC MEMORY CD8 T CELLS IN LYMPH NODES

Analysis of the distribution of the different populations of memory CD8 T lymphocytes is currently being performed in lymph nodes. Preliminary results indicate that the distribution of the HIV-specific memory CD8 T cell populations is similar between blood and lymph node. However, some differences have been observed with regard to the proportion of HIV-specific CCR7^{-} cells able to secrete IFN-γ. The results obtained seem to indicate that only about 30% of lymph node CCR7^{-} effectors cells are able to secrete IFN-γ compared to 60-70% of blood CCR7^{-} cells. It is unlikely that there is a selective accumulation of defective HIV-specific CD8 cells in lymph node. The hypothesis that we are testing is that despite the fact that the lymph node CCR7^{-} CD8 T cells are undistinguishable from a phenotypic standpoint from those in the blood, they are cells at a different stage of maturation and differentiation.

We have had also the opportunity to investigate the distribution of the different populations of CMV-specific memory CD8 T cells in lymph node. The preliminary data obtained are of interest. There are major differences in the distribution of CMV-specific memory CD8 T cells in blood and lymph node since the frequency of CMV-specific CD8 T cells in lymph node is at least 1 log lower compared to blood.

These differences in the in the anatomic distribution of HIV- and CMV-specific T cells can be explained by two recently published studies in mice (41, 42). These studies demonstrated that accumulation of virus-specific memory CD4 and CD8 T cells occurs in the target organs away from the lymphoid tissue. Since lung, retina and cervix are the target organs for CMV, our preliminary findings are consistent with the above studies in mice. Along the same line, the high frequency of HIV-specific memory CD8 T cells in lymph node can be explained by the fact that the lymphoid tissue is the target organ for HIV.

PHENOTYPIC CHARACTERIZATION OF VIRUS-SPECIFIC MEMORY CD4 T CELLS

The characterization of virus-specific memory CD4 T cells is more problematic than that of CD8 T cells for a number of reasons. Firstly, the technology of tetramer complex for MHC class II antigens is still in early phase of development. The reasons are multiple. There are a number of limitations including production of large stock of the reagents, rapid downregulation of the T cell receptor following staining of freshly isolated mononuclear cells (most of the data available are on cultured activated CD4 T cells) and, more importantly, the different class II antigen alleles are not widely distributed in the population as for example the HLA-A2 class I antigen. Secondly, with the exception of the long-term nonprogressors, HIV-infected subjects do not have detectable lymphoproliferative responses to HIV antigens. However, HIV-specific IFN-γ secreting cells are readily detected by flow cytometry. Therefore, the type of characterization that we have performed involves only the subsets of antigen-specific CD4 T cells able to secrete IFN-γ following appropriate stimulation. Furthermore, since the majority of subjects are co-infected with CMV, we have performed comparative analysis of the composition of the pools of HIV- and CMV-specific memory CD4 T cells. As for the virus-specific memory CD8 T cell populations, it has been assessed whether similar population of virus-specific CD4 T cells were identified based on the surface expression of CD45RA and CCR7. Preliminary data indicate that four populations of virus-specific CD4 T cells can be identified within the population of IFN-γ secreting cells: $CD45RA^+CCR7^+$, $CD45RA^-CCR7^+$, $CD45RA^-CCR7^-$, and $CD45RA^+CCR7^-$ (Figure 3).

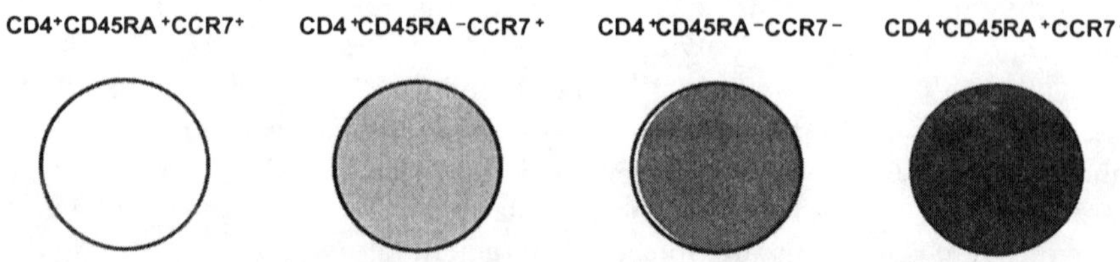

Figure 3: populations of HIV- specific CD4 T cells identified on the basis of the ability to mediate effector function , e.g. IFN-γ secretion

The virus-specific IFN-γ secreting cells are mostly contained within the $CCR7^-$ cell populations that also represent the majority of the virus-specific memory CD4 T cells. The identification of the four populations as for the CD8 T cells is of interest. In the previous study by Sallusto *et al.* (37), the $CD45RA^+CCR7^-$ CD4 population was not identified. A possible explanation for this discrepancy is that a superantigen and not a conventional antigen was used in the stimulation experiments of T cells.

FUNCTIONAL CHARACTERIZATION OF VIRUS-SPECIFIC MEMORY CD4 T CELLS

In order to delineate the antiviral-specific primary CD4 T cell-mediated immune response, we have assessed the ability of the different populations of CD4 T cells as defined by the expression and/or by the absence of CCR7 to secrete IFN-γ. Since a significant proportion of patients are also co-infected with CMV, we have had the opportunity to compare the HIV- and CMV-specific CD4 T-cell immune responses. The studies have been performed on cohort of HIV-infected patients with primary and chronic infection. Our results clearly demonstrate that both HIV-specific and CMV-specific $CD4^+$ IFN-γ secreting cells are contained within the $CCR7^-$ cell population. However, in patients experiencing primary HIV and CMV co-infections the percentage of HIV-specific $CD4^+$ IFN-γ secreting cells was significantly lower as compared to that of CMV-specific $CD4^+$ T cells observed in the four patients with CMV primary infection. These results indicated that there are major differences in the magnitude (at least one order of magnitude) of the primary CMV- vs. HIV-specific CD4 T cell responses. We have also analyzed whether there were differences between primary CMV- and HIV-specific CD4 T cell responses and those in patients with chronic CMV- and HIV-infections. These analyses showed that the percentage of CMV-specific CD4 T in the patients with CMV primary infection was significantly higher ($P=0.005$) compared to that observed in patients with chronic CMV infection. Of interest, there were no differences ($P=0.1$) in the percentage of HIV-specific CD4 T cells in patients with primary and chronic infections. These results indicate that a major expansion of virus-specific CD4 T cells is associated with primary CMV but not with HIV infection. The magnitude of primary HIV-specific CD4 T cell response is indeed not different from that observed during chronic infection. The differences in the magnitude of primary HIV- and CMV-specific immune responses may translate in a different effectiveness of the two antiviral responses in the control of virus replication. Another interesting observation from our studies is the different composition of the pool of HIV- and CMV-specific memory CD4 T cells. The HIV-specific IFN-γ secreting $CD45RA^+CCR7^-CD4^+$ cell population seem to be poorly represented and/or absent. If this latter population is, as for CD8 T cells, composed by terminally differentiated cells, an abnormality similar to that demonstrated for CD8 T cells can be proposed also for the maturation of HIV-specific memory CD4 T cells.

DISTRIBUTION OF VIRUS-SPECIFIC MEMORY CD4 T CELLS IN LYMPH NODES

Analysis of the distribution of the different populations of memory CD4 T lymphocytes in lymph nodes obtained from HIV-infected subjects enrolled in therapeutic trials has also been performed.

As already discussed above, the characterization of virus-specific memory CD4 T cells is limited to those cells able to secrete IFN-γ. At present, we can confirm that the differences in the distribution observed for HIV- and CMV-specific CD8 T cells have also been found in CD4 T cells. The percentage of HIV-specific CD4 T cells is similar between blood and lymph node while CMV-specific CD4 T cells accumulates mostly in blood. Based on the limited number of patients tested (n=7), we have no definitive data on the representation of the different populations of virus-specific CD4 T cells in lymph node as well as whether there are differences in the proportion of IFN-γ secreting cells within the lymph node and blood memory CD4 cell populations.

CONCLUSIONS

Our studies performed in HIV-infected subjects with chronic infection have proposed a novel differentiation pattern for memory CD8 T lymphocytes. Different populations of virus-specific memory CD8 T cells at different stages of differentiation and with different functional capacities have been identified based on the expression of the chemokine receptor CCR7 and the surface CD45RA molecule.

The different composition of the pool of memory CD8 T cells specific for HIV and CMV appears to be associated with a different ability to control these infections. It is therefore important to determine the composition of the pool of virus-specific memory CD8 T cells in order to correctly evaluate the potential efficacy of the immune response elicited. This issue is particularly important with regard to the evaluation of the immune response induced following vaccination. A number of vaccine approaches using recombinant viral vectors and/or DNA are being pursued in order to induce HIV-specific cellular immune responses. It is likely that to confer partial and/or complete protection from HIV, the ideal immune response generated by a putative vaccine has to be able to rapidly respond to HIV infection and to be long lasting. These two features of the immune response are going to be strictly dependent upon the type of memory CD4 and CD8 T cell immune responses that have been generated by the putative vaccine.

REFERENCES

1. Pantaleo G et al. Major expansion of $CD8^+$T cells with a predominant Vb usage during the primary immune response to HIV. *Nature*. **370**, 463-467 (1994)
2. Pantaleo G et al. The qualitative nature of the primary immune response to HIV infection is a prognosticator of disease progression independent of the initial level of plasma viremia. *Proc. NatlAcad. Sci. USA*. **94**, 254-258 (1997)
3. Koup RA et al. Temporal association of cellular immune response with the initial control of viremia in primary human immunodeficiency virus type 1 syndrome. *J. Virol*. **68**, 4650-4655 (1994)
4. Borrow P, Lewicki H, Hahn BH, Shaw GM, Oldstone MB. Virus-specific $CD8^+$ cytotoxic T-lymphocyte activity associated with control of viremia in primary human immunodeficiency virus type 1 infection. *J. Virol*. **68**, 6103-6110 (1994)
5. Schmitz JE et al. Control of viremia in simian immunodeficiency virus infection by $CD8^+$ lymphocytes. *Science*. **283**, 857-860 (1999)
6. Jin X et al. Dramatic rise in plasma viremia after CD8(+) T cell depletion in simian immunodeficiency virus-infected macaques. *J.Exp.Med*. **189**, 991-998 (1999)
7. Philips RE et al. Human immunodeficiency virus genetic variation that can escape cytotoxic T cell recognition. *Nature*. **354**, 453-459 (1991)
8. Evans DT et al. Virus-specific cytotxic T-lymphocyte responses select for amino-acid variation in simian immunodeficiency virus Env and Nef. *Nature Med*. **5**, 1270-1276 (1999)

9. Carrington M et al. *HLA* and HIV-1 : heterozygote advantage and *B*35-Cw*04* disadvantage. *Science*. **283**, 1748-1752 (1999)
10. Kaslow RA et al. Influence of combinations of human major histocompatibility complex genes on the course of HIV-1 infection. *Nature Med*. **2**, 405-411 (1996)
11. Rowland-Jones SL et al. Cytotoxic T cell responses to multiple conserved HIV epitopes in HIV-resistant prostitutes in Nairobi. *J.Clin.Invest*. **102**, 1758-1765 (1998)
12. Lifson JD et al. Containment of simian immunodeficiency virus infection : cellular immune responses and protection from rechallenge following transient postinoculation antiretroviral treatment. *J. Virol*. **74**, 2584-2593 (2000)
13. Putkonen P, Makitalo B, Bottiger D, Biberfeld G, Thorstensson R. Protection of human immunodeficiency virus type 2-exposed seronegative macaques from mucosal simian immunodeficiency virus transmission. *J. Virol*. **71**, 4981-4984 (1997)
14. Clerici M, et al. Detection of three distinct patterns of T helper cell dysfunction in asymptomatic, human immunodeficiency virus-seropositive patients. Independence of $CD4^+$ cell numbers and clinical staging. *J.Clin.Invest*. **84**, 1892-1899 (1989)
15. Rosenberg ES et al. Vigorous HIV-1-specific $CD4^+$ T-cell responses associated with control of viremia. *Science*. **278**, 1447-1450 (1997)
16. Rosenberg ES et al. Immune control of HIV-1 after early treatment of acute infection. *Nature*. **407**, 523-526 (2000)
17. Pitcher CJ et al. HIV-1-specific $CD4^+$ T cells are detectable in most individuals with active HIV-1 infection, but decline with prolonged viral suppression. *Nature Med*. **5**, 518-525 (1999)
18. Autran B et al. Positive effects of combined antiretroviral therapy on $CD4^+$ T cell homeostasis and function in advanced HIV disease. *Science*. **277**, 112-116 (1997)
19. Rizzardi GP, Tambussi G, Bart PA et al. Virological and immunological responses to HAART in asymptomatic therapy-naive HIV-1-infected subjects according to CD4 cell count. *AIDS*. **14**, 2257-2263 (2000)
20. Oxenius A et al. Early highly active antiretroviral therapy for acute HIV-1 infection preserves immune function of $CD8^+$ and $CD4^+$ T lymphocytes. *Proc.Natl Acad.Sci.USA*. **97**, 3382-3387 (2000)
21. Walker BD et al. HIV-specific cytotoxic T lymphocytes in seropositive individuals. *Nature*. **328**, 345-348 (1987)
22. Plata F et al. AIDS virus specific cytotoxic T lymphocytes in lung disorders. *Nature*. **328**, 348-351 (1987)
23. Ogg GS et al. Quantitation of HIV-1-specific cytotoxic T lymphocytes and plasma viral RNA load. *Science*. **279**, 2103-2106 (1998)
24. Jassoy C et al. Human immunodeficiency virus type 1-specific cytotoxic T lymphocytes release gamma interferon, tumor necrosis factor alpha (TNF-alpha), and TNF-beta when they encounter their target antigens. *J. Virol*. **67**, 2844-2852 (1993)
25. Meylan PR, Guattelli JC, Munis JR, Richman DD, Kornbluth RS. Mechanisms for the inhibition of HIV replication by interferons-α, -β, and $-\gamma$ in primary human macrophages. *Virology*, **193**, 138-148 (1993)
26. Emilie D, Maillot MC, Nicolas JF, Fior R, Galanaud P. Antagonistic effect of interferongamma on tat-induced transactivation of HIV long terminal repeat. *J. Biol.Chem*. **267**, 20565-20570 (1992)
27. Wagner L et al. β-Chemokines are released from HIV-1-specific cytolytic T-cell granules complexed to proteoglycans. *Nature*. **391**, 908-911 (1998)
28. Price DA et al. Antigen-specific release of β-chemokines by anti-HIV-1 cytotoxic T lymphocytes. *Curr.Biol*. **8**, 355-358 (1998)
29. Cocchi F et al. Identification of RANTES, MIP-1α, and MIP-1β as the major HIV-suppressive factors produced by $CD8^+$ T cells. *Science*. **270**, 1811-1815 (1995)
30. Mackewicz C, Levy JA. CD8+ cell anti-HIV activity : nonlytic suppression of virus replication. *AIDS Res.Hum.Retroviruses*. **8**, 1039-1050 (1992)
31. Levy JA, Mackewicz CE, Barker E. Controlling HIV pathogenesis : the role of noncytotoxic anti-HIV response of CD8+ T cells. *Immunol.Today*. **17**, 217-224 (1996)
32. Copeland K, McKay PJ, Rosenthal KL. Suppression of activation of the HIV LTR by CD8+ cells is not lentivirus specific. *AIDS Res.Hum.Retroviruses*. **11**, 1321-1325 (1995)
33. Appay V et al. HIV-specific $CD8^+$ T-cells produce antiviral cytokines but are impaired in cytolytic function. *J.Exp.Med*. **192**, 63-75 (2000)
34. Yang OO et al. Efficient lysis of human immunodeficiency virus type 1-infected cells by cytotoxic T lymphocytes. *J. Virol*. **70**, 5799-5806 (1996)
35. Zajac AJ et al. Viral immune evasion due to persistence of activated T cells without effector function. *J.Exp.Med*. **188**, 2205-2213 (1998)
36. Kalams SA, Walker BD. The critical need for CD4 help in maintaining effective cytotoxic T lymphocyte responses. *J.Exp.Med*. **188**, 2199-2204 (1998)
37. Sallusto F, et al. Two subsets of memory T lymphocytes with distinct homing potentials and effector functions. *Nature* **401**, 708-712 (1999)

38. Hamann D et al. Phenotypic and functional separation of memory and effector human CD8^{+} T cells. *J.Exp.Med.* **186**, 1407-1418 (1997)
39. Champagne P et al. Skewed maturation of memory HIV-1 specific CD8 T lymphocytes. *Nature* **410**, 106-111 (2001)
40. Callan MFC et al., Direct visualization of antigen-specific CD8+ T cells during primary immune response to Epstein-Barr Virus *in vivo*. *J.Exp.Med.* **187**, 1395-1402 (1998)
41. Reinhardt R et al. Visualizing the generation of memory CD4 T cells in the whole body. *Nature* **410**, 101-105 (2001)
42. Masopust D, et al. Preferential localization of effector memory cells in nonlymphoid tissue. *Science* **291**, 2413-2417 (2001)

HOMEOSTATIC PROLIFERATION BUT NOT THE GENERATION OF VIRUS SPECIFIC MEMORY CD8 T CELLS IS IMPAIRED IN THE ABSENCE OF IL-15 OR IL-15Rα

E. John Wherry,[1§] Todd C. Becker,[1§] David Boone,[2§] Murali-Krishna Kaja,[3] Averil Ma,[2] and Rafi Ahmed[1#*]

ABSTRACT

The generation and efficient maintenance of antigen specific memory T cells is essential for long-lasting immunological protection. Antigen specific memory CD8 T cells are known to be maintained via antigen-independent homeostatic proliferation. However, signals that drive memory T cell generation and/or influence the slow turnover of memory T cells are unknown. Recently, IL-15 has received attention for its potential effect on memory CD8 T cells. In this report we examine the role of IL-15 in the generation and maintenance of virus specific memory CD8 T cells using mice deficient in either IL-15 or the IL-15 receptor α chain. Both cytokine and receptor deficient mice mount a robust CD8 T cell response to infection with lymphocytic choriomeningitis virus (LCMV) that is initially only slightly lower than in control mice. Further, virus specific memory CD8 T cells are generated in both IL-15 -/- and IL-15Rα -/- mice. However, longitudinal analysis reveals a slow attrition of LCMV specific memory CD8 T cells in the absence of IL-15 signals. Indeed, direct examination of homeostatic proliferation reveals a severe defect in the turnover of antigen specific memory CD8 T cells in the absence of IL-15. Together these results suggest that IL-15 is not essential for the generation of memory CD8 T cells, but is required for homeostatic proliferation to maintain populations of memory cells over long periods of time.

[§]These authors contributed equally.

[#]To whom correspondence should be addressed: ra@microbio.emory.edu Phone: 404-727-4700; Fax: 404-727-3722.

[*1]Emory Vaccine Center and Department of Microbiology and Immunology, Emory University School of Medicine, 1510 Clifton Road, Atlanta, GA 30322; [2]Department of Medicine, University of Chicago, Chicago, IL; [3]Department of Immunology, University of Washington, Seattle, WA

INTRODUCTION

The formation and maintenance of memory CD8 T cells following antigenic stimulation is essential for long-lasting immunity. Several characteristics define T cell memory. These include a substantial increased in the precursory frequency of epitope specific T cells, an ability to perform effector functions more rapidly than naïve cells and the ability to persist for long periods of time in the absence of antigen. In particular, understanding the long-term maintenance of memory CD8 T cells is of major interest. Memory CD8 T cells are known to be maintained at relatively constant numbers by homeostatic proliferation in the absence of antigen.[1-3] However, the signals the drive this homeostatic maintenance as well as the generation of memory CD8 T cells from a pool of effectors are not well understood. Recent evidence suggests that IL-15, an IL-2-like cytokine that shares the IL-2 receptor β and γ chain, but uses a unique receptor α chain, could be a particularly important signal for memory CD8 T cells.[4-7]

Two distinct, but non-mutually exclusive, mechanisms exist for how IL-15 may influence CD8 T cell memory. First, IL-15 could impact the initiation of an antigen-specific CD8 T cell response and the subsequent generation of memory CD8 T cells. Alternatively or in addition, IL-15 could drive the homeostatic survival and/or turnover of the memory pool. IL-15 is known to be induced by type 1 interferons,[6,8] a common feature of the early response to many viral infections,[9] suggesting that IL-15 could be available during the early generation of a CD8 T cell response. In fact, increasing the levels of IL-15 can enhance CD8 T cell responses to several infectious agents.[5,10-13] However, it is not currently known if IL-15 is essential during priming for the efficient induction of CD8 T cell responses or whether IL-15 signals early influence the initial generation of CD8 T cell memory.

A role for IL-15 in the maintenance and/or turnover of the memory CD8 T cell population has also been implied. The administration of exogenous IL-15, or IL-15 inducers selectively stimulates the division of CD44^{hi} "memory phenotype" CD8 T cells[6,14] and the overexpression of IL-15 leads to an expansion of memory phenotype CD8 T cells that gradually progress to leukemia.[15] Two groups have generated mice defective in IL-15 signals by genetically ablating either the IL-15Rα chain or IL-15 itself.[4,5] In both cases knockout mice have a substantial reduction of CD8 T cells of a memory phenotype (CD44^{Hi} CD8), consistent with a role for IL-15 in either memory CD8 T cell generation or maintenance. Though these studies implicate IL-15 in proliferation of CD44^{Hi} CD8 T cells, the antigenic stimuli and signals that result in the generation of these cells is not known. It is not clear if these CD44^{Hi} CD8 T cells from naïve mice represent T cells that experience the same type of activation and display the same characteristics as antigen specific CD8 T cells induced following a viral infection. To date no study has addressed whether IL-15 is required to generate virus specific memory CD8 T cells or is necessary for the turnover of memory CD8 T cells of a defined specificity.

We investigated the role of IL-15 in the generation of virus-specific CD8 T cells, and in the homeostatic maintenance of antigen-specific CD8 T cell memory. Our data demonstrate that mice deficient in either IL-15 or the IL-15Rα chain can mount robust CD8 T cell responses following lymphocytic choriomeningitis virus (LCMV) infection. LCMV specific memory CD8 T cells are efficiently generated in an IL-15 deficient environment. However, memory CD8 T cell numbers undergo slow attrition in the absence of IL-15 signals. Strikingly, homeostatic turnover of LCMV specific memory CD8 T cells is nearly completely lost in the absence of IL-15, but T cell numbers only slowly decline. Thus, IL-15 is not absolutely required for the initial generation of an effective virus specific CD8 T cell response and the establishment of a pool of antigen specific memory CD8 T cells. However, IL-15 signals appear crucial for homeostatic proliferation of memory CD8 T cells and in its absence the long-term antigen specific CD8 T cell memory slowly declines.

MATERIALS AND METHODS

Mice

Four- to six-week-old female C57Bl/6 were purchased from The Jackson Laboratory (Bar Harbor, ME). The generation and description of IL-15Rα$^{-/-}$ and IL-15$^{-/-}$ have been previously described.[4,5] IL-15$^{-/-}$ mice were obtained from Dr. Michael Caligiuri (Ohio State University, Columbus, OH). Both IL-15α$^{-/-}$ and IL-15$^{-/-}$ and littermate controls were fully backcrossed to B6 (>10 generations).

Viral Infections

In all cases mice were infected i.p. with $2x10^5$ PFU of the Armstrong strain of LCMV. All LCMV stocks were plaque purified on Vero cells and grown in BHK-21 cells as described.[16]

Cell Surface, Intracellular Staining, Flow Cytometry and FACS Analysis

Preparation of cells and staining has been described.[17]

Antibodies, MHC Tetramers, and CFSE Staining

All antibodies were purchased from Pharmingen (San Diego, CA). The construction and purification of the MHC class I/peptide tetramers has been previously described.[17] Cells were labeled with CFSE (Molecular Probes, Eugene, OR) as described.[3]

RESULTS AND DISCUSSION

IL-15$^{-/-}$ and IL-15Rα$^{-/-}$ Mice Generate Robust CD8 T Cell Responses to LCMV Challenge, but LCMV Specific T Cell Numbers Decline Slowly

Previous reports have demonstrated that mice lacking either IL-15 or IL-15Rα have reduced CD8 T cell numbers, particularly those of a CD44^{hi} memory phenotype,[4,5] suggesting compromised antigen-specific CD8 T cell memory in these mice. To investigate the induction and maintenance of virus specific memory CD8 T cells in the absence of IL-15 signals IL-15-/- mice and littermate controls were infected with LCMV. LCMV specific CD8 T cells responses were monitored longitudinally in the PBMC of individual mice. Figure 1A shows staining with MHC class I/peptide tetramers for 4 LCMV CD8 T cell epitopes from representative mice. At the peak of the response, day 8, the frequency of tetramer positive cells is similar between control and IL-15-/- mice, when expressed as a percent of CD8 T cells (Figure 1A), or as a percent of total PBMCs (data not shown). This demonstrates that potent virus specific CD8 T cell responses can be generated in the absence of IL-15 (Figure 1A). Similar generation of LCMV specific CD8 T cell responses was observed in IL-15rα -/- mice (data not shown). Indeed, functional virus-specific responses were not dramatically impacted by the absence of IL-15 signals since effector CD8 T cells from +/- and -/- mice were equally capable of IFN-γ production and virus was effectively cleared from -/- mice (see below and data not shown). Importantly, the persistence of tetramer positive cells at days 30, 50 and 140 demonstrates that memory T cells were generated in the absence of IL-15 (Figure 1). However, it became evident during longitudinal tracking that LCMV specific memory T cells were inefficiently maintained in IL-15 -/- animals (Figure 1A and B). While initially similar, the response in IL-15-/- mice falls to below 50% of that in controls by 140 days post infection and the number antigen specific CD8 T cells/10^6 PBMCs drops markedly for each LCMV epitope in IL-15-/- mice (Figure 1B). For example, GP33-specific responses fall from 60-70% of controls at day 8 post infection to approximately 30% of controls by 2 months post-infection in IL-15 -/-mice. Similar results were observed in IL-15 receptor alpha chain deficient (IL-15rα -/-) mice (data not shown). Overall IL-15 -/- and IL-15Rα -/- mice mount a substantial antiviral CD8 T cell response. However, the magnitude of the LCMV specific memory population in the absence of IL-15 signals declines over time.

To evaluate whether the frequency of antigen-specific cells in the PBMC accurately reflects the number of LCMV specific memory CD8 T cells in vivo, lymphocytes from lymphoid and non-lymphoid organs were isolated and stained with LCMV specific MHC tetramers. LCMV immune IL-15-/- and IL-15Rα -/- mice were sacrificed 80 and 180 days post infection respectively and NP396 and GP33 specific cells were enumerated in the spleen, bone marrow, PBMC and liver (Figure 2). The distribution of antigen specific memory T cells among lymphoid tissues was similar in IL-15 and IL-15rα deficient and control animals. LCMV specific CD8 T cell numbers in IL-15-/- were reduced 1.4 to 4.2-fold in all organs examined compared to +/- control mice (Figure 2B). Similar results were also observed for IL-15Rα-/- mice at 180 days post infection (Figure 2C). In this case, NP396 and GP33 specific memory CD8 T cell numbers were reduced between 1.6 and 2 fold in the spleens and livers. These results suggest that the waning memory CD8 T cell numbers in the longitudinal PBMC analysis was not simply due to a redistribution of CD8 T cells to other locations. Importantly, LCMV specific CD4 responses were similar in both IL-15 -/- and IL-15Rα -/- and +/- control mice (data not shown).

Virus-specific memory CD8 T cells generated in the absence of IL-15 signals were qualitatively similar to those induced in the presence of IL-15. First, IFN-γ production by memory CD8 T cells from IL-15 +/- and -/- mice was assessed following a 5-hour stimulation with a peptide corresponding to a dominant LCMV epitope. Figure 3 shows MHC tetramer staining along with intracellular cytokine staining for IFN-γ and TNF-α. The frequencies obtained by IFN-γ production were similar to those obtained by MHC tetramer staining, indicating that all antigen specific cells from IL-15 -/- mice were capable of rapidly synthesizing the effector cytokine IFN-γ in this short term assay. Cells from IL-15+/- and -/- mice were also similar with respect to TNF-α production (Figure 3). Finally, cells were analyzed for expression of surface markers associated with memory CD8 T cell phenotype. T cells generated in IL-15-/- mice expressed equivalent levels of CD44 and CD122, the IL-2/15Rβ chain (Figure 3B). The latter is significant, since this is thought to allow memory CD8 T cells to respond to IL-15.

Together these results demonstrate that virus specific memory CD8 T cells can be generated in the absence of IL-15 signals. LCMV specific memory CD8 T cells are detected long after infection in IL-15 and IL-15Rα deficient mice and are phenotypically and functionally similar to memory T cells generated in +/- mice. However, LCMV specific CD8 T cell numbers decline over time in receptor and cytokine knockout mice, suggesting a defect in the maintenance of memory CD8 T cells.

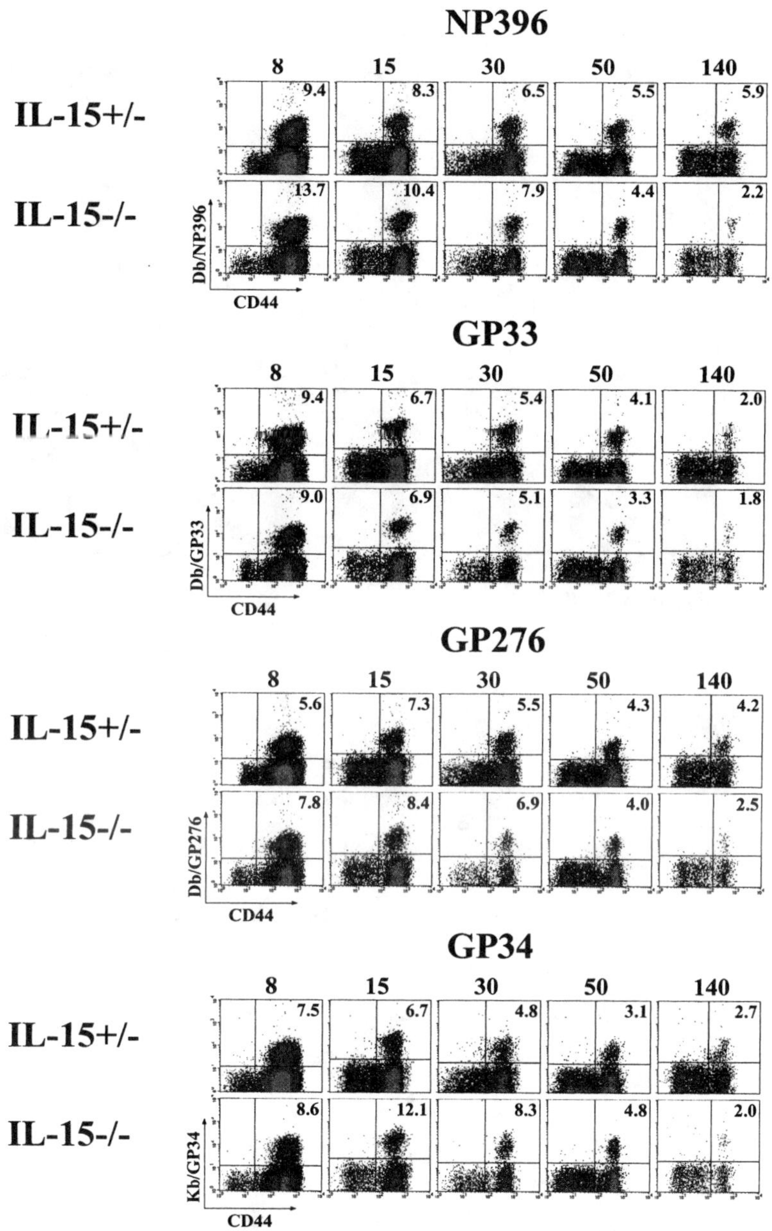

Figure 1. IL-15 -/- mice mount a substantial CD8 T cell response to LCMV infection but memory CD8 T cell numbers gradually decline. (A) Longitudinal PBMC analysis was performed by serially bleeding individual mice on the indicated days post infection with LCMV. PBMCs were stained with MHC tetramers of 4 LCMV epitopes (Db/NP396, Db/GP33, Db/GP276, and Kb/GP34). Plots are gated on CD8+ T cells, and the numbers represent the number of MHC tetramer binding cells as a percentage of CD8+ T cells. Staining is representative of 4-7 mice/group. (B) The numbers of epitope-specific cells / 10^6 PBMCs in IL-15+/- and IL-15-/- mice. Staining was performed as in (A). Numbers are averages of 4-7 mice. Similar results were observed in IL-15Rα-/- mice (data not shown).

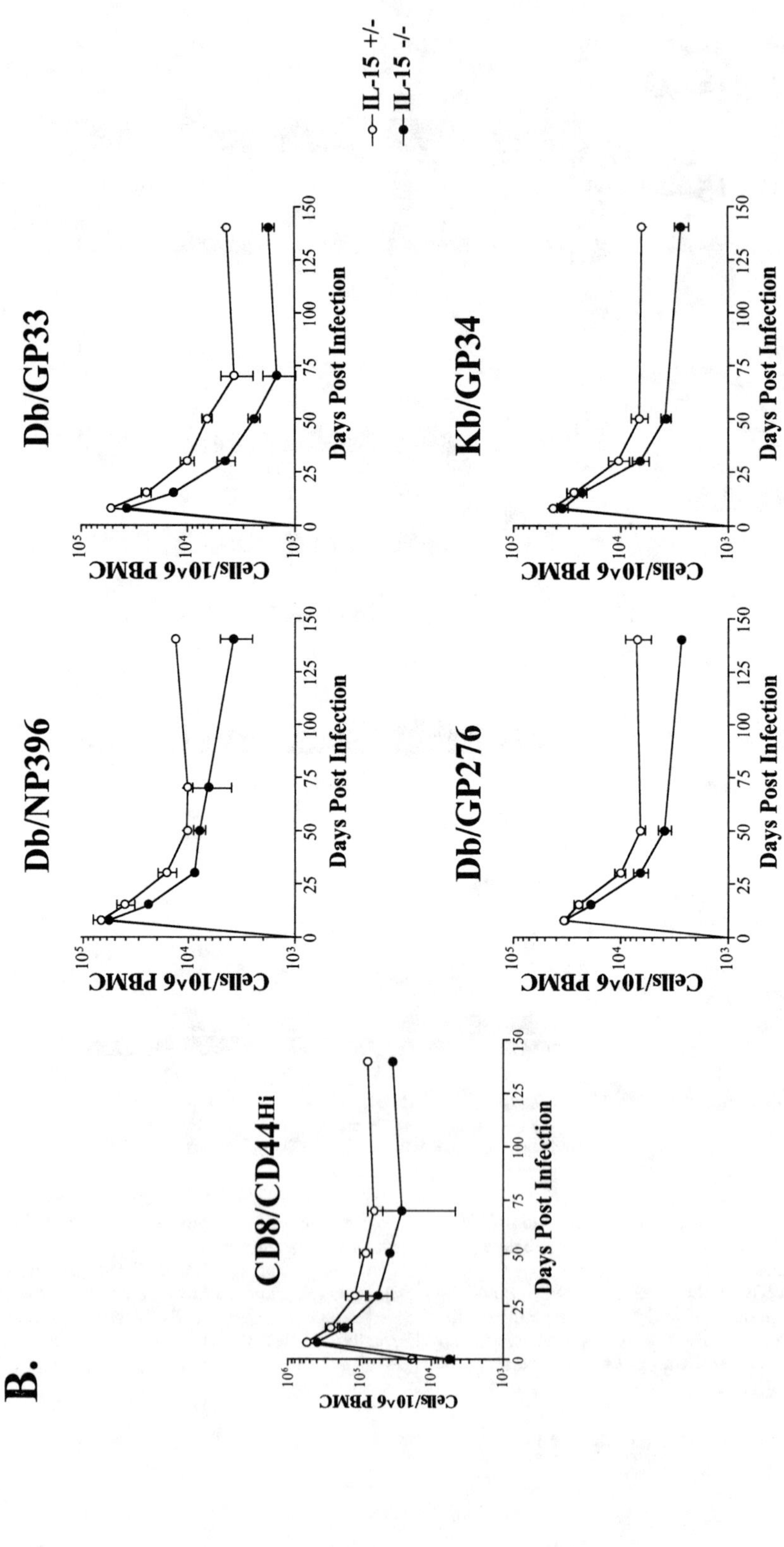

Figure 1. *(Continued)*

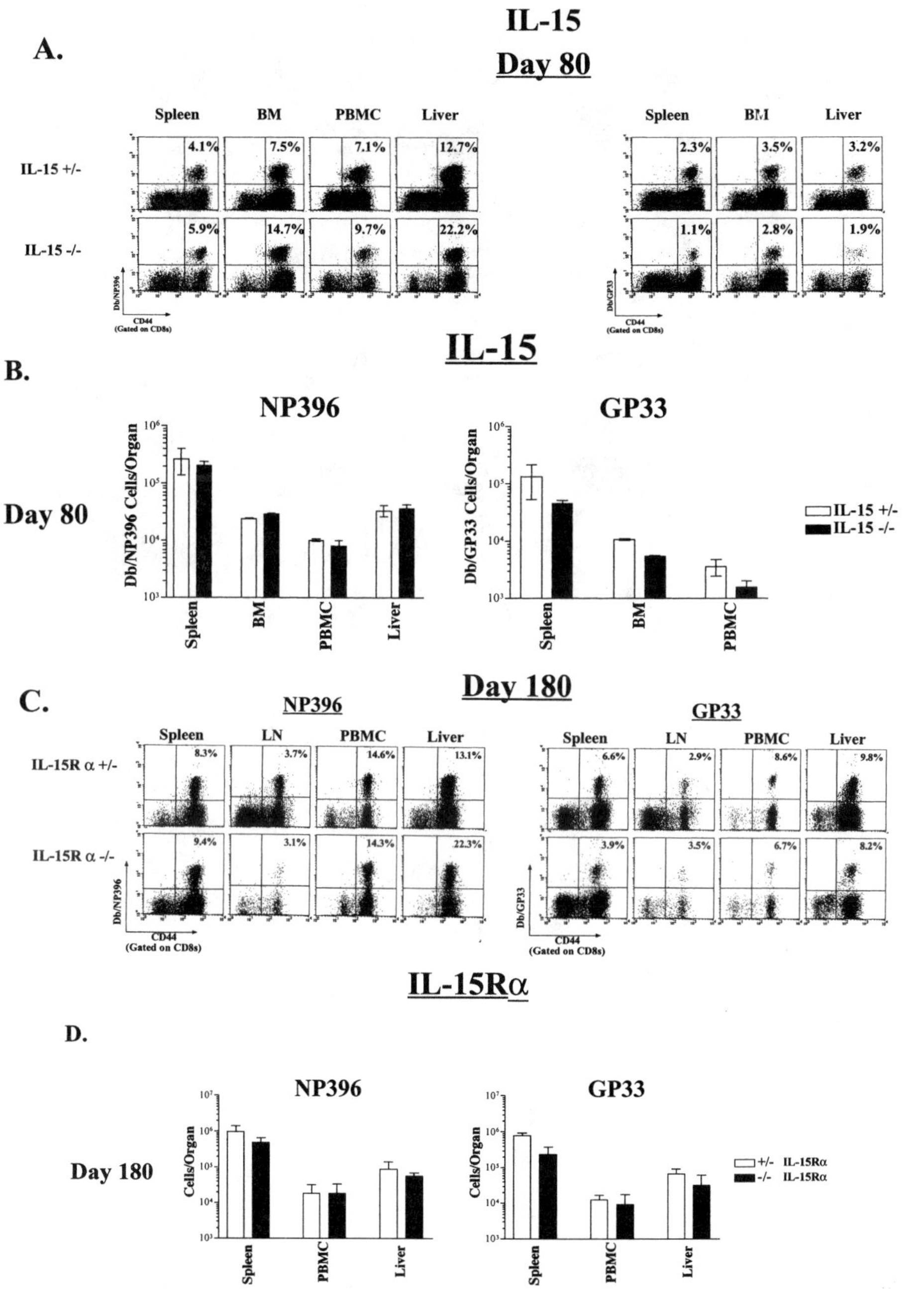

Figure 2. LCMV specific memory CD8 T cells are found in both lymphoid and non-lymphoid locations in both IL-15-/- and IL-15Rα-/-. Lymphocytes isolated from spleen, bone marrow (BM), PBMC, and liver of IL-15+/- and IL-15-/- (A and C) and IL-15Rα+/- and IL-15Rα-/- (B and D) were stained for NP396 and GP33specific CD8 T cells. Staining is representative of 2-5 mice/group. C and D. Total numbers of memory LCMV-specific CD8 T cells were determined for this indicated tissues. LCMV specific CD4 responses were similar in both IL-15+/- and IL-15-/- as well as IL-15Rα -/- and +/- control mice (data not shown).

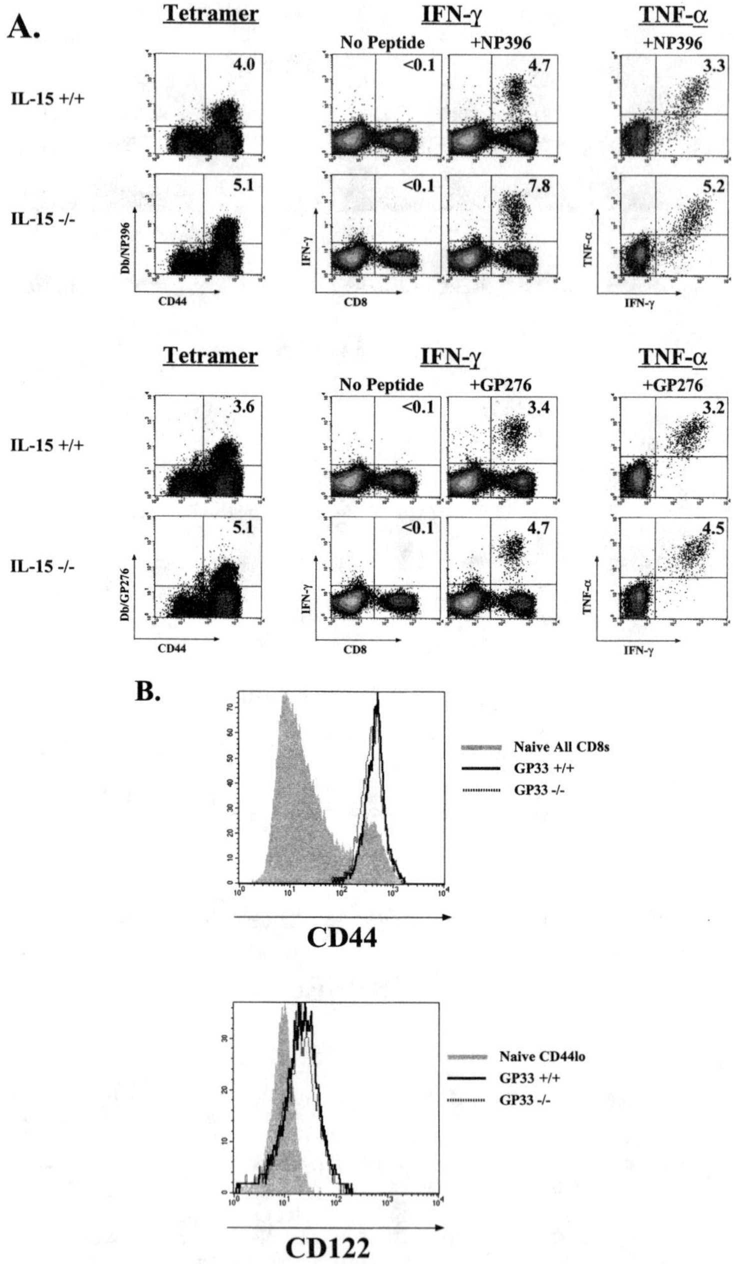

Figure 3. Antigen specific memory CD8 T cells in IL-15 -/- mice are functionally and phenotypically similar to memory CD8 T cells in +/+ mice. (A) Splenocytes from IL-15 -/- or +/+ control mice were prepared 40 days post-infection. Cells were stimulated in vitro for 5 hours with NP396 or GP276 peptides, then intracellular cytokine staining was performed for IFN-γ and TNF-α. The percentage of CD8 T cells producing either cytokine is essentially equivalent to the percentage of tetramer-binding cells in all cases. Tetramer staining and IFN-γ/TNF-α dual staining is shown gated on CD8 T cells. Numbers in the upper right represent the percent of CD8 T cells (B) LCMV-specific GP33 tetramer+ CD8 T cells from the spleens of IL-15 -/- mice or littermate controls were stained for CD44 and CD122 at 40 days post-infection. The dark line indicates GP33 tetramer+ cells from a +/+ mouse while the dashed line represents GP33 tetramer+ CD8 T cells from an IL-15 -/- mouse. Naïve controls are shown in light grey; all CD8s from a naïve mouse for the CD44 plot and $CD44^{Lo}$CD8s for the CD122 plot.

Memory CD8 T Cells are Unable to Undergo Homeostatic Proliferation in An IL-15 -/- Environment

Based on the relatively gradual decline in antigen-specific memory cell numbers in the absence of IL-15 signals, we next evaluated the ability of memory CD8 T cells to undergo proliferative renewal in the absence of IL-15. To examine memory CD8 T cell turnover, homeostatic proliferation was evaluated over a one-month period by adoptively transferring CFSE-labeled memory CD8 T cells into either IL-15 competent or IL-15 deficient hosts. Approximately 140 days after infection with LCMV memory CD8 T cells from B6 mice were isolated, labeled with CFSE, and adoptively transferred into intact, non-irradiated, B6 or IL-15 -/- hosts. Homeostatic proliferation of these memory T cell populations was assessed by the loss of CFSE fluorescence 30 days post transfer. When LCMV specific CD8 T cells generated in B6 animals were labeled and adoptively transferred to normal B6 mice, the NP396 and GP33 specific memory CD8 T cell populations were maintained and underwent normal homeostatic proliferation (Figure 4). A large proportion of the transferred NP396 and GP33 specific CD8 T cells had divided up to 4 times. In striking contrast, when LCMV specific CD8 T cells from B6 mice were transferred into IL-15-/- mice little division was detected (Figure 4). The turnover of total $CD44^{hi}$ CD8 T cells from LCMV immune B6 mice was also nearly ablated when adoptively transferred into IL-15 -/- hosts while normal turnover was observed in wildtype recipients. It is unlikely that IL-15 is produced to significant levels by T cells or other transferred cell population in these experiments since when wildtype CD8 T cells were transferred into into IL-15 -/- mice very little memory T cell turnover resulted. These transfer experiments demonstrate that in an IL-15-/- environment memory CD8 T cells, both true LCMV-specific tetramer+ memory CD8 T cells and total $CD44^{hi}$ CD8 T cells, are unable to undergo efficient homeostatic proliferation. Thus, IL-15 or an IL-15-derived signal is essential for normal homeostatic turnover of memory CD8 T cells.

Memory CD8 T cells generated in an IL-15 deficient environment might adopt an IL-15 independent phenotype. To examine whether memory CD8 T cells from IL-15 -/- mice were capable of normal homeostatic proliferation a similar experiment was performed. Memory cells from LCMV immune IL-15-/- mice were CFSE-labeled and adoptively transferred into IL-15-/- hosts. Under these conditions virtually no proliferation was detectable at one month post transfer (Figure 4 right panels). Thus, the memory cells generated in IL-15 deficient animals were not an adapted population capable of IL-15-independent homeostatic proliferation. However, memory CD8 T cells from IL-15-/- mice underwent substantial proliferation when transferred to normal B6 recipients, indicating that memory cells generated in IL-15 deficient mice are still competent to respond to the signals for homeostatic proliferation, but are unable to do so in an IL-15 deficient environment.

CONCLUDING REMARKS

There are two main findings in the present study. First, IL-15 signals are not absolutely required for the generation of an antiviral CD8 T cell response in vivo. Second, IL-15 is essential for homeostatic proliferation of virus-specific memory CD8 T cells since no turnover of CFSE-labeled memory CD8 T cells was observed in IL-15 -/- mice.

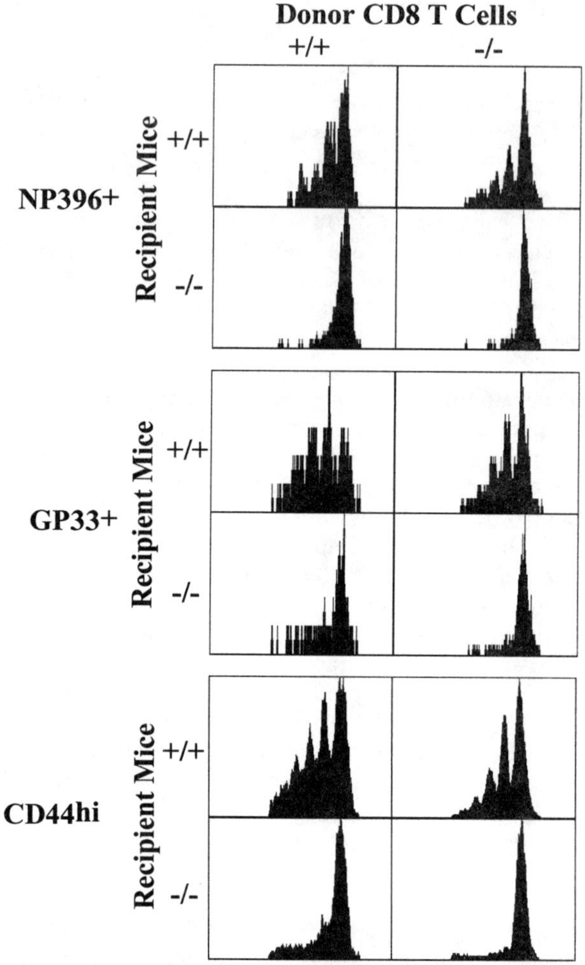

Figure 4. Memory CD8 T cells do not undergo homeostatic proliferation, but undivided cells are maintained in an IL-15 -/- environment. Splenocytes from LCMV immune mice were labeled with CFSE and transferred into naïve recipients. 30 days post-transfer, spleens were removed from recipient animals and cells were stained using MHC class I tetramers to identify antigen specific memory CD8 T cells. Transferred antigen-specific cells proliferated in IL-15 +/+ but not -/- mice over a 30-day period. In the right column, the same experiment was performed using splenocytes from immune IL-15 -/- mice and similar results were observed. Homeostatic proliferation of CD4 T cells occurred and was normal in all environments (data not shown).

Effective LCMV specific responses can be generated in both IL-15 and IL-15Rα deficient mice. In both cases a substantial population of virus specific memory T cells is generated suggesting that the signals important for the formation of memory CD8 T cells are not dramatically impaired in the absence of IL-15 or its high affinity receptor. It should be noted that control of LCMV is highly dependent on CD8 T cells.[18] During infections with other pathogens that are sensitive to NK cells, NK T cells, or γδ T cells, which are absent in IL-15 -/- or IL-15Rα -/- mice,[4,5] the impact of an IL-15 deficiency may be dramatic. For example, IL-15 -/- mice succumb to infection with a dose of vaccinia virus easily controlled by wild type mice.[5] The use of LCMV in the current study illustrates that despite a defect in other cell types a potent LCMV specific CD8 T cell response is generated without IL-15 signals. Rather, the slow attrition of memory T cells over time suggests that the main role of IL-15 for CD8 T cells may be in homeostatic turnover or maintenance. In fact, the data presented in this study provide convincing evidence that antigen specific memory CD8 T cells require IL-15 for homeostatic proliferation. It remains to be determined precisely how IL-15 causes homeostatic proliferation. Recent studies demonstrate that IL-15 may not act directly on CD8 T cells, but rather via another bone marrow-derived cell population.[19] Indeed, it has been reported that dendritic cells function may be altered in the absence of IL-15 signals.[20] This may allow for a strict level of control of homeostatic proliferation of CD8 T cells via dendritic cells since IL-15 is thought to be produced by many cells types in multiple tissues.[21] Future studies will be necessary to dissect the components involved in IL-15-dependent memory T cell turnover.

Our results suggest that the main role of IL-15 for CD8 T cells is in the maintenance of the memory pool. Initial responses and generation of a pool of memory CD8 T cells can occur without IL-15 or IL-15Rα. However, homeostatic proliferation is ablated in the absence of IL-15 or IL-15Rα and the size of the memory pool slowly declines. More detailed understanding of the interaction between IL-15 and the memory CD8 T cell compartment may, in the future, allow selective maintenance or augmentation of antigen-specific memory T cell populations.

REFERENCES

1. Lau, L. L., Jamieson, B. D., Somasundaram, T., and Ahmed, R. (1994) *Nature* **369**(6482), 648-52.
2. Swain, S. L., Hu, H., and Huston, G. (1999) *Science* **286**(5443), 1381-3.
3. Murali-Krishna, K., Lau, L. L., Sambhara, S., Lemonnier, F., Altman, J., and Ahmed, R. (1999) *Science* **286**(5443), 1377-81.
4. Lodolce, J. P., Boone, D. L., Chai, S., Swain, R. E., Dassopoulos, T., Trettin, S., and Ma, A. (1998) *Immunity* **9**(5), 669-76.
5. Kennedy, M. K., Glaccum, M., Brown, S. N., Butz, E. A., Viney, J. L., Embers, M., Matsuki, N., Charrier, K., Sedger, L., Willis, C. R., Brasel, K., Morrissey, P. J., Stocking, K., Schuh, J. C., Joyce, S., and Peschon, J. J. (2000) *J Exp Med* **191**(5), 771-80.
6. Zhang, X., Sun, S., Hwang, I., Tough, D. F., and Sprent, J. (1998) *Immunity* **8**(5), 591-9.
7. Ku, C. C., Murakami, M., Sakamoto, A., Kappler, J., and Marrack, P. (2000) *Science* **288**(5466), 675-8.
8. Mattei, F., Schiavoni, G., Belardelli, F., and Tough, D. F. (2001) *J Immunol* **167**(3), 1179-87.
9. Biron, C. A. (2001) *Immunity* **14**(6), 661-4.
10. Umemura, M., Nishimura, H., Hirose, K., Matsuguchi, T., and Yoshikai, Y. (2001) *J Immunol* **167**(2), 946-56.
11. Yajima, T., Nishimura, H., Ishimitsu, R., Watase, T., Busch, D. H., Pamer, E. G., Kuwano, H., and Yoshikai, Y. (2002) *J Immunol* **168**(3), 1198-203.
12. Khan, I. A., and Casciotti, L. (1999) *J Immunol* **163**(8), 4503-9.
13. Khan, I. A., and Kasper, L. H. (1996) *J Immunol* **157**(5), 2103-8.
14. Tough, D. F., Borrow, P., and Sprent, J. (1996) *Science* **272**(5270), 1947-50.
15. Fehniger, T. A., Suzuki, K., Ponnappan, A., VanDeusen, J. B., Cooper, M. A., Florea, S. M., Freud, A. G., Robinson, M. L., Durbin, J., and Caligiuri, M. A. (2001) *J Exp Med* **193**(2), 219-31.
16. Ahmed, R., Salmi, A., Butler, L. D., Chiller, J. M., and Oldstone, M. B. (1984) *J Exp Med* **160**(2), 521-40.
17. Murali-Krishna, K., Altman, J. D., Suresh, M., Sourdive, D. J., Zajac, A. J., Miller, J. D., Slansky, J., and Ahmed, R. (1998) *Immunity* **8**(2), 177-87.
18. Buchmeier, M. J., and Zajac, A. J. (1999) in *Persistent Viral Infections.* (Ahmed R., C. I. S. Y., ed), pp. 575-605., John Wiley & Sons, Chichester, UK
19. Lodolce, J. P., Burkett, P. R., Boone, D. L., Chien, M., and Ma, A. (2001) *J Exp Med* **194**(8), 1187-94.
20. Ohteki, T., Suzue, K., Maki, C., Ota, T., and Koyasu, S. (2001) *Nat Immunol* **2**(12), 1138-43.
21. Giri, J. G., Kumaki, S., Ahdieh, M., Friend, D. J., Loomis, A., Shanebeck, K., DuBose, R., Cosman, D., Park, L. S., and Anderson, D. M. (1995) *Embo J* **14**(15), 3654-63.

IRREVERSIBLE MARKING OF DENDRITIC CELLS IN VIVO

For Contributed Volumes

Sanjay Garg, Alp Oran, Charles Maris and Joshy Jacob*

1. INTRODUCTION

Dendritic cells (DC) are bone marrow derived antigen presenting cells that function as sentinels of immune system (Akbari *et al.*, 1999; Bancherea, 2000; Banchereau and Steinman, 1998; Cutler *et al.*, 2001; Mellman and Steinman, 2001). In the peripheral tissues such as skin, they exist as immature DC - cells that are highly efficient at antigen uptake via receptor mediated and non-receptor mediated mechanisms. Upon activation by tissue damage, CpG DNA, pathogens or inflammatory cytokines the immature DC transform into mature DC, a cell type optimized for antigen presentation and T cell activation. DC efficiently process the ingested antigens and present the resultant peptides on the cell surface in the context of class I and class II MHC. Mature DC up regulate expression of class II MHC and activation markers such as CD54, CD58, CD80 and CD86. Mature DC down regulate chemokine receptors like CCR1, CCR5, and CCR6 that keep them in the non-lymphoid tissues and up regulate expression of CCR7, which enables the them to migrate to the draining lymph nodes (LN). In the draining lymph nodes, the antigen bearing mature DC move to T cell area and activate quiescent naïve CD4 and CD8 T cells; DC are the most potent initiators of primary immune responses - they are two orders of magnitude more effective as antigen presenting cells than B-lymphocytes and macrophages. This pattern of antigen capture in the peripheral non-lymphoid tissues and migration into the draining lymph nodes induces activation and clonal expansion of rare antigen-specific T cells.

Migration of activated DC to the draining lymph nodes has been studied using two experimental approaches to tag migrating DC – (a) skin painting with contact sensitizer fluorescein isothiocyanate (FITC) and tracking FITC positive DC in draining LNs (Macatonia *et al.*, 1987) and (b) genetic immunization with plasmid DNA encoding a reporter such as bacterial β-galactosidase on the abdominal epidermis and tracking

* Sanjay Garg, Alp Oran, Charles Maris, Joshy Jacob, Vaccine Research Center, Emory University, 954 Gatewood Road, Atlanta, GA 30329.

tagged DC in the draining lymph nodes (Porgador *et al.*, 1998). Both of these approaches have shown that the frequency of antigen-bearing DC that migrate from the peripheral tissues to the draining LN is quite small. Germain and colleagues (Porgador *et al.*, 1998) studied DC migration by gene gun immunization with a cytomegalovirus enhancer/promoter driven β-galactosidase. They showed that majority of antigen-bearing DC were directly transfected with the plasmid DNA and by counting the number of the migrating β-galactosidase expressing DC in the draining LN they concluded that there are approximately 50-100 antigen-bearing DC per LN (Porgador *et al.*, 1998). A major disadvantage with tracking DC in this manner is that the ability to mark DC is dependent upon the promoter activity of β-gal encoding plasmid. Once the promoter activity of the plasmid DNA is extinguished, the DNA plasmid-bearing DC cannot be visualized. We have attempted to overcome this hurdle by permanently marking DC using cre recombinase mediated irreversible genetic recombination.

2. CRE LOX RECOMBINATION TO MARK DC

ROSA26 is a ubiquitous transcriptionally active locus in the mouse genome, which was originally identified by Dr. Soriano and colleagues by using a retroviral gene trapping strategy (Zambrowicz *et al.*, 1997). The ROSA promoter drives very high levels of gene expression in all the tissues tested, embryonic as well as adult (Zambrowicz *et al.*, 1997). Two independent groups have generated novel transgenic mouse strains to monitor Cre-mediated DNA recombination by modifying the ROSA locus (Mao *et al.*, 1999; Soriano, 1999). Cre recombinase is an enzyme made by P1 bacteriophage and it binds to its recognition sequence called loxP. Upon binding to tandem loxP sites, the Cre recombinase catalyzes a genetic recombination event between the loxP sites. Soriano and Mao *et al.* modified the ROSA locus by introducing a loxP-flanked or neomycin phospho transferase cDNA cassette and the β-gal reporter gene downstream of the ROSA promoter to generate ROSA26R mice (shown in Fig.1). Although the ROSA promoter is transcriptionally active in all cell types and at all developmental stages, the downstream dormant β-gal is not expressed because of the intervening loxP flanked neomycin cassette. β-gal can be expressed only after Cre recombinase mediated excisional recombination of the intervening loxP flanked neomycin cassette (Fig 1) and juxta positioning of the β-gal immediately downstream of the ROSA promoter. Once recombined, β-gal is constantly expressed for the life of the cell (Mao *et al.*, 1999; Soriano, 1999).

To mark DC in ROSA26R mice, we constructed a CMV promoter driven Cre recombinase encoding DNA vaccine (CMV-Cre). To target the recombinase to the nucleus we engineered into the Cre recombinase the nuclear localization signal sequence from simian virus 40 large T antigen. The plasmid DNA was coated on gold bullets and used for immunization. Cohorts of ROSA26R mice were immunized on the abdominal epidermis with a single gene gun immunization of 2μg CMV-Cre and the draining inguinal LNs were analyzed for β-gal expressing antigen presenting cells. At various time points following immunization, expression of β-gal was detected in the draining lymph nodes using the histochemical substrate X-gal (Porgador *et al.*, 1998). β-gal$^+$ cells were observed in the inguinal LNs of mice immunized with CMV-Cre plasmid but not control CMV-influenza hemagglutinin plasmid (Fig.2).

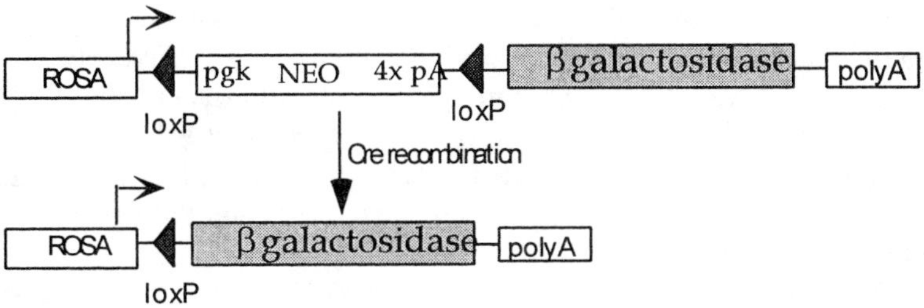

Figure 1. Transient expression of Cre recombinase leads to excisional recombination at the loxP sites and juxta positioning of the β gal reporter immediately downstream of the ubiquitous ROSA promoter. Once recombined the cell expresses β gal permanently.

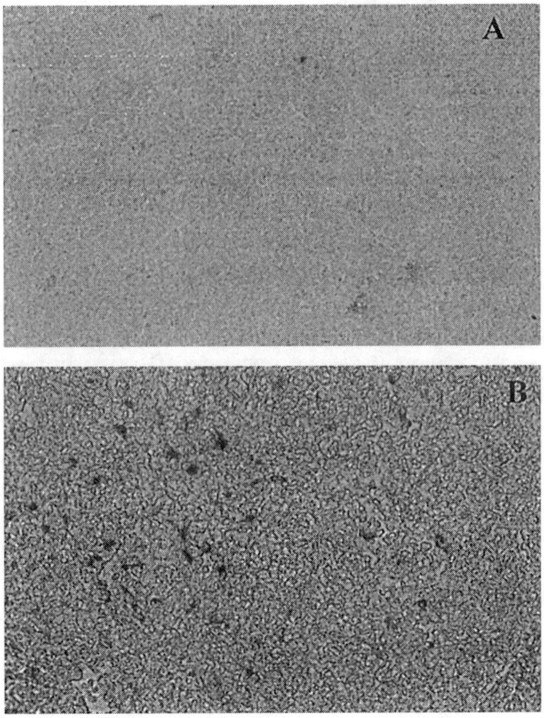

Figure 2. Genetic immunization of ROSA26R mice with CMV Cre plasmid DNA leads to irreversible marking. ROSA26R mice were gene gun immunized at the abdominal epidermis with gold bullets coated with either (A) CMV-influenza hemagglutinin or (B) CMV-Cre plasmid DNA. 48 hours post immunization, the draining LNs were isolated and processed for frozen sectioning. Eight micron thick frozen sections were fixed in paraformaldehyde and incubated overnight at 37°C with the substrate, X-gal. β gal^{+} DC are seen in mice immunized with CMV-Cre (B) but not control non-Cre plasmid (A).

Surprisingly the number of β-gal marked cells in the draining LNs following gene gun immunization with CMV-Cre plasmid DNA, was substantially higher than previous estimates. Germain and colleagues have estimated that there are 50-100 antigen-bearing DC per LN (Porgador *et al.*, 1998). In our irreversible tracking model we observed, on average, 70-100 β-gal marked cells per 8 μm frozen section. Are the β-gal marked cells found in the LN dendritic cells? To better phenotype the β-gal marked cells we used flow cytometry to track the β-gal$^+$ migrating cells by utilizing fluorescein-di-β-D-galactopyranoside (FDG), a substrate for β-gal (Fiering *et al.*, 1991; Nolan *et al.*, 1988). Upon hypotonic loading of FDG, cells that express β galactosidase hydrolyze the non-fluorescent FDG and release fluorescein into the cytoplasm and the fluorescent cells can be monitored by flow cytometry.

Cohorts of ROSA26R mice were gene gun immunized with gold bullets coated with CMV-Cre plasmid and at various time points the draining inguinal LNs were analyzed for β-gal$^+$ cells by flow cytometry. All of the β-gal$^+$ cells in the draining LN were exclusively CD11c$^+$ dendritic cells. None of the B cells, T cells, macrophages or other accessory cells expressed β-gal. This clearly demonstrates that the antigen presenting cells that migrate from the skin to the draining LNs following gene gun immunization are CD11c$^+$ dendritic cells. We determined the frequency of β-gal$^+$ CD11c$^+$ DC at various time points following immunization and found that the peak was at 2.5 days post gene gun immunization. At this time point, 10-12% of CD11c$^+$ DCs were β-gal$^+$. At its peak there are approximately $5x10^3$ antigen-bearing DCs per draining inguinal LN. β-gal$^+$ CD11c+ DC were observed at days 7 and 10 and even as late as day 14 post immunization but the numbers decreased with time.

In summary, we have shown that CD11c$^+$ dendritic cells can be irreversibly marked by cre-lox recombination in the abdominal skin epidermis and tracked in the draining lymph nodes. The number of antigen-bearing DC that migrate from the skin to the draining LN following gene gun immunization is approximately hundred fold higher that previous estimates. We show that the β-gal marked DC persist for approximately two weeks in the draining lymph nodes. Numerous reports in different animal models have demonstrated that genetic vaccination can generate effective immune responses against viruses, bacteria and parasites (Leitner *et al.*, 1999; Robinson, 1999; Steele *et al.*, 2001). The recent reports of prevention of AIDS in Rhesus Macaques by a multiprotein DNA vaccine have established without doubt the efficacy and feasibility of genetic immunization (Amara *et al.*, 2001). Our data provides a rational explanation for the efficacy of DNA vaccines - DNA immunization leads to priming of large numbers of antigen-bearing DCs that persists for extended periods of time (approximately two weeks in mice). Since dendritic cells are the most potent initiators of primary immune responses, this leads to robust T cell responses and protective immunity.

REFERENCES

Akbari, O., Panjwani, N., Garcia, S., Tascon, R., Lowrie, D., and Stockinger, B. (1999). DNA vaccination: transfection and activation of dendritic cells as key events for immunity, Journal of Experimental Medicine *189*, 169-78.

Amara, R. R., Villinger, F., Altman, J. D., Lydy, S. L., O'Neil, S. P., Staprans, S. I., Montefiori, D. C., Xu, Y., Herndon, J. G., Wyatt, L. S., *et al.* (2001). Control of a mucosal challenge and prevention of AIDS by a multiprotein DNA/MVA vaccine, Science *292*, 69-74.

Banchereau, J., Briere, F., Caux, C., Davoust, J., Lebecque, S., Liu, Y. J., Pulendran, B., and Palucka, K. (2000). Immunobiology of dendritic cells, Annual Review of Immunology *18*, 767-811.

Banchereau, J., and Steinman, R. M. (1998). Dendritic cells and the control of immunity, Nature *392*, 245-52.

Cutler, C. W., Jotwani, R., and Pulendran, B. (2001). Dendritic cells: immune saviors or Achilles' heel?, Infection & Immunity *69*, 4703-8.

Fiering, S. N., Roederer, M., Nolan, G. P., Micklem, D. R., Parks, D. R., and Herzenberg, L. A. (1991). Improved FACS-Gal: flow cytometric analysis and sorting of viable eukaryotic cells expressing reporter gene constructs, Cytometry *12*, 291-301.

Leitner, W. W., Ying, H., and Restifo, N. P. (1999). DNA and RNA-based vaccines: principles, progress and prospects, Vaccine *18*, 765-77.

Macatonia, S. E., Knight, S. C., Edwards, A. J., Griffiths, S., and Fryer, P. (1987). Localization of antigen on lymph node dendritic cells after exposure to the contact sensitizer fluorescein isothiocyanate. Functional and morphological studies, Journal of Experimental Medicine *166*, 1654-67.

Mao, X., Fujiwara, Y., and Orkin, S. H. (1999). Improved reporter strain for monitoring Cre recombinase-mediated DNA excisions in mice, Proc Natl Acad Sci U S A *96*, 5037-42.

Mellman, I., and Steinman, R. M. (2001). Dendritic cells: specialized and regulated antigen processing machines. Cell *106*, 255-8.

Nolan, G. P., Fiering, S., Nicolas, J. F., and Herzenberg, L. A. (1988). Fluorescence-activated cell analysis and sorting of viable mammalian cells based on beta-D-galactosidase activity after transduction of Escherichia coli lacZ, Proc Natl Acad Sci U S A *85*, 2603-7.

Porgador, A., Irvine, K. R., Iwasaki, A., Barber, B. H., Restifo, N. P., and Germain, R. N. (1998). Predominant role for directly transfected dendritic cells in antigen presentation to CD8+ T cells after gene gun immunization, Journal of Experimental Medicine *188*, 1075-82.

Robinson, H. L. (1999). DNA vaccines: basic mechanism and immune responses (Review), International Journal of Molecular Medicine *4*, 549-55.

Soriano, P. (1999). Generalized lacZ expression with the ROSA26 Cre reporter strain [letter], Nat Genet *21*, 70-1.

Steele, K. E., Stabler, K., and VanderZanden, L. (2001). Cutaneous DNA vaccination against Ebola virus by particle bombardment: histopathology and alteration of CD3-positive dendritic epidermal cells, Veterinary Pathology *38*, 203-15.

Zambrowicz, B. P., Imamoto, A., Fiering, S., Herzenberg, L. A., Kerr, W. G., and Soriano, P. (1997). Disruption of overlapping transcripts in the ROSA beta geo 26 gene trap strain leads to widespread expression of beta-galactosidase in mouse embryos and hematopoietic cells, Proc Natl Acad Sci U S A *94*, 3789-94.

TRACKING ARTERIAL SMOOTH MUSCLE-SPECIFIC T CELLS IN THE INFLAMED VASCULATURE

Burkhard Ludewig[1]

1. INTRODUCTION

Atherosclerotic disease and its major complications such as myocardial infarction and ischemic stroke, is still the major health problem in the Western World. Atherosclerosis is now regarded as a chronic inflammatory disease developing in response to injury in the vessel wall and is characterized by the infiltration of mononuclear lymphocytes into the intima, local expansion of vascular smooth muscle cells and accumulation of extracellular matrix.[1] Classical risk factors predisposing for atherosclerosis include hypercholesterolemia, hypertension, and diabetes mellitus. In addition, infections with different bacterial and viral pathogens have been associated with atherosclerotic disease. Initial vascular injury in atherosclerosis may result from the direct cytopathic effect of the pathogen replicating in the vascular wall or can be mediated by an immune response directed against antigens present in smooth muscle or endothelial cells. Chronic immune-mediated inflammatory responses in atherosclerotic processes may be antigen-specific and directed against (modified) autoantigens or pathogen-derived structures. Alternatively, metabolic and/or physical distress may induce non-specific activation of certain components of the immune system.

Classical primary vasculitis syndromes such as Takayasu arteritis or temporal arteritis, involve massive recruitment of lymphocytes and macrophages into the vascular wall, destruction of the medial layer with fibrosis, and proliferation of smooth muscle cells in the intima leading to neointima formation. Similar to atherosclerosis, the pathogenesis of systemic vasculitides probably involves infectious and/or autoimmune mechanisms. Importantly, inflammatory changes associated with systemic vasculitides may promote formation of atherosclerotic lesions.[2]

---

[1] Institute of Experimental Immunology, Department of Pathology, University of Zürich, Schmelzbergstr. 12, CH-8091, Zürich, Switzerland.

2. INFLAMMATORY RESPONSES IN THE VASCULAR WALL: INFECTION OR AUTOIMMUNITY?

A number of potential mechanisms have been implicated in the development inflammatory reactions in the vascular system. Microbial infection and autoimmune reactions are thought to possess the potential to elicit and maintain vascular inflammation in systemic vasculitides. For example, Wegener's granulomatosis has been linked with human parvovirus B19[3] and Staphylococcus aureus infection.[4] Likewise, human parvovirus B19 was suggested to be involved in Kawasaki disease[5] and polyarteritis nodosa.[6] These and other studies (reviewed in Mandell and Calabrese 1998)[7] suggest that several infectious agents may cause vascular pathologies, often in addition to the main clinical manifestation. Furthermore the number of infectious agents involved in the pathogenesis of atherosclerosis support the notion that multiple infections with the same or with different infectious agents possessing a tropism for cells of the vasculature may be required for the initiation and perpetuation of vascular inflammatory reactions.[8]

Pathogen-induced inflammation often precipitates genuine autoimmune responses against antigens present in the target tissue. Destruction of cells in the vascular wall either by the direct cytolytic activity of the infectious agent or by immune mechanisms directed against the pathogen may lead to release of self-antigens and subsequent activation of transient autoimmune responses. The pathological consequences of autoimmunity usually become evident when autoimmune responses directed against self antigens become chronic. Frequent or persistent infections of the vascular wall may favor autoimmune disease by increasing the amounts of released self antigens, generation of cytokine-mediated bystander activation of self-reactive lymphocytes, or by activation of self-reactive lymphocytes via antigenic determinants shared between the pathogen and the host, a mechanism that has been termed “molecular mimicry”.

B cell-mediated autoimmune reactions against endothelial cells are frequent in all systemic vasculitides as outlined in a recent review.[9] Furthermore, anti-β2 glycoprotein I antibodies directed against endothelial cells have been found in atherosclerosis patients[10] and T cells from mice primed with β2 glycoprotein were able to enhance atherosclerotic processes in genetically susceptible LDL-receptor-deficient mice.[11] An example for molecular mimicry leading to vascular autoimmunity is the mycobacterial heat shock protein 65 (mHSP65). Immunization of rabbits[12] or mice[13] with mHSP65 induced atherosclerotic alterations in arteries and human carotid atherosclerosis appears to be associated with elevated antibody titers against mHSP65.[14] Oxidatively modified low density lipoprotein (oxLDL) is a further autoantigen that is possibly involved in immunopathological alterations in the vessel wall. Patients with advanced atherosclerotic lesions showed elevated levels of circulating antibodies to oxLDL[15] and T cells derived from human atherosclerotic lesions have been shown to recognize oxLDL.[16] Interestingly, immunization with oxLDL and native LDL reduced the severity of atherosclerotic lesions in hypercholesterolemic LDLR-deficient mice.[17] Taken together, both infection-associated and lipid imbalance-induced autoimmune processes are most likely important pathogenic factors in the atherosclerotic process.

3. IMPACT OF HYPERCHOLESTEROLEMIA ON VASCULAR INFLAMMATION AND GENERAL IMMUNE RESPONSIVENESS

Hypercholesterolemia is one of the major risk factors in atherosclerosis. Therapeutic intervention using lipid-lowering agents significantly reduces mortality rates in patients with coronary heart disease.[18] Interestingly, drugs like statins also effect general immune-responsiveness. For example, statins may act as direct inhibitors of induction of MHC-II expression on endothelial cells by IFN-gamma and thereby suppress MHC-II-mediated T-cell activation.[19]

Cholesterol metabolism may directly impact at various points on the responsiveness of the immune system such as predisposing the microvasculature to intense leukocyte-endothelial cell adhesion in response to inflammatory stimuli[20] or by increasing macrophage chemotaxis.[21] However, high lipoprotein levels in plasma diminish systemic cytokine responses[22,23] and impaired anti-bacterial immune responses were shown in genetically hypercholesterolemic apolipoprotein E-deficient mice.[24,25] Similarly, hypercholesterolemic mice lacking the low density lipoprotein receptor are highly susceptible to disseminated *Candida albicans* infection[26] indicating that beside local stimulatory effects in vascular inflammatory responses, hypercholesterolemia may also exert negative effects on general immune responsiveness.

Although the role of hypercholesterolemia in the pathogenesis of atherosclerosis has been studied extensively in experimental mouse models,[27-29] a thorough analysis of T cell reactivity in these mice in response to a viral infection has only recently been described.[30] We used the well-characterized lymphocytic choriomeningits virus infection which induces immunopathological disease in a dose, age and MHC-dependent fashion.[32] ApoE- and LDLR-deficient mice were infected with the hepatotropic LCMV WE strain and MHC class I tetramers complexed with defined viral epitopes were used to follow activation and peripheral recruitment of virus-specific CTL. In addition, antiviral T cell effector function was followed by cytotoxicity and cytokine production assays. Our results showed that hypercholesterolemia may substantially impair antiviral cellular immune responses leading to delayed viral clearance from from the target organs. Hypercholesterolemia may thus disrupt the usually well-maintained pathogen-host equilibrium, ultimately leading to enhanced immunopathological disease.

Our findings support the notion that altered immune-reactivity favors virus-induced vascular immunopathology.[32],[33] Furthermore, mice with impaired IFN-gamma responses are more susceptible to infection with murine CMV[34] or gammaherpesvirus 68[35] and develop progressive chronic arterial inflammation. Similarly, *C. pneumoniae* infects preferentially pre-damaged atheromatous areas of the aorta in severely hypercholesterolemic ApoE$^{-/-}$ or LDLR$^{-/-}$ mice, but only very rarely healthy aortas of normocholesterolemic C57BL/6 mice[36] or aortas of hypercholesterolemic mice before development of cholesterol-induced lesions.[37] It is therefore likely that the frequent association of *C. pneumoniae* and human CMV infection with atherosclerotic disease is due, at least in part, to altered immune-responsiveness and the impaired pathogen control. Importantly, herpesvirus infection[32,38] can further alter cholesterol metabolism and chronic *C pneumoniae* infection was shown to enhance the effect of the metabolic syndrome on the risk of coronary heart disease.[39] Thus, self-perpetuating immunopathological disease circuits may develop when chronic hypercholesterolemia-mediated immunosuppression impairs the usually well-balanced host-pathogen equilibrium.

4. LINKING IMMUNE-MEDIATED ARTERIAL INFLAMMATION AND CHOLESTEROL-INDUCED ATHEROSCLEROSIS

In studies on the pathogenesis of pathogen-induced immunopathological disease, it is usually difficult to distinguish between direct pathogen-mediated tissue damage and the immune-mediated inflammatory response. Uncoupling microbial replication and anti-microbial immune responses can be achieved by using mycobacterial HSP.[12,13] However, HSP is a rather "general" immune activator that may also induce or aggravate other autoimmune diseases suggesting that anti-HSP responses may not be exclusively linked with vascular immunopathology.

In order to study immune-mediated inflammation restricted to the vascular system, a transgenic mouse model of inducible cardiovascular immunopathology was established. SM-LacZ transgenic mice express the microbial β-galactosidase antigen under the control of a smooth muscle-specific promotor in cardiomyocytes of the right heart and in arterial smooth muscle cells.[40] Although the transgene is widely expressed in the vascular system, T cells ignore the peripheral antigen and immunization with DC presenting β-galactosidase peptide elicited distinct cardiovascular immunopathology with arteritis, myocarditis and eventual dilated cardiomyopathy.[41] It is interesting to note that the regenerational capacity of the target organ (heart vs. arteries) determined the clinical outcome of the DC-induced cardiovascular immunopathology. After short-term repetitive priming of SM-LacZ mice with DC presenting β-galactosidase peptide, SM-LacZ mice developed strong lymphocytic infiltration in lung arteries and in the right heart. However, cessation of chronic self-antigen delivery by DC to secondary lymphoid organs after 10 - 12 days resulted in resolution of the arterial lesions in the lung arteries with only mild histiocytic infiltrates around the vessels indicating the high capacity of the arterial wall to regenerate after inflammatory injuries. However, in this chronic phase, despite cessation of immunization with β-galactosidase-loaded DC, SM-LacZ mice showed a severe loss of the poorly regenerating heart tissue and fibrosis that eventually led to dilated cardiomyopathy.[42] These results showed that autoimmune vascular damage may resolve if an antigen is not presented in lymphoid tissues any longer, supporting the notion that sustained antigen presentation in secondary lymphoid organs is probably an important prerequisite for chronic arterial inflammation.

By crossing SM-LacZ mice on the hypercholesterolemic ApoE$^{-/-}$ background[27] we were able to assess (a) the influence of hypercholesterolemia in the development of immune-mediated arterial inflammation and (b) to evaluate the significance of immune-mediated arteritis in the chronological process of cholesterol-induced atherosclerosis.[42] We found that hypercholesterolemia in SM-LacZxApoE$^{-/-}$ mice selectively enhanced and perpetuated arterial inflammation. Furthermore, arterial inflammation significantly increased the susceptibility of the arterial wall to cholesterol-induced atherosclerosis. The inflammatory reaction directed against the microbial antigen seemed to dominate over effects of hypercholesterolemia early in the course of vascular disease and the established lesions resolved within 2 – 3 wk in the absence of hypercholesterolemia. Importantly, the coexistence of hypercholesterolemia prevented recovery from arterial inflammation.

Mutual detrimental effects of hypercholesterolemia and vascular immune responses have been shown in other experimental systems. Zhou et al.[43] have shown that the number of IFN-gamma-producing T-helper 1 cells infiltrating atherosclerotic lesions decreases under severe hypercholesterolemic conditions and that the subsequent T-helper 1 to T-helper 2 switch results in the formation of IgG1 autoantibodies to oxidized LDL. Furthermore, C. pneumoniae infection was shown to aggravate diet-induced atherosclerosis in normal C57BL/6 mice[44] and the transfer of CD4 T cells from apoE-deficient mice into immunodeficient apoE(-/-)/scid/scid mice increased the severity of atherosclerosis.[45] Overall, these data together with the above mentioned studies support the notion that the mutual perpetuation of arteritis and cholesterol-induced atherosclerosis might eventually lead to a vicious circle of chronic vascular injury.

ACKNOWLEDGEMENTS

This work was supported by the Swiss National Science Foundation and the Kanton Zürich.

REFERENCES

1. R. Ross. 1999. Atherosclerosis—an inflammatory disease. *N.Engl.J.Med.* 340:115-126.
2. F. Numano. 2000. Vasa vasoritis, vasculitis and atherosclerosis. *Int.J.Cardiol.* 75:S1-S8
3. S. Nikkari, J. Mertsola, H. Korvenranta, R. Vainionpaa, and P. Toivanen. 1994. Wegener's granulomatosis and parvovirus B19 infection. *Arthritis Rheum.* 37:1707-1708.
4. C.A. Stegeman, J.W. Tervaert, W.J. Sluiter, W.L. Manson, P.E. de Jong, and C.G. Kallenberg. 1994. Association of chronic nasal carriage of Staphylococcus aureus and higher relapse rates in Wegener granulomatosis. *Ann.Intern.Med.* 120:12-17.
5. G. Nigro, M. Zerbini, A. Krzysztofiak, G. Gentilomi, M.A. Porcaro, T. Mango, and M. Musiani. 1994. Active or recent parvovirus B19 infection in children with Kawasaki disease. *Lancet* 343:1260-1261.
6. M. Leruez-Ville, A. Lauge, F. Morinet, L. Guillevin, and P. Deny. 1994. Polyarteritis nodosa and parvovirus B19. *Lancet* 344:263-264.
7. B.F. Mandell and L.H. Calabrese. 1998. Infections and systemic vasculitis. *Curr.Opin.Rheumatol.* 10:51-57.
8. B. Chiu, E. Viira, W. Tucker, and I.W. Fong. 1997. Chlamydia pneumoniae, cytomegalovirus, and herpes simplex virus in atherosclerosis of the carotid artery. *Circulation* 96:2144-2148.
9. S. Praprotnik, M. Blank, P.L. Meroni, B. Rozman, A. Eldor, and Y. Shoenfeld. 2001. Classification of anti-endothelial cell antibodies into antibodies against microvascular and macrovascular endothelial cells: the pathogenic and diagnostic implications. *Arthritis Rheum.* 44:1484-1494.
10. Y. Sherer, A. Tenenbaum, S. Praprotnik, J. Shemesh, M. Blank, E.Z. Fisman, D. Harats, J. George, Y. Levy, J.B. Peter, M. Motro, and Y. Shoenfeld. 2001. Coronary artery disease but not coronary calcification is associated with elevated levels of cardiolipin, beta-2-glycoprotein-I, and oxidized LDL antibodies. *Cardiology* 95:20-24.
11. J. George, D. Harats, B. Gilburd, A. Afek, A. Shaish, J. Kopolovic, and Y. Shoenfeld. 2000. Adoptive transfer of beta(2)-glycoprotein I-reactive lymphocytes enhances early atherosclerosis in LDL receptor-deficient mice. *Circulation* 102:1822-1827.
12. Q. Xu, H. Dietrich, H.J. Steiner, A.M. Gown, B. Schoel, G. Mikuz, S.H. Kaufmann, and G. Wick. 1992. Induction of arteriosclerosis in normocholesterolemic rabbits by immunization with heat shock protein 65. *Arterioscler.Thromb.* 12:789-799.
13. J. George, Y. Shoenfeld, A. Afek, B. Gilburd, P. Keren, A. Shaish, J. Kopolovic, G. Wick, and D. Harats. 1999. Enhanced fatty streak formation in C57BL/6J mice by immunization with heat shock protein-65. *Arterioscler.Thromb.Vasc.Biol.* 19:505-510.
14. Q. Xu, J. Willeit, M. Marosi, R. Kleindienst, F. Oberhollenzer, S. Kiechl, T. Stulnig, G. Luef, and G. Wick. 1993. Association of serum antibodies to heat-shock protein 65 with carotid atherosclerosis. *Lancet* 341:255-259.
15. J.T. Salonen, S. Yla-Herttuala, R. Yamamoto, S. Butler, H. Korpela, R. Salonen, K. Nyyssonen, W. Palinski, and J.L. Witztum. 1992. Autoantibody against oxidised LDL and progression of carotid atherosclerosis. *Lancet* 339:883-887.
16. S. Stemme, B. Faber, J. Holm, O. Wiklund, J.L. Witztum, and G.K. Hansson. 1995. T lymphocytes from human atherosclerotic plaques recognize oxidized low density lipoprotein. *Proc.Natl.Acad.Sci.U.S.A.* 92:3893-3897.
17. S. Freigang, S. Horkko, E. Miller, J.L. Witztum, and W. Palinski. 1998. Immunization of LDL receptor-deficient mice with homologous malondialdehyde-modified and native LDL reduces progression of atherosclerosis by mechanisms other than induction of high titers of antibodies to oxidative neoepitopes. *Arterioscler.Thromb.Vasc.Biol.* 18:1972-1982.
18. A.M.J. Gotto. 1998. Assessing the benefits of lipid-lowering therapy. *Am.J.Cardiol.* 82:2M-4M.
19. B. Kwak, F. Mulhaupt, S. Myit, and F. Mach. 2000. Statins as a newly recognized type of immunomodulator. *Nat.Med.* 6:1399-1402.
20. D.D. Henninger, M.E. Gerritsen, and D.N. Granger. 1997. Low-density lipoprotein receptor knockout mice exhibit exaggerated microvascular responses to inflammatory stimuli. *Circ.Res.* 81:274-281.
21. M. Navab, S.S. Imes, S.Y. Hama, G.P. Hough, L.A. Ross, R.W. Bork, A.J. Valente, J.A. Berliner, D.C. Drinkwater, and H. Laks. 1991. Monocyte transmigration induced by modification of low density

lipoprotein in cocultures of human aortic wall cells is due to induction of monocyte chemotactic protein 1 synthesis and is abolished by high density lipoprotein. *J.Clin.Invest.* 88:2039-2046.
22. D.M. Levine, T.S. Parker, T.M. Donnelly, A. Walsh, and A.L. Rubin. 1993. In vivo protection against endotoxin by plasma high density lipoprotein. *Proc.Natl.Acad.Sci.U.S.A.* 90:12040-12044.
23. M.G. Netea, P.N. Demacker, B.J. Kullberg, O.C. Boerman, I. Verschueren, A.F. Stalenhoef, and J.W. van der Meer. 1996. Low-density lipoprotein receptor-deficient mice are protected against lethal endotoxemia and severe gram-negative infections. *J.Clin.Invest.* 97:1366-1372.
24. S.E. Roselaar and A. Daugherty. 1998. Apolipoprotein E-deficient mice have impaired innate immune responses to Listeria monocytogenes in vivo. *J.Lipid Res* 39:1740-1743.
25. N. de Bont, M.G. Netea, P.N. Demacker, I. Verschueren, B.J. Kullberg, K.W. van Dijk, J.W. van der Meer, and A.F. Stalenhoef. 1999. Apolipoprotein E knock-out mice are highly susceptible to endotoxemia and Klebsiella pneumoniae infection. *J.Lipid Res.* 40:680-685.
26. M.G. Netea, P.N. Demacker, N. de Bont, O.C. Boerman, A.F. Stalenhoef, J.W. van der Meer, and B.J. Kullberg. 1997. Hyperlipoproteinemia enhances susceptibility to acute disseminated Candida albicans infection in low-density-lipoprotein-receptor-deficient mice. *Infect.Immun.* 65:2663-2667.
27. S.H. Zhang, R.L. Reddick, J.A. Piedrahita, and N. Maeda. 1992. Spontaneous hypercholesterolemia and arterial lesions in mice lacking apolipoprotein E. *Science* 258:468-471.
28. S. Ishibashi, J.L. Goldstein, M.S. Brown, J. Herz, and D.K. Burns. 1994. Massive xanthomatosis and atherosclerosis in cholesterol-fed low density lipoprotein receptor-negative mice. *J.Clin.Invest.* 93:1885-1893.
29. J.L. Breslow. 1996. Mouse models of atherosclerosis. *Science* 272:685-688.
30. B. Ludewig, M. Jaggi, T. Dumrese, K. Brduscha-Riem, B. Odermatt, H. Hengartner, and R.M. Zinkernagel. 2001. Hypercholesterolemia exacerbates virus-induced immunopathologic liver disease via suppression of antiviral cytotoxic T cell responses. *J.Immunol.* 166:3369-3376.
31. R.M Zinkernagel, E. Haenseler, T. Leist, A. Cerny, H. Hengartner, A. Althage. 1986. T cell-mediated hepatitis in mice infected with lymphocytic choriomeningitis virus. Liver cell destruction by H-2 class I-restricted virus-specific cytotoxic T cells as a physiological correlate of the 51Cr-release assay? *J. Exp. Med.* 164:1075-1092.
32. K. Berencsi, V. Endresz, D. Klurfeld, L. Kari, D. Kritchevsky, and E. Gonczol. 1998. Early atherosclerotic plaques in the aorta following cytomegalovirus infection of mice. *Cell Adhes.Commun.* 5:39-47.
33. M.C. Persoons, M.J. Daemen, J.H. Bruning, and C.A. Bruggeman. 1994. Active cytomegalovirus infection of arterial smooth muscle cells in immunocompromised rats. A clue to herpesvirus-associated atherogenesis? *Circ.Res.* 75:214-220.
34. R.M. Presti, J.L. Pollock, A.J. Dal Canto, A.K. O'Guin, and H.W.4. Virgin. 1998. Interferon gamma regulates acute and latent murine cytomegalovirus infection and chronic disease of the great vessels. *J.Exp.Med.* 188:577-588.
35. K.E. Weck, A.J. Dal Canto, J.D. Gould, A.K. O'Guin, K.A. Roth, J.E. Saffitz, S.H. Speck, and H.W. Virgin. 1997. Murine gamma-herpesvirus 68 causes severe large-vessel arteritis in mice lacking interferon-gamma responsiveness: a new model for virus- induced vascular disease. *Nat.Med.* 3:1346-1353.
36. T.C. Moazed, C. Kuo, J.T. Grayston, and L.A. Campbell. 1997. Murine models of Chlamydia pneumoniae infection and atherosclerosis. *J.Infect.Dis.* 175:883-890.
37. H. Hu, G.N. Pierce, and G. Zhong. 1999. The atherogenic effects of chlamydia are dependent on serum cholesterol and specific to Chlamydia pneumoniae. *J.Clin.Invest.* 103:747-753.
38. C.G. Fabricant, D.P. Hajjar, C.R. Minick, and J. Fabricant. 1981. Herpesvirus infection enhances cholesterol and cholesteryl ester accumulation in cultured arterial smooth muscle cells. *Am.J.Pathol.* 105:176-184.
39. M. Leinonen and P. Saikku. 1999. Interaction of Chlamydia pneumoniae infection with other risk factors of atherosclerosis. *Am.Heart J.* 138:S504-S506
40. H. Moessler, M. Mericskay, Z. Li, S. Nagl, D. Paulin, and J.V. Small. 1996. The SM 22 promotor directs tissue-specific expression in arterial but not in venous or visceral smooth muscle cells in transgenic mice. *Development* 122:2415-2425.
41. B. Ludewig, A.F. Ochsenbein, B. Odermatt, D. Paulin, H. Hengartner, and R.M. Zinkernagel. 2000. Immunotherapy with dendritic cells directed against tumor antigens shared with normal host cells results in severe autoimmune disease. *J.Exp.Med.2000* 191:795-804.
42. B. Ludewig, S. Freigang, M. Jaggi, M.O. Kurrer, Y.C. Pei, L. Vlk, B. Odermatt, R.M. Zinkernagel, and H. Hengartner. 2000. Linking immune-mediated arterial inflammation and cholesterol-induced atherosclerosis in a transgenic mouse model. *Proc.Natl.Acad.Sci.U.S.A.* 97:12752-12757.
43. X. Zhou, G. Paulsson, S. Stemme, and G.K. Hansson. 1998. Hypercholesterolemia is associated with a T helper (Th) 1/Th2 switch of the autoimmune response in atherosclerotic apo E- knockout mice. *J.Clin.Invest.* 101:1717-1725.

44. E. Blessing, L.A. Campbell, M.E. Rosenfeld, N. Chough, and C. Kuo. 2001. Chlamydia pneumoniae infection accelerates hyperlipidemia induced atherosclerotic lesion development in C57BL/6J mice. *Atherosclerosis* 158:13-17.
45. X. Zhou, A. Nicoletti, R. Elhage, and G.K. Hansson. 2000. Transfer of CD4(+) T cells aggravates atherosclerosis in immunodeficient apolipoprotein E knockout mice. *Circulation* 102:2919-2922.

REGULATION OF T CELL MIGRATION THROUGH FORMATION OF IMMUNOLOGICAL SYNAPSES
The Stop Signal Hypothesis

Michael L. Dustin*

1. INTRODUCTION

The antigen specific immune response is initiated by molecular interactions in the nanometer scale gap between the T cell and antigen-presenting cell (APC). This specialized cell-cell junction has been aptly described as an immunological synapse (Norcross, 1984; Paul and Seder, 1994). The immunological synapse is the site of key signaling events that set thresholds for T cell activation. If these thresholds are appropriately set the T cell is defended against a wide array of pathogens and toxins. If these thresholds are incorrectly set the host is susceptible to pathogen escape, at one extreme, and destructive autoimmunity at the other extreme. The immunological synapse is the site of a sensitive molecular search where as few as 10-100 agonist MHC-peptide complexes are sufficient to activate a T cell (Peterson, et al., 1999). The immunological synapse is also the product of an extensive cellular search that brings the T cell and APC together in the secondary lymphoid tissues. The interaction of T cells agonist MHC-peptide complexes on an appropriate surface generates a robust stop signal (Dustin, et al., 1997; Dustin, et al., 1996; Negulescu, et al., 1996). The common characteristic of the stop signal is that the TCR signal appears to control the polarity of the T cell cytoskeleton and maintains a balance of integrin mediated traction forces to hold the T cell in place- an apparently delicate balance (Dustin, et al., 1997). Subsequent studies on the molecular interactions in the immunological synapse reveal a complex signaling landscape in which molecular topology and dynamic assemblies of cytoskeletal molecules and adhesion molecules create distinct supramolecular activation clusters (SMACs) in a bull's eye pattern (Grakoui, et al., 1999;Monks, et al., 1998). The T cell receptor (TCR) and CD28 occupy the center of the bull's eye, whereas the adhesion molecules LFA-1 and CD2 occupy the surrounding ring (Bromley, et al., 2001; Grakoui, et al., 1999; Monks, et al., 1998). The immunological synapse is formed through a distinct intermediate in which the

*Program in Molecular Pathogenesis and Department of Pathology, Skirball Institute of Molecular Medicine, New York University School of Medicine, 540 First Avenue, 2nd Floor, Labs 4 and 5, New York, NY 10016.

bulk of TCR interactions are located in a peripheral ring surrounding a central region of LFA-1 interactions (Grakoui, et al., 1999; Lee, et al., 2002). Others have observed more disordered early patterns (Krummel, et al., 2000), but there is general agreement that formation of the central antigen cluster requires several minutes and involves progressive movement of TCR clusters toward the center (Grakoui, et al., 1999; Krummel, et al., 2000). It is well established that both the TCR and adhesion molecules are involved in signal transduction processes and the molecular movements may be strongly influenced by signaling (Wülfing, et al., 1998). This suggests that two kinds of signal integration are taking place in the T cell- one based on molecular proximity at the interface where surface receptors can directly cooperate to assemble the immunological synapse and the other based on merging of signals from spatially distinct signaling complexes with integration taking place in the cytosol and nucleus. In this chapter I will describe current information on the molecular organization of the immunological synapse, a hypothesis for the role of the immunological synapse in regulation of T cell migration and the role of adhesion molecules in distinct signaling processes that create and then follow from this unique organization.

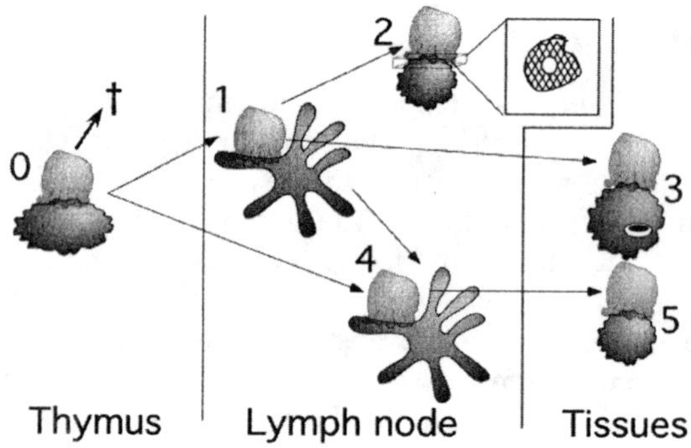

Figure 1. Immunological synapses- 0- Thymocyte/thymic epithelial cell, 1- CD4 (helper) T cell/dendritic cell, 2- CD4 T cell/B cell, 3- CD4 T cell/macrophage (with intracelllular bacteria), 4- CD8 (cytotoxic) T cell/licensed ($CD40^+$) dendritic cell, 5- Effector CD8 T cell/target cell.

A productive immune response involving cellular and humoral components requires the formation of a series of immunological synapses. Immunological synapses are a subset of informational synapses defined as specialized cell-cell junctions with a specific polarity and a microscopic pattern of molecular organization that serves sustained cell-cell communication. Fig. 1 shows 5 examples of immunological synapses in a primary immune response. Synapses between naïve CD4 T cells and dendritic cells (synapse 1) initiate the response by inducing CD4 T cell expansion and differentiation and by licensing the dendritic cell for activation of naïve CD8 T cells. Synapses between effector CD4 T cells and B cells help antibody production (synapse 2). Synapses between effector CD4 T cells and macrophages activate the macrophage for destruction of phagocytosed bacteria (synapse 3). Synapses between licensed dentritic cells and naïve CD8 T cells (synapse 4) induce differentiation of the naïve CD8 T cells to effector cells that can form synapses with target cells to trigger target cell apoptosis (synapse 5). Prior to this process the immature CD4 and CD8 positive lymphocyte undergoes selection in the thymus through formation of yet another set of synapses (synapse 0). The CD4 T cell-

B cell synapse is best characterized (Krummel, et al., 2000; Monks, et al., 1998; Wülfing, et al., 2000; Wülfing, et al., 1998) and some information is available on natural killer cells synapses (Davis, et al., 1999). It has also been suggested that B cells may form synapses with cells presenting intact antigen (Batista, et al., 2001). Immunological synapses between CD4 T cells and supported planar bilayers containing the LFA-1 ligand ICAM-1 and the MHC-peptide complex resemble the CD4 T cell-B cell synapses and provide a very high resolution view of events mediated by live T cells (Grakoui, et al., 1999; Johnson, et al., 2000). Recent views of T cell-macrophage synapses reveal a slower version of the synapse formation process than observed with supported planar bilayers or B cell tumors (Lee, et al., 2002). Clearly there are significant gaps in knowledge, but we can begin to appreciate the organization of signaling molecules in synapses based on the current examples.

The CD4 T cell-B cell immunological synapse can be divided into the central and peripheral SMACs, the cSMAC and the pSMAC (Monks, et al., 1998). The TCR and the signaling molecule protein kinase C-θ are molecular markers of the cSMAC while LFA-1 and the cytoskeletal protein talin are molecular markers of the pSMAC. The formation of the immunological synapse is correlated with full T cell activation including cytokine production and proliferation (Bromley, et al., 2001; Grakoui, et al., 1999; Monks, et al., 1998). The pSMAC and cSMAC do not exist independently of each other, however, it is useful to consider the distinct signaling processes associated with antigen receptor and integrin signaling with respect to the goals of cytokine gene transcription, cytokine responses, cell cycle progression, and cell growth and survival.

2. SIGNALING IN THE IMMUNOLOGICAL SYNAPSE

In order to be fully activated the T cell must sustain signaling for several hours. The time required to activate naïve T cells with MHC-peptide complexes on plastic is about 24 hours (Iezzi, et al., 1998), but this time appears to be reduced to 4-6 hours through use of natural APC (Lee, et al., 2002). Immunological synapse formation takes place over a fraction of this time (5-30 minutes), whereas mature synapses are clearly stable over the time frame identified by Lee et al of 4-6 hours. Both the timing of proliferation and the duration of the synapse are strikingly different than the short apparent duration of tyrosine phosphorylation that is the best understood mode of TCR signaling (Dustin and Chan, 2000; Dustin and Cooper, 2000; Lee, et al., 2002; Weiss and Littman, 1994). The TCR signaling pathway controls the activation of three tyrosine kinase families, the src, syk and itk families, and a lipid kinase phosphatidylinositol-3'-kinase (PI-3-K). These kinases are assembled together around a tyrosine phosphorylated scaffold formed by the TCR associated CD3γ,δ, ε and ζ subunits and the Linker of Activated T cells (LAT). Critical down steam pathways include activation of the mitogen activated protein kinase and Jun kinase pathways and NFκB and NFAT transcription factors that are required for full T cell activation. The tyrosine kinase cascade was throught to be transient and Lee et al confirmed this in the synapse since they could not detect Lck and ZAP-70 activation after 15 min and 30 min, respectively, in T cell-macrophage immunological synapses. There is evidence from both the sustained TCR engagement in the immunological synapse and the continual requirement for MHC-peptide complexes to sustain T cell activation to strongly support the notion that the TCR is required to maintain the synapse for longer than 30 min. Furthermore, cells forming immunological synapses on planar bilayers display elevated cytoplasmic Ca^{2+} for more than an hour. Why prolonged TCR mediated signal are required is not clear, but may be related to the need to transcribe new proteins that have to be modified by the TCR signaling process, the requirements for passing certain cell-cycle checkpoints and possibly also to the time required for chromatin remodeling to activate cytokine gene expression (Avni and Rao, 2000). We would propose that the scaffolding of the immunological synapse is assembled through a radical process that requires a burst of tyrosine phosphorylation to fully rearrange the cytoskeleton and assemble the scaffold. Once this is accomplished the requirement for TCR signaling may be greatly reduced and the efficiency of signaling in the assembled synapse may be high enough to operate at low levels that we do not readily detect with the current imaging technology. The synaptic scaffold may also integrate other signaling processes such as costimulation through CD28 and perhaps other receptors such as members of the TNF receptor family (Bromley, et al., 2001).

The ability of the T cell to suppress TCR mediated death signaling is carried out by co-stimulatory receptors like CD28 and ICOS (Genot, et al., 2000; Noel, et al., 1996; Tafuri, et al., 2001). These survival receptors augment activation of PI-3-K and the generation of phosphatidylinositol-3,4,5-trisphosphate augments recruitment and activation of the pleckstrin homology (PH) domain containing kinases like Itk and Akt (also known as protein kinase B) (Burgering and Coffer, 1995). Akt is a serine/throenine kinase that activates or inhibits downstream molecules to promote T cell survival (Scanga, et al., 2000). CD28 looks like an adhesion molecule that might be expected to activate PI-3-K in response to T cell contact with any ligand expressing cell, but this may not be the case in vivo. CD28 engagement is tightly controlled by TCR engagement to an extent that unlike adhesion molecules LFA-1 and CD2. CD28 does not appear to augment TCR engagement (Viola and Lanzavecchia, 1996), but rather it appears to be the other way around (Bromley, et al., 2001), that TCR mediated formation of the immunological synapse induces interaction of CD28 with CD80 or CD86, if these ligands are present on the APC. Thus, it is appropriate to classify CD28 in a separate category from LFA-1 and CD2. It remains to be seen whether ICOS and other co-stimulatory receptors are similarly regulated.

The immumological synapse coordinates adhesion, signaling and cytoskeleton. Immunological synapse formation requires an intact actin cytoskeleton and it has been suggested that it is a classical case of tension generation by myosin II contraction of an actin filament network (Dustin and Cooper, 2000; Rozdzial, et al., 1995). An alternative view of synapse formation has been presented based on the physics of membrane interfaces, molecular topology and TCR interaction kinetics (Qi, et al., 2001). Whether cytoplasmic transport or membrane interfacial properties are dominant, the trigger for this process is signaling through the TCR and LFA-1 with cross-talk between these spatially distinct signaling regions in the synapse (Dustin and Chan, 2000) (Fig. 2). TCR signaling rapidly activates adhesion through changes in LFA-1 activity (Dustin and Springer, 1989; van Kooyk, et al., 1989). The mechanism of this process is not known, but candidates include the branches of the mitogen activated protein kinase pathway and the regulation of actin cytoskeleton through Syk family tyrosine kinases, the Vav exchange factor (Miranti, et al., 1998), and the adapter protein ADAP (fyb/SLAP-130) (Griffiths, et al., 2001; Peterson, et al., 2001). While the details of this signaling pathway are not known it may also involve activation of the small G protein Rap1 (Katagiri, et al., 2000; Reedquist, et al., 2000; Suga, et al., 2001). Knock out of the SLP:-76 associated adapter ADAP has a strong phenotype in which CD3 mediated T cell activation and inside out signaling are impaired. In the ADAP knock out all early TCR signaling appears normal (Griffiths, et al., 2001; Peterson, et al., 2001). Therefore, the ADAP knock-out raises an interesting point about the nature of the longer lived sustained signals from the TCR. One potential message of the ADAP knock out may be that early TCR signals are used to set up an integrin dependent sustained signal through downstream effectors of ADAP. Like other adapter molecules, ADAP has no enzymatic activation and appears to link the SLP-76 adapter to members of the ENA/VASP family of cytoskeletal protiens. Physiological TCR engagement and activation of adhesion pathways are likely to be totally interdependent processes that cannot be separated, but are compartmentalized (Fig. 2).

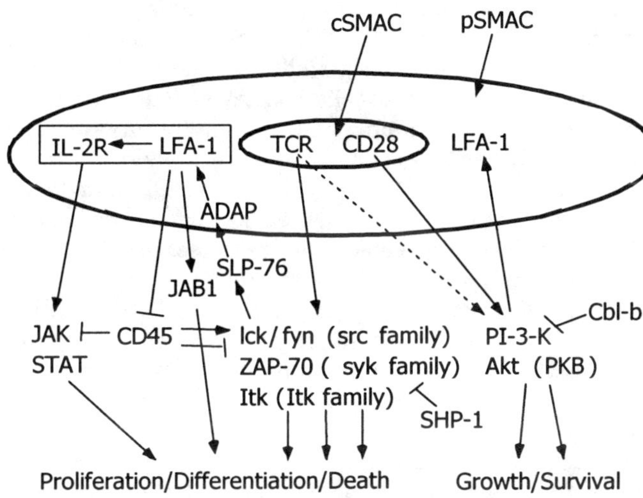

Figure 2. Compartmentalization of the signaling in the immunological synapse. Simplified summary of positive and negative signaling in the mature immunological synapse with a central cluster of TCR and CD28 (cSMAC) and a ring of LFA-1 (pSMAC). Arrows are positive effects and perpendicular lines are negative effects. Signaling from both the cSMAC and pSMAC are required for full T cell activation.

Signaling from the adhesion ring is also likely to play a key role in cell cycle progression. The best-characterized biochemical pathway is based on activation of Jun/Fos transcriptional co-activator JAB1 by LFA-1 (Bianchi, et al., 2000). Sustained LFA-1 engagement in conjunction with IL-2 receptor engagement also down regulates the cell cycle inhibitor p27kip, thus enabling movement from G1 to S phases of the cell cycle (Geginat, et al., 1999). Another interesting property of the LFA-1 ring is that it is relatively depleted of the transmembrane tyrosine phosphatase CD45 (Johnson, et al., 2000). CD45 has recently been shown to act as a Janus kinase phosphatase and thus a negative regulator of cytokine signaling pathways (Irie-Sasaki, et al., 2001). In order to enter the cell cycle the T cell must integrate TCR, survival and cytokine signals. It is attractive to speculate that the LFA-1 ring may be a site of active cytokine signaling both due to the depletion of CD45 and by analogy to focal adhesion complexes that also have a well known role in sustaining growth factor signals (Miyamoto, et al., 1995).

3. INHIBITORY SIGNALING

The immunological synapse is also laced with inhibitory signaling mechanisms that contribute to setting signaling thresholds at appropriate levels. Signaling for activation of immature T cells may operate near the theoretical limit of single molecule sensitivity. In order to prevent autoimmunity the sensitivity of mature T cells is less than the sensitivity of thymocytes. This amounts to a biochemical safety margin where mature T cell activation requires 50-100 MHC-peptide complexes to achieve full activation (Peterson, et al., 1999). Thymocytes are also more sensitive to self-MHC-peptide complexes than mature T cells. However, it has recently been found that self-MHC peptide complexes can play a supporting role in recognition of very small numbers of agonist MHC-peptide complexes (Wulfing, et al., 2002). There are a number of biochemical mechanisms that contribute to the attenuated sensitivity of mature T cells. One mechanism is the recruitment of tyrosine phosphatases like SHP-1. Strong recruitment of SHP-1 can extinguish signaling (Binstadt, et al., 1996), while moderate recruitment is likely to be involved in setting thresholds in both T cells and thymocytes (Johnson, et al., 1999). Another axis of regulation is based on ubiquitin addition by ring finger family member E3 ubiquitin ligases like c-cbl and cbl-b (Levkowitz, et al., 1999). One distinct regulatory network is based on cbl-b inhibition of PI-3-K (Fang, et al., 2000). In the absence of cbl-b the TCR mediated recruitment of PI-3-K is sufficient to fully activate mature T cells without a requirement for CD28 engagement. In the presence of cbl-b, the addition of ubiquitin to the p85 subunit inhibits PI-3-K and imposes a requirement for additional PI-3-K activation through CD28 to activate effectors like Akt to achieve full T cell activation and survival. Ubiquitin addition may inhibit a pathway by proteosome-mediated degradation and/or by steric inhibition in the absence of degradation (Fang, et al., 2000). Negative regulators of inositol lipid phosphorylation also impinge on this pathway. PTEN is a phosphatidylinositol-3'-phosphatase that counteracts PI-3-K and regulates signaling thresholds (Wang, et al., 2000). For example, the T cell line Jurkat lacks PTEN, a fact that may account for Jurkat's general high sensitivity to activating signals (Shan, et al., 2000). PTEN also has a role in controlling the sensitivity of T cells to FAS mediated apoptosis, a major mediator of antigen induced cell death (Di Cristofano, et al., 1999). SHIP is phosphatidylinositol 5' phosphatase that generates PI-3,4-P_2. Akt has a PH domain that preferentially binds PI-3,4,5-P_3 and action of SHIP reduces Akt recruitment and activation, but generates new binding sites for negative regulators that have PH domains interacting preferentially with PI-3,4-P_2. Thus, the formation of the immunological synapse may provide a mechanism to balance positive and negative biochemical pathways with TCR and CD28 engagement to achieve activation at an appropriate level of TCR engagement. The activation of T cells through serial engagement in a collagen gel may place less emphasis on sustaining this balance. In this situation T cells my carry a finite biochemical memory of earlier encounters with APC. This memory may be encoded in slowly reversible processes like degradation of cell cycle inhibitors and in chromatin remodeling.

Negative signaling molecules may be recruited to the immunological synapse by specific receptors. One of the most notable examples is CTLA-4, an alternative receptor for CD80 and CD86. CTLA-4 induces a poorly defined negative signal that may involve the specific recruitment of SHP-1 to the cSMAC and may also involve simple competition with CD28. CTLA-4 is very weakly expressed on naïve T cells, but is upregulated on activated T cells within a few hours and reaches high levels of protein in the cell by 24 hours after antigen stimulation. The cytoplasmic domain of CTLA-4 contains a strong internalization motif that rapidly diverts it from the Golgi apparatus and cell surface to a late endosome like compartment. CTLA-4 has a short half-life in T cells suggesting that the function of endocytosis is to degrade CTLA-4 and to maintain a low steady state density on the cell surface (Egen and Allison, 2002). In response to strong antigenic stimulation CTLA-4 is recruited to the interface in a ligand dependent manner.

One mode for this is that phosphorylation of the tyrosine based internalization motif may block its function and allow CTLA-4 to remain the surface longer. It has been proposed that strong agonist signaling through the TCR recruits CTLA-4 to the interface to attenuate these very strong signals and to even the playing field between weaker agonists and underrepresented agonist and the most highly expressed agonist MHC-peptide complexes. This may help account for the observation that T cell responses to MHC-peptide complexes generated from a single antigen can expand similar numbers of T cell clones recognizing MHC-peptide complexes that differ in number of the APC by several orders of magnitude (Velazquez, et al., 2001).

A novel mode of inhibition of immunological synapse formation has been suggested by studies of N-linked glycosylation. Deficiency in enzymes that regulate production of galectin-1 binding sites on T cells suggest that the interaction of the bivalent lectin galectin-1 has an inhibitory effect on T cell activation through a retarding effect on immunological synapse formation (Demetriou, et al., 2001). Restriction of lateral mobility has been proposed as a mechanism for regulating a number of adhesive processes including LFA-1 regulation (Kucik, et al., 1996). The report of Demitriou et al is novel in that the molecular mechanism of restricted mobility is not a cytoskeletal interaction, but an extracellular lectin associated with the thymic and lymph node microenvironment.

4. THE STOP SIGNAL HYPOTHESIS

The stop signal delivered by TCR mediated signaling is a very early event in immunological synapse formation. Cahalan and colleagues proposed that elevation of cytoplasmic Ca^{2+} is necessary and sufficient to stop T cell migration in response to APC with agonist MHC-peptide complexes (Negulescu, et al., 1996). I found that for ex vivo activated T cells from transgenic mice that cytoplasmic Ca^{2+} elevation was not necessary for the stop signal (Dustin, et al., 1997). Increases in cytoplasmic Ca^{2+} induce a "cringe" response in leukocytes that is known to stop migration (Pasternak and Elson, 1985). However, the lymphocytes stopped on surfaces coated with MHC-peptide complexes and adhesion molecules do not appear to be in a sustained cringe response, but rather appear to undergo spikes of cytoplasmic Ca^{2+} increase that punctuate longer periods of low Ca^{2+}. Between these Ca^{2+} spikes the T cells do not migrate, but appear to strike a dynamic balance of traction forces between active lamellapodia that are extended in all around the central region defined by the synapse. This balance of forces allows the T cell to remain in place while displaying a highly active cytoskeleton that may be important for other aspects of sustained signaling (Valitutti, et al., 1995). TCR signaling triggers dramatic changes in integrin activity, particularly in ex vivo resting T cells (Dustin and Springer, 1989). Therefore, Ca^{2+} mobilization alone appears unable to address the basic mechanism of the stop signal.

As described above, inside out signaling is a process through which TCR signaling triggers a dramatic increase in the avidity of LFA- and other integrins. Studies with naïve T cells demonstrate that this increase in LFA-1 avidity results in an 8-fold increase in the number of LFA-1/ICAM-1 interactions in the contact area (Bromley and Dustin, in preparation). While high levels of adhesion counteract cell movement in general, migrating cells can have many adhesive interactions assuming that they have strong polarization (Palecek, et al., 1997). Studies with previously activated T cells demonstrate that migrating and stopped cells can display similar high numbers of LFA-1/ICAM-1 interactions. Thus, high LFA-1 avidity alone does not explain the stop signal.

Polarity is the other major element of the stop signal (Dustin, et al., 1997). The polarity of T cells is controlled by molecules like CDC42, which in mutated form interferes with polarization of the MTOC toward the antigen presenting cells (Stowers, et al., 1995). While polarization was not necessary for IL-2 production by hybridomas, it is unlikely that polarization or migration are important under the conditions of most in vitro experiments where APC are plentiful and ubiquitous in the culture and T cell can be activated through serial encounters (Gunzer, et al., 2000). The downstream effectors of CDC42 in control of T cell polarity are not clear, but it is likely that microtubule motors anchored to the SMACs may play a role in polarization of this structure. It is not known if molecules like ADAP have any role in regulation of polarity in addition to their role in regulation of integrin function.

The stop signal is clearly not an absolute event following effective TCR engagement. Environmental factors may play an important role in this process. For example, specific chemokines when presented as soluble gradients have the ability to override the antigen stop signal (Bromley, et al., 2000). This was demonstrated with the CCR7 ligand CCL21 and CCL19 and the CXCR3 ligand IP-10. Other very potent chemokines were unable to distract T cells from surfaces bearing ICAM-1 and MHC-peptide complexes. On the other hand CXCL12, a highly efficacious CXCR4 ligand, and CCL3, a potent CCR5 ligand, were unable to force T cells to bypass MHC-peptide complexes. This suggests a hierarchy between chemokines and TCR signals where some chemokines are dominant over antigen and others are subordinate. The control of TCR signaling through location may be a powerful mechanism to link T cell responses to environmental factors. It is possible that the serial encounter process described by Gunzer et al is driven by an appropriately titrated mixture of TCR and chemokine signaling (Gunzer, et al., 2000).

5. CONCLUSIONS

The key signaling features of the immunological synapse that distinguish it from earlier studied phenomena such as antibody mediated capping is the requirement to coordinate cell-cell adhesion and the signaling process. This is clearly a highly choreographed process in which the segregation of integrin engagement from TCR engagement necessitates specific mechanisms for communication between the two systems. It is likely that the actin cytoskeleton plays a substantial role in bridging these distinct singling systems at the membrane. Therefore, some LFA-1 and TCR signaling is integrated at the membrane. Other signals, like those mediated by JAB1 are likely to be integrated in the cell nucleus. The stop signal that is a co-requisite for immunological synapse formation may have an important role in environmental regulation of T cell responses, with chemokine gradients playing a key role deciding whether a T cell remains in a lymph node to be activated or exits the lymph node. Immunological synapse formation provides an important example of spatial organization of adhesion and signaling machinery that is an excellent model for large scale organization of signaling complexes and a highly prized target for drug discovery.

6. ACKNOWLEDGEMENTS

This work was supported by the NIH and the Irene Diamond Foundation. I thank my lab and collaborators for their contributions to the immunological synapse model and stop signal hypothesis.

REFERENCES

Avni, O. and Rao, A., 2000, T cell differentiation: a mechanistic view, *Curr Opin Immunol,* **12:**654.

Batista, F. D., Iber, D. and Neuberger, M. S., 2001, B cells acquire antigen from target cells after synapse formation, *Nature,* **411:**489.

Bianchi, E., Denti, S., Granata, A., Bossi, G., Geginat, J., Villa, A., Rogge, L. and Pardi, R., 2000, Integrin LFA-1 interacts with the transcriptional co-activator JAB1 to modulate AP-1 activity, *Nature,* **404:**617.

Binstadt, B. A., Brumbaugh, K. M., Dick, C. J., Scharenberg, A. M., Williams, B. L., Colonna, M., Lanier, L. L., Kinet, J. P., Abraham, R. T. and Leibson, P. J., 1996, Sequential involvement of Lck and SHP-1 with MHC-recognizing receptors on NK cells inhibits FcR-initiated tyrosine kinase activation, *Immunity,* **5:**629.

Bromley, S. K., Iaboni, A., Davis, S. J., Whitty, A., Green, J. M., Shaw, A. S., Weiss, A. and Dustin, M. L., 2001, The immunological synapse and CD28-CD80 interactions, *Nat Immunol,* **2:**1159.

Bromley, S. K., Peterson, D. A., Gunn, M. D. and Dustin, M. L., 2000, Cutting Edge: Hierarchy of Chemokine Receptor and TCR Signals Regulating T Cell Migration and Proliferation, *J Immunol,* **165:**15.

Burgering, B. M. and Coffer, P. J., 1995, Protein kinase B (c-Akt) in phosphatidylinositol-3-OH kinase signal transduction, *Nature,* **376:**599.

Davis, D. M., Chiu, I., Fassett, M., Cohen, G. B., Mandelboim, O. and Strominger, J. L., 1999, The human natural killer cell immune synapse, *Proc Natl Acad Sci U S A,* **96:**15062.

Demetriou, M., Granovsky, M., Quaggin, S. and Dennis, J. W., 2001, Negative regulation of T-cell activation and autoimmunity by Mgat5 N-glycosylation, *Nature,* **409:**733.

Di Cristofano, A., Kotsi, P., Peng, Y. F., Cordon-Cardo, C., Elkon, K. B. and Pandolfi, P. P., 1999, Impaired Fas response and autoimmunity in Pten+/- mice, *Science,* **285:**2122.

Dustin, M. L., Bromely, S. K., Kan, Z., Peterson, D. A. and Unanue, E. R., 1997, Antigen receptor engagement delivers a stop signal to migrating T lymphocytes, *Proc.Natl.Acad.Sci. USA,* **94:**3909.

Dustin, M. L. and Chan, A. C., 2000, Signaling takes shape in the immune system, *Cell,* **103:**283.

Dustin, M. L. and Cooper, J. A., 2000, The immunological synapse and the actin cytoskeleton: molecular hardware for T cell signaling, *Nature Immunology,* **1:**23.

Dustin, M. L., Miller, J. M., Ranganath, S., Vignali, D. A. A., Viner, N. J., Nelson, C. A. and Unanue, E. R., 1996, T cell receptor mediated adhesion of T cell hybridomas to planar bilayers containing purified MHC class II/peptide complexes and receptor shedding during detachment, *J. Immunol.,* **157:**2014.

Dustin, M. L. and Springer, T. A., 1989, T cell receptor cross-linking transiently stimulates adhesiveness through LFA-1, *Nature,* **341:**619.

Egen, J. G. and Allison, J. P., 2002, Cytotoxic T Lymphocyte Antigen-4 Accumulation in the Immunological Synapse Is Regulated by TCR Signal Strength, *Immunity,* **16:**23.

Fang, D., Wang, H. Y., Fang, N., Altman, Y., Elly, C. and Liu, Y. C., 2000, Cbl-b, a RING-type E3 ubiquitin ligase, targets phosphatidylinositol 3-kinase for ubiquitination in T cells, *J Biol Chem,* **21:**21.

Geginat, J., Bossi, G., Bender, J. R. and Pardi, R., 1999, Anchorage dependence of mitogen-induced G1 to S transition in primary T lymphocytes, *J Immunol,* **162:**5085.

Genot, E. M., Arrieumerlou, C., Ku, G., Burgering, B. M., Weiss, A. and Kramer, I. M., 2000, The T-cell receptor regulates Akt (protein kinase B) via a pathway involving Rac1 and phosphatidylinositide 3-kinase, *Mol Cell Biol,* **20:**5469.

Grakoui, A., Bromley, S. K., Sumen, C., Davis, M. M., Shaw, A. S., Allen, P. M. and Dustin, M. L., 1999, The immunological synapse: A molecular machine controlling T cell activation, *Science,* **285:**221.

Griffiths, E. K., Krawczyk, C., Kong, Y. Y., Raab, M., Hyduk, S. J., Bouchard, D., Chan, V. S., Kozieradzki, I., Oliveira-Dos-Santos, A. J., Wakeham, A., Ohashi, P. S., Cybulsky, M. I., Rudd, C. E. and Penninger, J. M., 2001, Positive regulation of T cell activation and integrin adhesion by the adapter Fyb/Slap, *Science,* **293:**2260.

Gunzer, M., Schafer, A., Borgmann, S., Grabbe, S., Zanker, K. S., Brocker, E. B., Kampgen, E. and Friedl, P., 2000, Antigen presentation in extracellular matrix: interactions of T cells with dendritic cells are dynamic, short lived, and sequential, *Immunity,* **13:**323.

Iezzi, G., Karjalainen, K. and Lanzavecchia, A., 1998, The duration of antigenic stimulation determines the fate of naive and effector T cells, *Immunity,* **8:**89.

Irie-Sasaki, J., Sasaki, T., Matsumoto, W., Opavsky, A., Cheng, M., Welstead, G., Griffiths, E., Krawczyk, C., Richardson, C. D., Aitken, K., Iscove, N., Koretzky, G., Johnson, P., Liu, P., Rothstein, D. M. and Penninger, J. M., 2001, CD45 is a JAK phosphatase and negatively regulates cytokine receptor signalling, *Nature,* **409:**349.

Johnson, K. G., Bromley, S. K., Dustin, M. L. and Thomas, M. L., 2000, A supramolecular basis for CD45 regulation during T cell activation, *Proc.Natl.Acad.Sci. USA,* **97:**10138.

Johnson, K. G., LeRoy, F. G., Borysiewicz, L. K. and Matthews, R. J., 1999, TCR signaling thresholds regulating T cell development and activation are dependent upon SHP-1, *J Immunol,* **162:**3802.
Katagiri, K., Hattori, M., Minato, N., Irie, S., Takatsu, K. and Kinashi, T., 2000, Rap1 is a potent activation signal for leukocyte function-associated antigen 1 distinct from protein kinase C and phosphatidylinositol-3-OH kinase, *Mol Cell Biol,* **20:**1956.
Krummel, M. F., Sjaastad, M. D., Wulfing, C. and Davis, M. M., 2000, Differential clustering of CD4 and CD3zeta during T cell recognition, *Science,* **289:**1349.
Kucik, D. F., Dustin, M. L., Miller, J. M. and Brown, E. J., 1996, Adhesion activating phorbol ester increases the mobility of leukocyte integrin LFA-1 in cultured lymphocytes, *J.Clin.Invest.,* **97:**2139.
Lee, K.-H., Holdorf, A. D., Dustin, M. L., Chan, A. C., Allen, P. M. and Shaw, A. S., 2002, Tyrosine kinase signaling precedes formation of the immunological synapse, *Science,* **In press:**
Levkowitz, G., Waterman, H., Ettenberg, S. A., Katz, M., Tsygankov, A. Y., Alroy, I., Lavi, S., Iwai, K., Reiss, Y., Ciechanover, A., Lipkowitz, S. and Yarden, Y., 1999, Ubiquitin ligase activity and tyrosine phosphorylation underlie suppression of growth factor signaling by c-Cbl/Sli-1, *Mol Cell,* **4:**1029.
Miranti, C. K., Leng, L., Maschberger, P., Brugge, J. S. and Shattil, S. J., 1998, Identification of a novel integrin signaling pathway involving the kinase Syk and the guanine nucleotide exchange factor Vav1, *Curr Biol,* **8:**1289.
Miyamoto, S., Teramoto, H., Coso, O. A., Gutkind, J. S., Burbelo, P. D., Akiyama, S. K. and Yamada, K. M., 1995, Integrin function: molecular hierarchies of cytoskeletal and signaling molecules, *J Cell Biol,* **131:**791.
Monks, C. R., Freiberg, B. A., Kupfer, H., Sciaky, N. and Kupfer, A., 1998, Three-dimensional segregation of supramolecular activation clusters in T cells, *Nature,* **395:**82.
Negulescu, P. A., Kraieva, T., Khan, A., Kerschbaum, H. H. and Cahalan, M. D., 1996, Polarity of T cell shape, motility, and sensitivity to antigen, *Immunity,* **4:**421.
Noel, P. J., Boise, L. H. and Thompson, C. B., 1996, Regulation of T cell activation by CD28 and CTLA4, *Adv Exp Med Biol,* **406:**209.
Norcross, M. A., 1984, A synaptic basis for T-lymphocyte activation, *Ann Immunol (Paris),* **135D:**113.
Palecek, S. P., Loftus, J. C., Ginsberg, M. H., Lauffenburger, D. A. and Horwitz, A. F., 1997, Integrin-ligand binding properties govern cell migration speed through cell-substratum adhesiveness, *Nature,* **385:**537.
Pasternak, C. and Elson, E. L., 1985, Lymphocyte mechanical response triggered by cross-linking surface receptors, *J.Cell Biol.,* **100:**860.
Paul, W. E. and Seder, R. A., 1994, Lymphocyte responses and cytokines, *Cell,* **76:**241.
Peterson, D. A., DiPaolo, R. J., Kanagawa, O. and Unanue, E. R., 1999, Negative selection of immature thymocytes by a few peptide-MHC complexes: differential sensitivity of immature and mature T cells, *J Immunol,* **162:**3117.
Peterson, E. J., Woods, M. L., Dmowski, S. A., Derimanov, G., Jordan, M. S., Wu, J. N., Myung, P. S., Liu, Q. H., Pribila, J. T., Freedman, B. D., Shimizu, Y. and Koretzky, G. A., 2001, Coupling of the TCR to integrin activation by Slap-130/Fyb, *Science,* **293:**2263.
Qi, S. Y., Groves, J. T. and Chakraborty, A. K., 2001, Synaptic pattern formation during cellular recognition, *Proc Natl Acad Sci U S A,* **98:**6548.
Reedquist, K. A., Ross, E., Koop, E. A., Wolthuis, R. M., Zwartkruis, F. J., van Kooyk, Y., Salmon, M., Buckley, C. D. and Bos, J. L., 2000, The small GTPase, Rap1, mediates CD31-induced integrin adhesion, *J Cell Biol,* **148:**1151.
Rozdzial, M. M., Malissen, B. and Finkel, T. H., 1995, Tyrosine-phosphorylated T cell receptor zeta chain associates with the actin cytoskeleton upon activation of mature T lymphocytes, *Immunity,* **3:**623.
Scanga, S. E., Ruel, L., Binari, R. C., Snow, B., Stambolic, V., Bouchard, D., Peters, M., Calvieri, B., Mak, T. W., Woodgett, J. R. and Manoukian, A. S., 2000, The conserved PI3'K/PTEN/Akt signaling pathway regulates both cell size and survival in Drosophila, *Oncogene,* **19:**3971.
Shan, X., Czar, M. J., Bunnell, S. C., Liu, P., Liu, Y., Schwartzberg, P. L. and Wange, R. L., 2000, Deficiency of PTEN in Jurkat T cells causes constitutive localization of Itk to the plasma membrane and hyperresponsiveness to CD3 stimulation, *Mol Cell Biol,* **20:**6945.
Stowers, L., Yelon, D., Berg, L. J. and Chant, J., 1995, Regulation of the polarization of T cells toward antigen-presenting cells by ras-related GTPase CDC42, *Proc.Natl.Acad.Sci.USA,* **92:**5027.
Suga, K., Katagiri, K., Kinashi, T., Harazaki, M., Iizuka, T., Hattori, M. and Minato, N., 2001, CD98 induces LFA-1-mediated cell adhesion in lymphoid cells via activation of Rap1, *FEBS Lett,* **489:**249.
Tafuri, A., Shahinian, A., Bladt, F., Yoshinaga, S. K., Jordana, M., Wakeham, A., Boucher, L.-M., Bouchard, D., Chan, V. S. F., Duncan, G., Odermatt, B., Ho, A., Itie, A., Horan, T., Whoriskey, J. S., Pawson, T., Penninger, J. M., Ohashi, P. S. and Mak, T. W., 2001, ICOS is essential for effective T-helper-cell respones, *Nature,* **409:**105.
Valitutti, S., Dessing, M., Aktories, K., Gallati, H. and Lanzavecchia, A., 1995, Sustained signalling leading to T cell activation results from prolonged T cell receptor occupancy. Role of T cell actin cytoskeleton, *J.Exp.Med.,* **181:**577.
van Kooyk, Y., van de Wiel-van Kemenade, P., Weder, P., Kuijpers, T. W. and Figdor, C. G., 1989, Enhancement of LFA-1-mediated cell adhesion by triggering through CD2 or CD3 on T lymphocytes, *Nature,* **342:**811.
Velazquez, C., DiPaolo, R. and Unanue, E. R., 2001, Quantitation of lysozyme peptides bound to class II MHC molecules indicates very large differences in levels of presentation, *J Immunol,* **166:**5488.
Viola, A. and Lanzavecchia, A., 1996, T cell activation determined by T cell receptor number and tunable thresholds, *Science,* **273:**104.

Wang, X., Gjorloff-Wingren, A., Saxena, M., Pathan, N., Reed, J. C. and Mustelin, T., 2000, The tumor suppressor PTEN regulates T cell survival and antigen receptor signaling by acting as a phosphatidylinositol 3-phosphatase, *J Immunol,* **164:**1934.

Weiss, A. and Littman, D. R., 1994, Signal transduction by lymphocyte antigen receptors, *Cell,* **76:**263.

Wülfing, C., Bauch, A., Crabtree, G. R. and Davis, M. M., 2000, The vav exchange factor is an essential regulator in actin-dependent receptor translocation to the lymphocyte-antigen-presenting cell interface, *Proc Natl Acad Sci U S A,* **97:**10150.

Wülfing, C., Sjaastad, M. D. and Davis, M. M., 1998, Visualizing the dynamics of T cell activation: intracellular adhesion molecule 1 migrates rapidly to the T cell/B cell interface and acts to sustain calcium levels, *Proc.Natl.Acad.Sci.USA,* **95:**6302.

Wülfing, C., Sjaastad, M. D. and Davis, M. M., 1998, Visualizing the dynamics of T cell activation: intracellular adhesion molecule 1 migrates rapidly to the T cell/B cell interface and acts to sustain calcium levels, *Proc Natl Acad Sci U S A,* **95:**6302.

Wulfing, C., Sumen, C., Sjaastad, M. D., Wu, L. C., Dustin, M. L. and Davis, M. M., 2002, Costimulation and endogenous MHC ligands contribute to T cell recognition, *Nat Immunol,* **3:**42.

TWO-PHOTON IMAGING IN INTACT LYMPHOID TISSUE

Sindy H. Wei, Mark J. Miller, Michael D. Cahalan, and Ian Parker*

1. INTRODUCTION

For imaging living tissue, the two-photon technique presents several advantages over conventional confocal fluorescence imaging, including reduced photobleaching, minimal phototoxicity, and the ability to resolve detail at greater depths. We have established a two-photon imaging method that permits five-dimensional (*x*, *y*, *z*, *t*, emission wavelength) *ex vivo* imaging in intact murine lymphoid organs of viable T lymphocytes, B lymphocytes, and antigen-presenting cells (APC) up to a depth of about 300 μm. Three-dimensional reconstruction of lymph nodes revealed B cells in shallow, discrete subcapsular follicles, T cells in interfollicular regions, reticular fiber networks, and migratory dendritic cells. Additionally, we have imaged lymphocyte activity within the periarteriolar lymphocyte sheath in the intact spleen. In both lymph node and spleen, distinct modes of B and T cell movement may reflect specific antigen search programs. In this paper we review the capabilities of our experimental system, and hope to demonstrate the importance of live tissue imaging studies at the cellular level.

1.1. Why Tissue Imaging is Important

Lymphocyte motility and T cell-APC interactions have been studied extensively using *in vitro* techniques, and by following morphological changes over time. Although the use of extracellular substrates and matrices attempts to simulate physiological tissue environments, there still remains the question of whether the behavior of cells observed outside of their native environment is relevant to cellular functions such as trafficking, chemotaxis, antigen presentation, activation, proliferation, and effector functions in the context of the intact immune system.

---

* Sindy H. Wei, Mark J. Miller, Michael D. Cahalan, Department of Physiology and Biophysics, University of California, Irvine, California 92697-4561, USA. Ian Parker, Department of Neurobiology and Behavior, University of California, Irvine, California 92697-4550, USA.

Lymphocytes do not passively flow between the circulatory and lymphatic systems, but require the development of motile characteristics in their perpetual search for antigen, and afterward, to act as effectors in both lymphatic and peripheral tissue. Although extravasation and homing have been thoroughly examined by *in vitro* and *in vivo* approaches, there exists only limited information on lymphocyte passage through, and antigen presentation within, lymphoid organs. This can be attributed to difficulties in replicating the myriad factors that affect motility of naïve lymphocytes, including the three-dimensional extracellular matrix (ECM), cytokine and chemokine gradients, and cell-cell interactions. Live tissue imaging with two-photon laser microscopy permits the study of dynamic lymphocyte behavior in the native environment, and may help to resolve questions regarding the role of the ECM and cell-cell interactions in modulating T cell motility *in vivo*.

1.2. Multi-photon Versus Confocal Imaging

In recent decades, conventional video and confocal microscopy methods have allowed researchers to study physiology at the single-cell level, and have inspired the rapid development of fluorophores and specialized indicators to probe intracellular structures, localization, signaling and function. Confocal microscopy was a drastic improvement over conventional fluorescence microscopy, and achieves high spatial and temporal resolution by rapidly scanning a laser spot across the sample and then selectively transmitting in-focus signals through a pinhole to acquire thin optical slices (Periasamy, 2001). Furthermore, serial acquisition of such optical slices through a sample in the Z-direction permits subsequent three-dimensional reconstruction of the sample.

However, serious concerns of viability arise when conventional confocal technique is used to study live specimens by continuous imaging over periods of time. The small confocal aperture rejects most of the fluorescence emitted from the sample, and thus requires high laser output or long scanning times to achieve a decent signal-to-noise ratio. This results in rapid photobleaching of fluorophores, and phototoxicity from free-radical formation that compromises tissue integrity and cellular function. Moreover, light scattering in biological tissues usually limits the depth to which confocal sections can be obtained to only a few tens of micrometres.

Many problems associated with live-cell imaging can be greatly mitigated by the use of the recently introduced technique of multi-photon fluorescence excitation. Two-photon microscopy is based on the principle that fluorophores can be excited by the near-simultaneous absorption of two low energy photons (longer wavelength), instead of a single photon of double the energy (shorter wavelength). Extremely high light intensities are required to produce two-photon excitation and, in order to avoid damage to the tissue, the excitation laser (e.g. mode-locked Ti:Sapphire laser) emits extremely short (ca. 100 fs) pulses yielding enormously high peak energies even though the overall laser power is low. The key point is that two-photon excitation occurs the second power of intensity and is thus essentially restricted to the focal spot formed by a microscope objective and, unlike confocal microscopy, no out-of-focus fluorescence is generated. By limiting excitation to the focal plane, two-photon microscopy eliminates the need for a pinhole to reject out-of-focus fluorescence, and greatly reduces photobleaching and phototoxicity (Denk et al., 1990). Moreover, multi-photon microscopy can image much deeper (hundreds of microns) into tissues than is possible with confocal microscopy, owing both

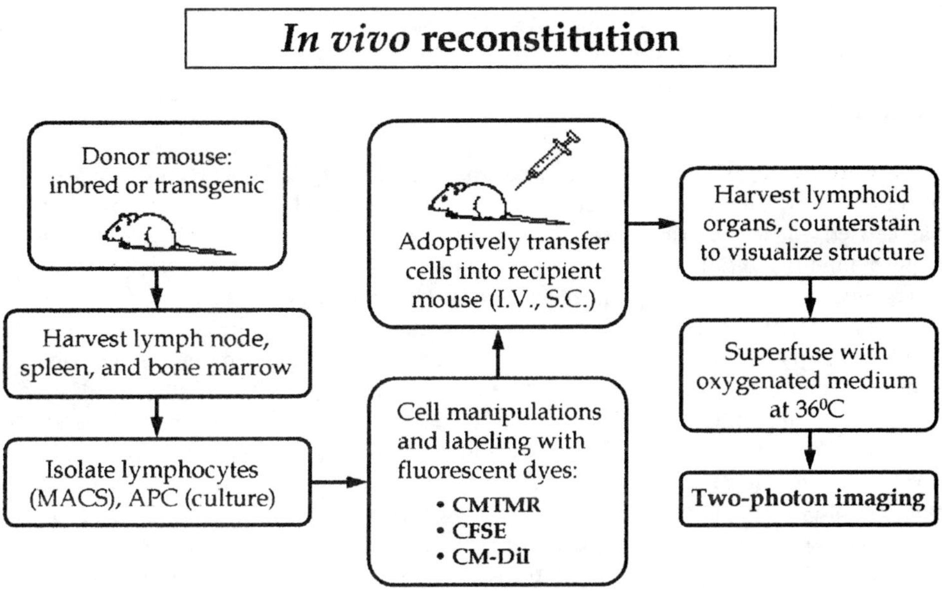

Figure 1. Schematic illustrating procedures for lymphocyte isolation, fluorescent labeling and adoptive transfer prior to two-photon imaging of excised lymph nodes.

to the lack of detection pinhole limitations and the use of infrared or near-infrared excitation wavelengths that pass through biological specimens with little absorption or scattering.

2. METHODS

Figure 1 describes the general paradigm that we have adopted to image lymphocytes and APC in intact lymphoid organs.

2.1. Adoptive Transfer of Fluorescently Labeled Cells

In order to image the behavior of living cells inside intact lymphoid organs we used adoptive transfer of isolated, purified subsets of T lymphocytes, B lymphocytes, and dendritic cells (or other APC) from BALB/c mice that were labeled *in vitro* with Cell Tracker dyes (Molecular Probes), following established methods (Ingulli et al., 1997). For motility studies we isolated BALB/c T and B cells from lymph node and spleen, respectively. After applying magnetic bead purification to >98% purity, 10-30 million naïve T and B cells were labeled with fluorescent CellTracker (Molecular Probes) dyes including CFSE (green) or CMTMR (orange-red) and were adoptively transferred into a recipient BALB/c mouse by tail vein injection. Transferred cells were allowed 6 hours to 5 days to home to peripheral lymph nodes or spleen prior to imaging. To visualize cells interacting with macroscopic components in their environment, we counterstained lymph nodes to label structures including reticular fibers by incubating intact lymphoid organs with fluorescent dyes, including the CellTrackers (Miller et al., in press).

2.2. Imaging

Five-dimensional imaging was performed under physiological conditions using a custom-built two-photon microscopy system (Nguyen et al., 2001), and image analysis was carried out as described previously (Miller et al., in press). This system is able to acquire high-resolution optical sections through a volume of 200 x 200 x 50 μm, corresponding to 2 nanoliters, at intervals of 10 seconds/frame. Acquisitions at intervals of greater than 20-30 seconds do not yield high enough time resolution to track individual lymphocyte paths. With this *in vivo* reconstitution system, we have observed significantly increased T and B lymphocyte velocities (up to 35 μm/min for T cells) in lymph node in contrast to observations made on 2D substrates, as well as random migration. Both B and T lymphocytes exhibit a high degree of temperature dependence, developing maximal motility near physiological temperature (approximately 36°C). For motility studies it is crucial to implement steady temperature control, as slight cooling of 2-4°C can decrease velocities drastically (Miller et al., in press). Fluorescence intensity did not decrease appreciably due to photobleaching for at least one hour, while rapidly crawling lymphocytes were observed for at least 6 hours.

3. RESULTS AND DISCUSSION

3.1. Time-Lapse Imaging of T cells in Different Regions of the Lymph Node

Even though the unseen factors within lymphoid tissue are likely to play significant roles in the homing and motility patterns of lymphocytes, we found evidence for intrinsic B or T cell antigen search programs that are independent of the local environment. T cell regions are highly fibrillar in nature, whereas B cell regions express their own distinct set of cytokines, chemokines, and ECM. As the lymph node is highly compartmentalized, T cells rarely wander into B cell regions, and vice versa. Assuming that lymphocytes employ a "random walk" mode of movement, analysis of T and B cell motility patterns within their proper location in the lymph node show the motility coefficient of T cells to be five- to six-fold greater than that of B cells (Miller et al., in press). Panels 1-6 in Figure 2 show sequential time-lapse images (20 s per frame) within a lymph node, and illustrate the motion of a T cell (marked by arrowhead) that crosses into a B cell zone within the primary follicle. Even though the chemical, cellular, and structural environment of the follicle differs dramatically from that of the interfollicular space, the T cell intruder appears to maintain its basic motility characteristics. These include a high motility coefficient, periodic changes in the shape index (a long, stretched-out form alternating with rounded morphology), and lunging motility. The lone T cell can be compared with B cells (triangular morphology) within the follicle, and other T cells outside of the follicle. In looking at the cells within the tissue, volume imaging is the only method to be certain that a given cell is in a particular spatial relationship with other cells. This will be important in examining regional variations of cell behavior, defining the extent of contact between T cells and APC, and tracking cell proliferation.

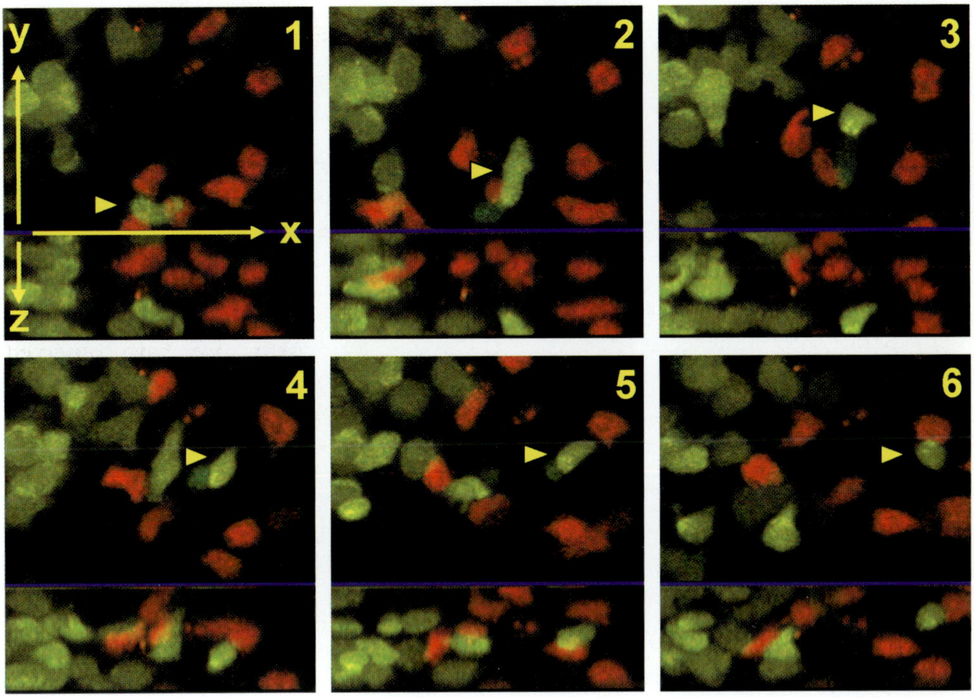

Figure 2. T cell behavior within a B cell follicle. For each time point, Z-sections through the imaging volume were reconstructed in 3-D and compressed to create an *x-y* and *x-z* view (separated by a blue line). To ensure that the T cell (arrowhead) was passing through, but not above or in front of the B cell region, we must compare the location of the T cell relative to B cells (red) and other T cells (green) in both the *x-y* and *x-z* directions.

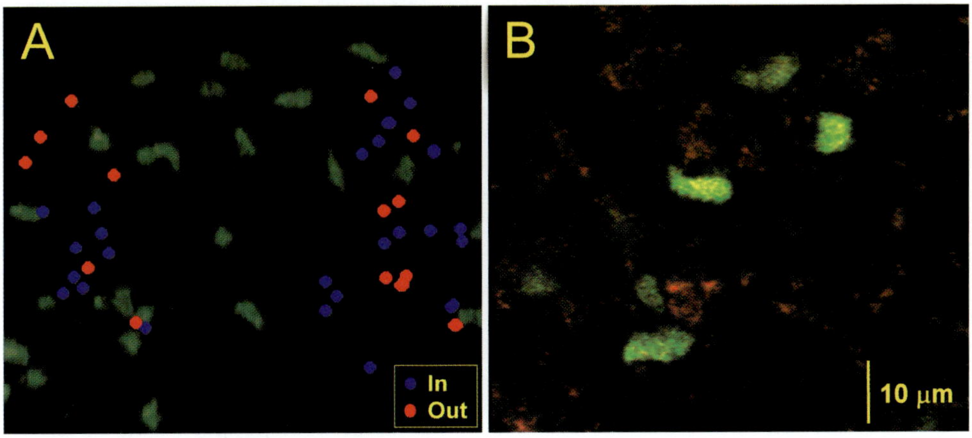

Figure 3. Imaging of lymphocytes within mouse lymph node and spleen. (A) Analysis of an imaging field was done to determine if T cell movement in the Z-direction occurs in specific regions. Blue and red dots represent individual migratory events into and out of the imaging volume, respectively. Clusters of dots depict areas where Z-movement appears to be favored, and may correlate to putative "conduits." (B) Fluorescently-labeled T cells (green) exhibit polarized, motile behavior among reticular fibers (red) in the spleen.

3.2. Conduits for *Z*-Axis Motion in the Lymph Node

We acquired stacks of images through a volume of tissue at 10 seconds/frame, and then compressed the fluorescent signals in all Z-stacks so that all lymphocytes within that imaging volume are represented. Tracking of lymphocyte motility in a video in the plane parallel to the lymph node capsule demonstrated the presence of distinct regions within the tissue structure that allow rapid movement of lymphocytes perpendicular to the capsule (Fig. 3A). There appear to be two modes of Z-movement: slow drift, and a fast dive into or out of the imaging volume. Blue and red dots mark locations where T cells moved into or out of the imaging volume, respectively, in a direction perpendicular to the capsule. In this field, two regions of active motion in the *z*-axis are discernable, while other regions appear to favor movement in the *x-y* plane parallel to the capsule. These putative conduits give us some idea of how lymphocytes may transit through this highly compartmentalized organ in search for antigen.

3.3 Imaging in Spleen

Similar to murine lymph node, the spleen is comprised of a dense collagen capsule with a reticular fiber network. Figure 3B demonstrates the use of two-photon imaging within the spleen; polarized, motile lymphocytes (green) are revealed within a network of reticular fibers (red). In the spleen, lymphocytes reside within the white pulp, more specifically in the periarteriolar lymphocyte sheath (PALS), while erythrocytes constitute the major cell population within the splenic cords and sinuses of the red pulp. As in the lymph node, the lymphatic tissue of the PALS is organized into diffuse T cell and nodular B cell regions.

4. CONCLUSIONS

We have shown that our imaging system provides a window into the dynamic behavior of living lymphocytes within intact lymphoid tissue, and reveals lymphocyte behaviors fundamental to the initiation of the immune response. Two-photon, five-dimensional laser microscopy allows the determination of location, structure, motility pattern, and the timing and dynamics of antigen presentation. We have applied this technique to isolated lymph nodes and spleen, which provide a much closer approximation to the true physiological environment than was possible in earlier experiments using artificial two- and three-dimensional substrates. Nevertheless, organ removal from the circulation may disrupt necessary chemical gradients that affect homing, cell motility, and behavior. To address this concern, we are presently developing an intravital imaging approach.

The imaging methodologies we describe should greatly enhance studies of immunological synapse formation, antigen-stimulated Ca^{2+} signaling and gene expression *in vivo*. We hope that this technique will stimulate the development of *in vivo* or genetically-encoded indicators that will be stable over the time course of an immune response. In conjunction with existing animal models for autoimmune disease and sites of tissue injury and inflammation, we plan to extend cell-imaging techniques to elucidate mechanisms of tissue damage as well as the action of candidate immunomodulatory agents.

REFERENCES

Denk, W., Strickler, J. H., and Webb, W. W., 1990, Two-photon laser scanning fluorescence microscopy, *Science* **248**:73.

Ingulli, E., Mondino, A., Khoruts, A., and Jenkins, M. K., 1997, In vivo detection of dendritic cell antigen presentation to CD4+ T cells, *J. Exp. Med.* **185**:2133.

Miller, M. J., Wei, S. H., Parker, I., and Cahalan, M. D., Two-photon imaging of lymphocyte motility and antigen response in intact lymph node, *Science*, in press.

Nguyen, Q. T., Callamaras, N., Hsieh, C., and Parker, I., 2001, Construction of a two-photon microscope for video-rate Ca^{2+} imaging, *Cell Calcium* **30**:383.

Periasamy, A., 2001, *Methods in Cellular Imaging*, Oxford University Press, Oxford, pp. 66-67.

INDEX